Ten Mathematical Essays on
Approximation in Analysis and Topology

Ten Mathematical Essays on Approximation in Analysis and Topology

Edited by

J. Ferrera
J. López-Gómez
F.R. Ruiz del Portal

Universidad Complutense de Madrid
28040 Madrid
Spain

2005
ELSEVIER

Amsterdam – Boston – Heidelberg – London – New York – Oxford
Paris – San Diego – San Francisco – Singapore – Sydney – Tokyo

ELSEVIER B.V.
Radarweg 29
P.O. Box 211, 1000 AE Amsterdam
The Netherlands

ELSEVIER Inc.
525 B Street, Suite 1900
San Diego, CA 92101-4495
USA

ELSEVIER Ltd.
The Boulevard, Langford Lane
Kidlington, Oxford OX5 1GB
UK

ELSEVIER Ltd.
84 Theobalds Road
London WC1X 8RR
UK

First edition 2005

Library of Congress Cataloging in Publication Data
A catalog record is available from the Library of Congress.

British Library Cataloguing in Publication Data
A catalogue record is available from the British Library.

ISBN: 0-444-51861-4

♾ The paper used in this publication meets the requirements of ANSI/NISO Z39.48-1992 (Permanence of Paper).
Printed in The Netherlands.

Contents

Eigenvalues and Perturbed Domains

J.K. Hale

Monotone Approximations and Rapid Convergence

V. Lakshmikantham

Preface

During the past century, the impact of mathematics on humanity has been more tremendous than ever since Galileo's agonizing fight against the old establishment and the revolution which physics experienced after Newton's subsequent synthesis.

At the beginning of the last century, mathematical ideas and techniques were spread to theoretical and applied physics by the influence of two of the greatest mathematicians of all times, D. Hilbert and H. Poincaré, being then at the zenith of their careers. Their ability to establish very deep — at first glance often hidden — connections between a priori separated branches of science convinced physicists to adopt and work with the most powerful existing mathematical tools. Whereas the 20th century really was the century of physics, mathematics enjoyed a well deserved reputation from its very beginning, so facilitating the huge impact it had subsequently on humanity. This reputation has been crucial for the tremendous development of science and technology. Although mathematics supported the development of weapons of mass destruction, it simultaneously promoted the advancement of computers and high technology, without which the substantial improvement of the living conditions humanity as a whole has experienced, could not have been realized. In no previous time the world has seen such a spectacular growth of scientific knowledge as during the last century, with mathematics playing a central role in most scientific and technological successes. But, not surprisingly, the extraordinary development of mathematics — and science politics — created an impressive number of difficulties which are to be solved in order to facilitate further real advance in mathematics. These concerns are not new as is witnessed by the following excerpts from a speech of R. Courant, given in 1962 in Göttingen on the occasion of Hilbert's hundredth birthday: [1]

Although mathematics has played an important role for more than two thousand years, it is still subject to changes of fashion and, above all, to departures from tradition. In the present era of over-active industrialization of science, propaganda, and the explosive manipulation of the social and personal basis of science, I believe that we find ourselves in such a period of danger. In our time of mass media, the call for reform, as a result of propaganda, can just as easily lead to a narrowing and choking as to a liberating of

---

[1] We follow the translation of C. Reid in *Hilbert* (Springer, New York, 1996, p. 220).

mathematical knowledge. That applies, not only to research in the universities, but also to the instruction in the schools. The danger is that the combined forces so press in the direction of abstraction that only that side of the great Hilbertian tradition is carried on.

Living mathematics rests on the fluctuation between the antithetical powers of intuition and logic, the individuality of 'grounded' problems and the generality of far-reaching abstractions. We ourselves must prevent the development being forced to only one pole of the life-giving antithesis.

Mathematics must be cherished and strengthened as a unified, vital branch in the broad river of science; it dares not trickle away in the sand.

Hilbert has shown us through his impressive example that such dangers are easily preventable, that there is no gap between pure and applied mathematics, and that between mathematics and science as a whole a fruitful community can be established. I am therefore convinced that Hilbert's contagious optimism even today retains its vitality for mathematics, which will succeed only through the spirit of Hilbert.

Indeed, one of the main difficulties we are faced with is to treat and to filter the huge amount of information at our disposal, originating in the explosive growth of human knowledge. What is the best way to deal with such an abundance of information? How to filter the new results in order to ascertain that the most novel and interesting discoveries and relations between the different areas of science and, in particular, within and with mathematics, do find the place they deserve? Certainly, as in former times, sometimes it may be difficult to publish really new results lying outside of the fashionable main stream research. This is an indication for a severe fragmentation of mathematics — and the community of mathematicians —, as well as for the failure of the media to scrutinize and propagate mathematical discoveries. As envisioned by R. Courant, science politics and influential groups may severely affect the quality of mathematics as well.

Not tackling these serious problems might lead to an increasing waste of enthusiasm and energy of a large number of highly specialized researchers who, in practice, stay almost completely ignorant of important mathematical progress, due to the strong fragmentation of the field. Undoubtedly, the enormous amount of information can add to this fragmentation, whose real cost in practical terms seems very difficult to estimate. Quite surprisingly, although mathematical methods for scrutinizing quality are, a priori, stronger than those used in other scientific disciplines, like physics, chemistry and biology, the fragmentation of mathematics is substantially higher than that of these disciplines. Why? Could it be possible that our stronger scrutiny methods facilitate tendencies of established networks to propagate almost exclusively main stream work of their associated groups instead of ensuring independence and quality of mathematical production? Certainly a challenging question to contemplate!

Another serious difficulty lies in the high number of mathematicians still having a genuine 19th century idea of mathematics. Actually, instead of encouraging and supporting the outset to new mathematical frontiers and real innovation, most of the world leading mathematical associations tend to reward almost exclusively the ability of mathematicians to solve open problems coming from the past — occasionally highly marginal ones —, instead of looking ahead and encouraging experts to face the new challenges of the globalized world in which we are living.

It is a significant irony that the list of Hilbert's Problems did not predict what Hilbert would work on by himself during the subsequent two decades, as there is no word about functional analysis or integral equations, and, most surprisingly, there is almost nothing on topology, which Poincaré had founded and which L. E. J. Brouwer revolutionized a decade after Hilbert's celebrated speech at Paris.

In his book *The Structure of Scientific Revolutions* (Chicago University Press, 1962), Thomas S. Kuhn advocated the existence of two different types of science: *normal science* and *revolutionary science*. In any scientific discipline, normal science follows a number of well established rules, concepts and methodologies, accepted by all researchers in the field. Occasionally, within normal science there arise unexpected discoveries inconsistent with the established paradigms. These discoveries generate high tension that increases in intensity until a scientific revolution is accepted to emerge beneath the surface of such an unorthodox finding. Then a new paradigm naturally arises under which all experts start to do normal science again. Normal science expands within existing paradigms, while revolutionary science sees a necessity to update the old established paradigms. Of course, too strong scrutiny within a strongly fragmented scientific discipline is far from facilitating the diffusion of unorthodox thinking, and so, is hampering progress.

Although these thoughts might be rather contentious, besides being irrelevant for the advance of mathematics, they were at the roots of this book — at the very beginning a rather fuzzy project. We decided to collect a series of papers covering wide areas of mathematics that have advanced along separate paths around the common theme approximation theory, in an attempt to bring together those fragmented areas, and, simultaneously, to facilitate the development of further connections between them. As a result, this book collects ten mathematical essays around approximation theory, understood in a broad sense, within the context of one of those mathematical fields that has enjoyed a most remarkable growth during the past century: Nonlinear Analysis. Since the pioneering results of H. Poincaré and L. E. J. Brouwer, Nonlinear Analysis defines itself as that part of mathematics dealing with the existence, structure, and multiplicity of solutions of equations in general topological spaces, Banach spaces in particular. Its main impetus comes from the fact that a huge number of mathematical problems — pure and applied ones — admit such an abstract formulation.

The mathematical analysis of the general properties of the set of solutions to abstract equations, that are preserved by continuous deformations, was an important impetus for the development of Topology, and, simultaneously, stimulated the extraordinary development that Functional Analysis — linear and non linear — experienced during the past century. Undoubtedly, the classification of the resulting local algebraic manifolds motivated the creation of Algebraic Geometry. The implementation of iterative schemes and continuation methods to compute solution manifolds is on the basis of modern Numerical Analysis. The study of the sensitivity of the states with respect to initial values and parameters triggered the theory of Dynamical Systems. The determination of the correct functional spaces to analyze Partial Differential Equations was a milestone in the development of Functional Analysis since the pioneering results of Hilbert and Banach.

Combining topological, analytical, algebraic and geometrical tools has proved to be imperative for finding the hidden structures of the solution sets in broad classes of nonlinear differential equations. Aside from the enormous interest of nonlinear differential equations per se, and for modelling a wide variety of real world problems, these equations have always provided an extraordinary testing ground for abstract mathematical techniques. Simultaneously, new mathematical tools have been generated by their study and novel ideas emerged that extrapolate to and have an impact on other areas of mathematics, economics and science. This methodology has tremendously facilitated the rise of scientific revolutions since Galileo and Newton, passing through the synthesis of Poincaré, Volterra and Hilbert, until today.

Linear results are the basis upon which the *local nonlinear theory* is constructed, while topology provides us with the most *universal global properties* of nonlinear problems. In complete agreement with I. Stewart in *The Problems of Mathematics* (Oxford Univ. Press, 1987, p. 156), we corroborate that:

The current success of topology ... owes little to Poincaré's abilities at crystal-gazing, but an enormous amount to his mathematical imagination and good taste. Topology is a success precisely because it forgot the details of those original problems, and instead concentrated on their deeper structure. This deep structure occurs in any scheme of mathematical investigation that makes use of continuity. In consequence, topology touches on almost everything. Of course, the extent to which this fact helps depends on the question you want to answer.

In the papers collected in this volume the authors discuss central problems of their respective research fields which are closely related to approximation theory, understood in the broadest possible sense, connecting topology, analysis, and applications to nonlinear differential equations. These jewelry pieces of *mathematical intra-history* will be a delight for many forthcoming generations of mathematicians, who will occupy themselves with these fascinating and important subjects. Some authors kindly outlined, as requested by the editors, the path that led to some of their most celebrated results, as well as the personal circumstances of their discoveries.

Before concluding this preface, we express our deepest gratitude to all who have generously contributed to the production of this volume, which could not have taken form without their friendship and unselfish efforts. Our warmest acknowledgements to all of them!

Madrid, January 2005.

The Editors

Ten Mathematical Essays on Approximation in Analysis and Topology
J. Ferrera, J. López-Gómez, F. R. Ruiz del Portal, Editors

1

Maximum Principles and Principal Eigenvalues

H. Amann

Institut für Mathematik, Universität Zürich,
Winterthurerstr. 190, CH–8057 Zürich, Switzerland

Abstract

In this paper well-known maximum principles are extended to second order cooperative linear elliptic systems with cooperative boundary conditions in strong, weak, and very weak settings. In addition, interrelations between maximum principles and principal eigenvalues are studied in detail, as well as continuity properties of principal eigenvalues under domain perturbations.

Key words: Maximum principles, principal eigenvalues, cooperative systems, cooperative boundary conditions, weak and very weak solutions, domain perturbations

1. Introduction

It is the main purpose of this paper to study maximum principles for linear second order cooperative elliptic systems under general linear first order cooperative boundary conditions. We are particularly interested in weak settings, in view of applications to nonlinear systems in situations where higher regularity either cannot be expected or does not constitute a convenient frame to deal with such problems.

Maximum principles for cooperative systems have already been discussed by several authors under various assumptions (cf. [20], [22], [32], [36], [43], [56], [60], [63], [71], [77]). However, in all these references, with the exception of [63], the case of Dirichlet boundary conditions is studied only. Furthermore, in almost all cases maximum principles in the strong sense are considered, that is, for C^2 functions, or, at least, for W_q^2 functions where q is sufficiently large.

It is well-known that maximum principles are of great importance for the study of existence and qualitative properties of nonlinear equations. For example, one of the most useful techniques in the theory of second order scalar elliptic (and parabolic) boundary value problems, the method of sub- and supersolutions, is based on maximum principles (cf. [1], [62], [64], [67]). This is true for systems as well, as has already been observed in [1, Sections 5 and 10] and has since been worked out by several authors under various hypotheses (cf. [59], [62], [66], and the references therein). However, in all those papers either Dirichlet conditions are considered only or, if Neumann boundary conditions are studied at all, it is assumed that either the boundary conditions decouple, a rather particular situation (e.g., [40], [41]), or that very strong regularity conditions are satisfied (e.g., [62]). It is one of the advantages of our work that our maximum principles allow, among other things, comparison theorems for semilinear problems with nonlinear boundary conditions, the latter depending on all components of the unknown vector function, in a weak setting.

The validity of maximum principles is closely related to the existence of a principal eigenvalue, that is, of a least real eigenvalue determining the position of the smallest closed right half plane containing the spectrum. This eigenvalue plays a predominant rôle in the qualitative study of nonlinear boundary value problems via bifurcation theory and in the method of sub- and supersolutions (cf. [37], [51], [53], [54], [57], [58], and the references therein). Consequently, we investigate in some detail questions of existence and continuous dependence on the data of the principal eigenvalue.

It should be noted that our results on maximum principles in weak settings are new, even in the scalar case. The same is true for our continuity results for the principal eigenvalue, since we allow perturbations of the Robin boundary as well.

To give a flavor of the content of this paper we describe now some of our results in a simple setting. Here we restrict ourselves to a 2×2 system with the diagonal Laplace operator as principal part. The general case is studied in the main body of this work.

Throughout this paper Ω is a C^2 domain in $\mathbb{R}^n$, where $n \geq 1$, with a nonempty compact boundary Γ. We denote by $\nu := (\nu^1, \dots, \nu^n)$ the outer unit normal on Γ.

However, to illustrate some of the main results by means of prototypical examples, we assume throughout the rest of this introduction that Ω is bounded.

Let u be a superharmonic distribution in Ω, which means

$$u \in \mathcal{D}'(\Omega), \qquad \langle -\Delta\varphi, u \rangle \geq 0 \quad \text{for all } \varphi \in \mathcal{D}(\Omega) \text{ with } \varphi \geq 0. \tag{1}$$

Then it is known that u is a regular distribution, in other words: $u \in L_{1,\mathrm{loc}}(\Omega)$. If, moreover, for some point-wise defined representative $\widetilde{u}$ of u,

$$\liminf_{\substack{x \in \Omega \\ x \to y}} \widetilde{u}(x) \geq 0 \qquad \text{for all } y \in \Gamma, \tag{2}$$

then $u \geq 0$, that is, $u(x) \geq 0$ for a.a. $x \in \Omega$ (e.g., [30, Propositions II.4.20 and II.4.21]). It is clear that from (1) alone nothing can be said about the boundary behavior of $u \in L_{1,\mathrm{loc}}(\Omega)$ since every test function $\varphi \in \mathcal{D}(\Omega)$ vanishes near Γ. Thus (1), without the additional information of (2), does not imply that $u \geq 0$.

The situation is different if we require the validity of the inequalities in (1) for a larger class of test functions and a little more regularity for u. For this, given $q \in (1, \infty)$, we put

$$W^2_{q',\gamma}(\Omega, \mathbb{R}) := \{ u \in W^2_{q'}(\Omega, \mathbb{R}) \; ; \; \gamma u = 0 \},$$

where

$$\frac{1}{q} + \frac{1}{q'} = 1$$

and γ is the trace operator. We also denote by $\langle \cdot, \cdot \rangle$ the usual L_q duality pairing. Then it is a consequence of our much more general results that the following *very weak maximum principle* (rather: minimum principle) is valid:

$$\left.\begin{array}{l} u \in L_q(\Omega, \mathbb{R}) \text{ for some } q \in (1, \infty), \\[2mm] \langle -\Delta v, u \rangle \geq 0 \text{ for all } v \in W^2_{q',\gamma}(\Omega, \mathbb{R}) \text{ with } v \geq 0 \end{array}\right\} \quad \text{imply } u \geq 0. \quad (3)$$

Very weak maximum principles are of importance in nonlinear problems involving low regularity data, for example (e.g., [12]). The maximum principles studied below are valid for cooperative systems also. To illustrate this we consider the model system $(\mathcal{A}, \mathcal{B})$ on Ω defined as follows: we put $u := (u^1, u^2)$ and assume that there are two decompositions of Γ:

$$\Gamma = \Gamma^1_0 \cup \Gamma^1_1 = \Gamma^2_0 \cup \Gamma^2_1 \qquad \Gamma^1_0 \cap \Gamma^1_1 = \Gamma^2_0 \cap \Gamma^2_1 = \varnothing,$$

such that Γ^1_0 and Γ^2_0 are open, hence closed, submanifolds of Γ. Then we define

$$\mathcal{A}u = (\mathcal{A}^1 u, \mathcal{A}^2 u)$$

by

$$\mathcal{A}^1 u := -\Delta u^1 + a^{11} u^1 + a^{12} u^2,$$
$$\mathcal{A}^2 u := -\Delta u^2 + a^{21} u^1 + a^{22} u^2,$$

and

$$\mathcal{B}u = (\mathcal{B}^1 u, \mathcal{B}^2 u)$$

by

$$\mathcal{B}^1 u := \begin{cases} u^1 & \text{on } \Gamma^1_0, \\ \partial_\nu u^1 + b^{11} u^1 + b^{12} u^2 & \text{on } \Gamma^1_1 \end{cases}$$

and

$$\mathcal{B}^2 u := \begin{cases} u^2 & \text{on } \Gamma^2_0, \\ \partial_\nu u^2 + b^{21} u^1 + b^{22} u^2 & \text{on } \Gamma^2_1, \end{cases}$$

where we assume that

$$a^{rs} \in L_\infty(\Omega, \mathbb{R}), \qquad b^{rs} \in C^{1-}(\Gamma, \mathbb{R}), \qquad r, s \in \{1, 2\},$$

with the hypothesis of 'cooperativity':

$$a^{12} \leq 0, \qquad a^{21} \leq 0, \qquad b^{12} \leq 0, \qquad b^{21} \leq 0.$$

As usual, C^{1-} means 'Lipschitz continuous'.

Denoting by χ^r the characteristic function of Γ_1^r and introducing matrix notation,

$$\chi := \begin{bmatrix} \chi^1 & 0 \\ 0 & \chi^2 \end{bmatrix}, \qquad a := \begin{bmatrix} a^{11} & a^{12} \\ a^{21} & a^{22} \end{bmatrix}, \qquad b := \begin{bmatrix} b^{11} & b^{12} \\ b^{21} & b^{22} \end{bmatrix},$$

we can rewrite this system in the concise from

$$\begin{aligned} \mathcal{A}u &= -\Delta u + au, \\ \mathcal{B}u &= \chi(\partial_\nu u + bu) + (1 - \chi)u. \end{aligned} \tag{4}$$

Of course, the boundary operator is to be understood in the sense of traces.

We also define the formally adjoint problem $(\mathcal{A}^\sharp, \mathcal{B}^\sharp)$ by

$$\begin{aligned} \mathcal{A}^\sharp v &:= -\Delta v + a^\top v, \\ \mathcal{B}^\sharp v &:= \chi(\partial_\nu v + b^\top v) + (1 - \chi)v, \end{aligned}$$

where $a^\top$ is the transposed of a, etc.

We endow all spaces of functions with their natural point- and component-wise defined order.

First we consider the eigenvalue problem

$$\mathcal{A}u = \lambda u \ \text{ in } \Omega, \qquad \mathcal{B}u = 0 \ \text{ on } \Gamma. \tag{5}$$

It will be shown that every eigenfunction of $(\mathcal{A}, \mathcal{B})$, that is, of (5), associated with any eigenvalue λ is regular in the sense that it belongs to

$$W_{\infty-}^2(\Omega, \mathbb{C}^2) := \bigcap_{1<q<\infty} W_q^2(\Omega, \mathbb{C}^2).$$

Furthermore, our results guarantee that $(\mathcal{A}, \mathcal{B})$, that is, problem (5), possesses a least real eigenvalue, the *principal eigenvalue*, $\lambda_0(\mathcal{A}, \mathcal{B})$, *of* $(\mathcal{A}, \mathcal{B})$, and it is associated with a positive eigenfunction.

Although there may exist other real eigenvalues of $(\mathcal{A}, \mathcal{B})$ possessing positive eigenfunctions, the principal eigenvalue characterizes the validity of the maximum principle for $(\mathcal{A}, \mathcal{B})$. More precisely, for $1 < q < \infty$ we put

$$W_{q',\mathcal{B}^\sharp}^2(\Omega, \mathbb{R}^2) := \{ v \in W_{q'}^2(\Omega, \mathbb{R}^2) \,;\, \mathcal{B}^\sharp v = 0 \}.$$

We say that $(\mathcal{A}, \mathcal{B})$ satisfies the *very weak maximum principle* if

$$\left. \begin{aligned} &u \in L_q(\Omega, \mathbb{R}^2) \text{ for some } q \in (1, \infty), \\ &\langle \mathcal{A}^\sharp v, u \rangle \geq 0 \text{ for all } v \in W_{q',\mathcal{B}^\sharp}^2(\Omega, \mathbb{R}^2) \text{ with } v \geq 0 \end{aligned} \right\} \quad \text{imply } u \geq 0. \tag{6}$$

Theorem 1. $(\mathcal{A}, \mathcal{B})$ *satisfies the very weak maximum principle iff* $\lambda_0(\mathcal{A}, \mathcal{B}) > 0$.

Since the principal eigenvalue of the Dirichlet Laplacean is positive, Theorem 1 is an extension of (3) to system (4).

As already mentioned, this theorem is new, even in the well-studied scalar case (where obvious analogues of the theorems of this introduction are valid). Indeed, to the best of our knowledge the very weak maximum principle has not been observed so far. Also note that there is no restriction on q, besides $1 < q < \infty$.

In order to guarantee that the principle eigenvalue is the only one with a positive eigenfunction we have to impose an additional condition. For this, the pair (a, b), more precisely: $(a, \chi b \chi)$, is said to be *irreducible* if

$$b^{12} | \Gamma_1^1 \cap \Gamma_1^2 = 0 \text{ implies } a^{12} \neq 0$$

and

$$b^{21} | \Gamma_1^1 \cap \Gamma_1^2 = 0 \text{ implies } a^{21} \neq 0,$$

putting

$$b^{12} | \varnothing := b^{21} | \varnothing := 0.$$

Note that these conditions can be rewritten as

$$a^{12} \neq 0 \quad \text{if} \quad \chi^1 b^{12} \chi^2 = 0,$$
$$a^{21} \neq 0 \quad \text{if} \quad \chi^2 b^{21} \chi^1 = 0.$$

For example, the pair (a, b) with

$$a = \begin{bmatrix} a^{11} & 0 \\ a^{21} & a^{22} \end{bmatrix}, \qquad b = \begin{bmatrix} b^{11} & b^{12} \\ 0 & b^{22} \end{bmatrix}$$

is irreducible if $a^{21} \neq 0$ and $b^{12} | \Gamma_1^1 \cap \Gamma_1^2 \neq 0$. Then the following improvement over the mere existence of a principal eigenvalue with a positive eigenfunction is valid.

Theorem 2. *Let (a, b) be irreducible. Then $\lambda_0(\mathcal{A}, \mathcal{B})$ is a simple eigenvalue of $(\mathcal{A}, \mathcal{B})$ and the only one with a positive eigenfunction.*

We refer to the main body of this paper for a precise definition of the simplicity of an eigenvalue of $(\mathcal{A}, \mathcal{B})$ and for further properties of $\lambda_0(\mathcal{A}, \mathcal{B})$ and the associated eigenfunction.

If (a, b) is irreducible then we obtain another useful characterization of the positivity of the principal eigenvalue. For this we say that u is a *very weak strict supersolution for* $(\mathcal{A}, \mathcal{B})$, provided

$$u \in L_q(\Omega, \mathbb{R}^2) \text{ for some } q \in (1, \infty) \text{ and}$$

$$\langle \mathcal{A}^\sharp v, u \rangle \geq 0 \text{ for all } v \in W^2_{q', \mathcal{B}^\sharp} \text{ with } v \geq 0,$$

with a strict inequality sign for at least one v.

It follows from Green's formula that u is a very weak strict supersolution for $(\mathcal{A}, \mathcal{B})$ if $u \in W_q^2(\Omega, \mathbb{R}^2)$ for some $q \in (1, \infty)$ and $(\mathcal{A}u, \mathcal{B}v) > 0$, meaning, of course, that $\mathcal{A}u \geq 0$ in Ω, $\mathcal{B}u \geq 0$ on Γ, and $(\mathcal{A}u, \mathcal{B}u) \neq (0, 0)$.

Theorem 3. *Let (a, b) be irreducible. Then $(\mathcal{A}, \mathcal{B})$ satisfies the very weak maximum principle iff there exists a positive very weak strict supersolution for $(\mathcal{A}, \mathcal{B})$.*

Theorems 2 and 3 (and their more general versions presented in Sections 6 and 7) generalize considerably the results of [56] and [71]. Indeed, besides of the fact that those authors consider only Dirichlet boundary conditions, our regularity hypotheses are substantially weaker than theirs. In particular, in Theorem 3 we are dealing with very weak supersolutions only.

In the following section we formulate the hypotheses used throughout most of this paper and give a precise formulation of the differential operators under consideration. In Section 3 we fix some general notations and describe the boundary spaces for our systems.

Our main results — very weak, weak, and strong maximum principles and their interrelations as well as monotonicity and continuity properties of the principal eigenvalue — are contained in Sections 4–11, where only the more elementary proofs are given. The somewhat deeper statements as well as additional results are proved in Sections 15–18.

In Section 12 we collect some functional analytical tools, and in Section 13 we recall the version of the maximum principle for scalar equations from which we derive our results for systems. Section 14 contains the fundamental solvability results for nonhomogeneous problems in the strong, weak, and very weak setting.

For all these results we impose enough regularity on the coefficients of the differential operators to guarantee that the assertions are independent of $q \in (1, \infty)$. In Section 19 we present weak maximum principles in W_q^1, assuming minimal q-dependent regularity only. They lead to comparison theorems for semilinear elliptic boundary value problems which are of importance in the study of such problems in situations where strong solutions do not exist.

We also show that the various realizations of cooperative elliptic systems generate positive analytic semigroups. These results have important implications for parabolic problems. Since this paper is already rather long we refrain from giving details.

2. Elliptic boundary value problems

In this section we give precise formulations of the elliptic problems under consideration and state the hypotheses used throughout the following, unless explicitly stated otherwise.

We assume that

$$N \in \mathbb{N}^\times := \mathbb{N} \setminus \{0\}.$$

The space of real $N \times N$ matrices, $a = [a^{rs}]$, is denoted by $\mathbb{R}^{N \times N}$, and $\mathbb{R}^{N \times N}_{\text{diag}}$ is the linear subspace of all diagonal matrices,

$$a = \text{diag}\,[a^1, \ldots, a^N].$$

We always use the summation convention with respect to j and k belonging to $\{1, \ldots, n\}$.

We also assume that

$$\left.\begin{array}{l} \bullet \quad a_{jk} = a_{kj} \in BUC^1(\Omega, \mathbb{R}^{N \times N}_{\text{diag}}),\ 1 \leq j, k \leq n; \\[4pt] \bullet \quad \boldsymbol{a}^r(x) := [a^r_{jk}(x)] \in \mathbb{R}^{n \times n} \text{ is positive definite for } 1 \leq r \leq N, \\[4pt] \qquad \text{uniformly with respect to } x \in \Omega; \\[4pt] \bullet \quad a_j \in W^1_\infty(\Omega, \mathbb{R}^{N \times N}_{\text{diag}}),\ 1 \leq j \leq n; \\[4pt] \bullet \quad a \in L_\infty(\Omega, \mathbb{R}^{N \times N}),\ b \in C^{1-}(\Gamma, \mathbb{R}^{N \times N}); \\[4pt] \bullet \quad -a \text{ and } -b \text{ are cooperative, that is, } a^{rs} \leq 0 \text{ and } b^{rs} \leq 0 \text{ for } r \neq s. \end{array}\right\} \quad (7)$$

Observe that there are no sign restrictions for the diagonal entries, neither for a nor for b.

We consider the elliptic differential operator $\mathcal{A}$ acting on $\mathbb{R}^N$-valued distributions

$$u := (u^1, \ldots, u^N)$$

on Ω, defined by

$$\mathcal{A}u := -\partial_j(a_{jk}\partial_k u) + a_j\partial_j u + au. \qquad (8)$$

Thus

$$\mathcal{A}u = (\mathcal{A}^1 u, \ldots, \mathcal{A}^N u),$$

where

$$\mathcal{A}^r u = -\partial_j(a^r_{jk}\partial_k u^r) + a^r_j\partial_j u^r + \sum_{s=1}^{N} a^{rs}u^s.$$

Note that $\mathcal{A}$ has diagonal principal and first order parts, but is coupled in its lowest order terms.

We fix

$$\chi \in C(\Gamma, \mathbb{R}^{N \times N}_{\text{diag}}) \qquad \text{with } \chi^r(y) \in \{0, 1\} \text{ for } y \in \Gamma \text{ and } 1 \leq r \leq N,$$

a **boundary identification map** for Ω. Hence

$$\Gamma^r_i := (\chi^r)^{-1}(i), \quad i \in \{0, 1\},$$

are open in Γ and disjoint with union Γ for $1 \leq r \leq N$. Then we define a boundary operator $\mathcal{B}$ by

$$\mathcal{B}u := \chi(\partial_\mu u + bu) + (1 - \chi)u, \qquad (9)$$

where

$$\partial_\mu := \text{diag}\,[\partial_{\mu^1}, \ldots, \partial_{\mu^N}]$$

with $\mu^r := \boldsymbol{a}^r\nu$ being the outer conormal with respect to $\boldsymbol{a}^r$. Of course, $\mathcal{B}$ is to be understood in the sense of traces. It follows that

$$\mathcal{B}u = (\mathcal{B}^1 u, \ldots, \mathcal{B}^N u),$$

where

$$\mathcal{B}^r u = \begin{cases} u^r & \text{on } \Gamma_0^r, \\ \partial_{\mu^r} u^r + \sum_{s=1}^N b^{rs} u^s & \text{on } \Gamma_1^r. \end{cases}$$

Thus $\mathcal{B}^r u$ is the Dirichlet boundary operator on Γ_0^r for the r-th component u^r of u, and a Neumann or Robin boundary operator for u^r on Γ_1^r. Note, however, that these boundary operators are coupled in their lowest order terms unless

$$\Gamma_0^1 = \cdots = \Gamma_0^N = \Gamma,$$

that is, unless $\mathcal{B} = \gamma$, the Dirichlet boundary operator. We express these facts by saying that $(\mathcal{A}, \mathcal{B})$ is a (weakly coupled second order) **cooperative elliptic boundary value problem** (on Ω).

If $n = 1$ then either $\Gamma = \{x_0\}$ or $\Gamma = \{x_0, x_1\}$ with $x_i \in \mathbb{R}$. Thus every space of $\mathbb{R}^N$ valued functions on Γ is naturally isomorphic to either $\mathbb{R}^N$ or $\mathbb{R}^N \times \mathbb{R}^N$, and all considerations of this paper apply with the obvious interpretations.

We put

$$\mathcal{A}^\sharp v := -\partial_j (a_{jk} \partial_k v + a_j v) + a^\top v$$

and

$$\mathcal{B}^\sharp v := \chi \left[\partial_\mu v + (\nu^j a_j + b^\top) v \right] + (1 - \chi) v.$$

Then $(\mathcal{A}^\sharp, \mathcal{B}^\sharp)$ is the elliptic boundary value problem formally adjoint to $(\mathcal{A}, \mathcal{B})$. We also put

$$\langle u, v \rangle := \int_\Omega u \cdot v \, dx, \qquad (u, v) \in L_{q'}(\Omega, \mathbb{R}^N) \times L_q(\Omega, \mathbb{R}^N). \qquad (10)$$

Similarly, $\langle \cdot, \cdot \rangle_\Gamma$ is the $L_q(\Gamma)$ duality pairing, obtained by replacing Ω and dx in (10) by Γ and $d\sigma$, respectively, $d\sigma$ being the volume measure of Γ. Then the Dirichlet form $\mathfrak{a}$ of $(\mathcal{A}, \mathcal{B})$ is defined by

$$\mathfrak{a}(v, u) := \langle \partial_j v, a_{jk} \partial_k u \rangle + \langle v, a_j \partial_j u + au \rangle + \langle \gamma v, \chi b \gamma u \rangle_\Gamma$$

for $(v, u) \in W_{q'}^1(\Omega, \mathbb{R}^N) \times W_q^1(\Omega, \mathbb{R}^N)$.

3. Notations and conventions

Henceforth, as long as Ω is kept fixed, we use the following simplified notation: if $E(\Omega, \mathbb{R}^N)$ is a vector subspace of

$$L_{1,\text{loc}} := L_{1,\text{loc}}(\Omega, \mathbb{R}^N)$$

then we denote it simply by E. For example,

$$W_q^k := W_q^k(\Omega, \mathbb{R}^N) \quad \text{for } k \in \mathbb{N}.$$

Similarly, we simply write $E(\Gamma)$ for $E(\Gamma, \mathbb{R}^N)$ if the latter is a vector subspace of

$$L_1(\Gamma) := L_1(\Gamma, \mathbb{R}^N).$$

We endow $L_{1,\text{loc}}(\Omega, \mathbb{R})$ with its natural 'point-wise' order induced by the positive cone

$$L_{1,\text{loc}}^+(\Omega, \mathbb{R}) := \{ u \in L_{1,\text{loc}}(\Omega, \mathbb{R}) \; ; \; u(x) \geq 0 \text{ a.a. } x \in \Omega \}. \tag{11}$$

Similarly, $L_1(\Gamma, \mathbb{R})$ is ordered by $L_1^+(\Gamma, \mathbb{R})$, the latter being defined in analogy to (11), but with respect to the volume measure $d\sigma$ of Γ.

If E is an ordered vector space then we write E^+ for its positive (proper) cone and $e > 0$ means $e \geq 0$ but $e \neq 0$, of course. If F is a linear subspace of E then F is given the induced order whose positive cone is $F^+ := E^+ \cap F$. Furthermore, if $E_1, \ldots, E_m$ are ordered vector spaces then $E_1 \times \cdots \times E_m$ is given the product order, that is, $(e_1, \ldots, e_m) \geq 0$ iff $e_i \geq 0$ for $1 \leq i \leq m$. Consequently, every vector subspace E of $L_{1,\text{loc}}$, or of $L_1(\Gamma)$, is an ordered vector space with respect to the naturally induced 'point-wise' product order. If E and F are ordered vector spaces then a linear map T from E into F is said to be positive, we write $T \geq 0$, if $T(E^+) \subset F^+$. Note, for example, that, consequently, $a \geq 0$ for $a \in L_\infty(\Omega, \mathbb{R}^{N \times N})$ means that $a^{rs}(x) \geq 0$ for $1 \leq r, s \leq N$ and a.a. $x \in \Omega$.

As a rule, in this paper all vector spaces are over the reals. However, if there occur, explicitly or implicitly, complex numbers in a given formula then it is always understood that the corresponding statement refers to the complexified version of that formula. For example, if A is a linear operator in a (real) Banach space then $\sigma(A)$, $\sigma_p(A)$, and $\rho(A)$ denote the spectrum, the point spectrum, and the resolvent set, respectively, of the complexification of A. For $\lambda \in \sigma_p(A)$ we denote by $N_A(\lambda)$ the algebraic eigenspace,

$$N_A(\lambda) := \bigcup_{k=1}^{\infty} \ker\big[(\lambda - A)^k\big],$$

of λ. Recall that $\dim\big(N_A(\lambda)\big)$ is the (algebraic) multiplicity of A, and λ is a simple eigenvalue if its multiplicity equals 1.

Let E and F be Banach spaces. Then $\mathcal{L}(E, F)$ is the Banach space of all bounded linear maps from E into F, and $\mathcal{L}(E) := \mathcal{L}(E, E)$. Moreover, $\mathcal{L}\text{is}(E, F)$ is the set of all isomorphisms in $\mathcal{L}(E, F)$. If

$$E \stackrel{d}{\hookrightarrow} F,$$

that is, E is continuously and densely embedded in F, then $\mathcal{H}(E, F)$ is the subset of all $A \in \mathcal{L}(E, F)$ such that $-A$, considered as a linear operator in F with domain E, generates a strongly continuous analytic semigroup, denoted by

$$\mathcal{U}_A := \{ U_A(t) \; ; \; t \geq 0 \},$$

on F, that is, in $\mathcal{L}(F)$. Given $A \in \mathcal{H}(E, F)$, there exists $\omega \in \mathbb{R}$ such that $[\text{Re } z > \omega]$ belongs to $\rho(-A)$, and the infimum of all such ω is the **spectral bound**, $s(-A)$, of A (e.g., [8, Section I.1.2]). We also set

$$\lambda_0(A) := -s(-A) = \inf\{ \text{Re } \lambda \; ; \; \lambda \in \sigma(A) \},$$

where $\inf(\varnothing) := \infty$. If E is an ordered Banach space (*OBS*) then the semigroup $\mathcal{U}_A$ is said to be positive if $U_A(t) \geq 0$ for $t \geq 0$.

We always assume that

$$1 < q < \infty.$$

For $|s| \leq 2$ we put

$$W_q^s(\Gamma_j) := \prod_{r=1}^{N} W_q^s(\Gamma_j^r), \qquad j = 0, 1,$$

with the understanding that $W_q^s(\varnothing) := \{0\}$. Then $W_q^s(\Gamma_j)$ is an ordered Banach space with its point-wise order if $s \geq 0$. If $s < 0$ then we endow $W_q^s(\Gamma_j)$ with the natural dual order whose positive cone is the dual (that is, polar) of $W_{q'}^{-s}(\Gamma_j)^+$. Then the injection maps

$$W_q^s(\Gamma_j) \overset{d}{\hookrightarrow} W_q^t(\Gamma_j), \qquad -2 \leq t < s \leq 2, \quad j = 0, 1,$$

are well-defined and positive. Finally, we put

$$\partial W_q^s := W_q^{s-1/q}(\Gamma_0) \oplus W_q^{s-1-1/q}(\Gamma_1), \qquad 0 \leq s \leq 2.$$

This means that we consider $W_q^{s-1}(\Gamma_0)$ and $W_q^{s-1-1/q}(\Gamma_1)$ as linear subspaces of $W_q^{s-1-1/q}(\Gamma)$, by extending the corresponding elements by zero over Γ, and endow their algebraic direct sum

$$\Sigma := W_q^{s-1/q}(\Gamma_0) + W_q^{s-1-1/q}(\Gamma_1)$$

with the unique topology for which

$$W_q^{s-1/q}(\Gamma_0) \times W_q^{s-1-1/q}(\Gamma_1) \to \Sigma, \qquad (g_0, g_1) \mapsto g_0 + g_1 \tag{12}$$

is a topological isomorphism. Then ∂W_q^s is an ordered Banach space with the unique order for which (12) and its inverse are positive.

4. Weak maximum principles

Using Sobolev embeddings, the trace theorem, and Hölder's inequality, it follows that

$$(\mathcal{A}, \mathcal{B}) \in \mathcal{L}(W_q^2, L_q \times \partial W_q^2). \tag{13}$$

Thus

$$W_{q,\mathcal{B}}^2 := \{ u \in W_q^2 \; ; \; \mathcal{B}u = 0 \}$$

is a closed linear subspace of W_q^2, hence a Banach space with

$$W_{q,\mathcal{B}}^2 \overset{d}{\hookrightarrow} L_q. \tag{14}$$

We denote by $A := A_{(q)}$ the L_q **realization** of $(\mathcal{A}, \mathcal{B})$, defined by $A := \mathcal{A} | W_{q,\mathcal{B}}^2$. Similarly, $A^\sharp$ is the $L_{q'}$ realization of $(\mathcal{A}^\sharp, \mathcal{B}^\sharp)$.

Theorem 4. $A \in \mathcal{H}(W_{q,\mathcal{B}}^2, L_q)$ and $A^\sharp \in \mathcal{H}(W_{q',\mathcal{B}^\sharp}^2, L_{q'})$ for $1 < q < \infty$. Moreover, $A^\sharp = A'$.

Proof. The first assertion is well-known (cf. [7, Theorem 4.1] and the references therein; also see [34]). Since $\gamma a_j \in C^\rho(\Gamma, \mathbb{R}^{N \times N})$ for every $\rho \in (0, 1)$, it follows from the preceding references that

$$A^\sharp \in \mathcal{H}(W^2_{q', \mathcal{B}^\sharp}, L_{q'}).$$

The last assertion is a consequence of Green's formula and the fact that A and $A^\sharp$ have a common point in their resolvent sets (even a half-plane, of course). □

Remarks 5.

(a) The assumption that $a_j \in W^1_\infty(\Omega, \mathbb{R}^{N \times N})$ guarantees that the assertions about $A^\sharp$ are valid. If we are only interested in

$$A \in \mathcal{H}(W^2_{q, \mathcal{B}}, L_q) \tag{15}$$

and not in the explicit representation of A' then it suffices to suppose that

$$a_j \in L_\infty(\Omega, \mathbb{R}^{N \times N}_{\text{diag}}), \qquad 1 \le j \le n. \tag{16}$$

It should be observed that, in either case, Theorem 4 and (15) hold for every $q \in (1, \infty)$, although $A_{(q)}$ depends on q, of course.

(b) It follows from (15) that $\lambda_0(A)$ is well-defined. □

We also set

$$W^1_{q, \mathcal{B}} := W^1_{q, (1-\chi)\gamma} := \{ u \in W^1_q ; (1 - \chi)\gamma u = 0 \}.$$

Note that

$$W^1_{q', \mathcal{B}^\sharp} = W^1_{q', (1-\chi)\gamma}.$$

In the following, $(\mathcal{A}, \mathcal{B})$ is said to satisfy the **very weak maximum principle** (in L_q) if, given any u such that

$$u \in L_q, \quad \langle A^\sharp v, u \rangle \ge 0 \text{ for all } v \in (W^2_{q', \mathcal{B}^\sharp})^+, \tag{17}$$

it follows that $u \ge 0$. It satisfies the **weak maximum principle** (in W^1_q) if it is a consequence of

$$u \in W^1_q, \quad \mathfrak{a}(v, u) \ge 0 \text{ for all } v \in (W^1_{q', (1-\chi)\gamma})^+, \quad (1 - \chi)\gamma u \ge 0 \tag{18}$$

that $u \ge 0$. Lastly, $(\mathcal{A}, \mathcal{B})$ satisfies the **maximum principle** or is **inverse positive** (on W^2_q) if

$$u \in W^2_q, \quad \mathcal{A}u \ge 0 \text{ in } \Omega, \quad \mathcal{B}u \ge 0 \text{ on } \Gamma$$

imply $u \ge 0$.

Theorem 6.

(1) *Consider the following assertions:*

(i) $(\mathcal{A}, \mathcal{B})$ *satisfies the very weak maximum principle in* L_q.
(ii) $(\mathcal{A}, \mathcal{B})$ *satisfies the weak maximum principle in* W^1_q.
(iii) $(\mathcal{A}, \mathcal{B})$ *is inverse positive on* W^2_q.

Then (i) *implies* (ii), *and* (ii) *implies* (iii). *If, in addition,* A *is surjective then all these assertions are equivalent.*

(2) *If* $\lambda_0 := \lambda_0(A) > 0$ *then* (A, B) *is inverse positive. Conversely, if* (A, B) *is inverse positive and* A *is surjective then* $\lambda_0 > 0$.

(3) *The semigroup* $\mathcal{U}_A$ *is positive.*

(4) *If* $\sigma(A) \neq \varnothing$ *then* $\lambda_0 \in \sigma(A)$.

The proof of this theorem is given in Section 16.

Remarks 7.

(a) Although our regularity assumptions guarantee that (15) holds for every $q \in (1, \infty)$, we do not know whether λ_0 is independent of q. This would follow from the spectral invariance of elliptic operators. However, the known results (see [13], [31], [45], [50], [70]) do not seem to apply to the present situation. It is also not known whether $\sigma(A) \neq \varnothing$, in general.

(b) Suppose that only the weaker assumption (16) is satisfied. Then Theorem 6 remains valid, provided we omit assertion (i) in (1). $\square$

We emphasize the fact that the results of this section are true under the mere assumption that Ω has a compact boundary. We are not aware of any related theorem valid for the case of exterior domains.

5. Nonhomogeneous problems

Of course, the validity of a maximum principle has implications on the solvability of nonhomogeneous elliptic boundary value problems. This is made precise in the present section.

We put
$$W_{q,\mathcal{B}}^{-2} := (W_{q',\mathcal{B}^\sharp}^2)', \quad W_{q,\mathcal{B}}^{-1} := W_{q,(1-\chi)\gamma}^{-1} := (W_{q',(1-\chi)\gamma}^1)'$$
with respect to the duality pairings $\langle \cdot, \cdot \rangle$ naturally induced by (10). It follows that
$$W_{q,\mathcal{B}}^2 \xhookrightarrow{d} W_{q,\mathcal{B}}^1 \xhookrightarrow{d} L_q \xhookrightarrow{d} W_{q,\mathcal{B}}^{-1} \xhookrightarrow{d} W_{q,\mathcal{B}}^{-2}. \tag{19}$$
We endow $W_{q,\mathcal{B}}^{-k}$ for $k \in \{1, 2\}$ with the natural dual order whose positive cone is the dual of the cone $(W_{q',\mathcal{B}^\sharp}^k)^+$. Then $W_{q,\mathcal{B}}^{-1}$ and $W_{q,\mathcal{B}}^{-2}$ are *OBS*s and each one of the injection maps in (19) is positive.

Next we consider the nonhomogeneous boundary value problem
$$Au = f \text{ in } \Omega, \quad Bu = g \text{ on } \Gamma. \tag{20}$$
A **(strong)** W_q^2 **solution** is a $u \in W_q^2$ with
$$(Au, Bu) = (f, g) \quad \text{in } L_q \times \partial^2 W_q.$$

By a **(weak)** W_q^1 **solution** we mean a $u \in W_q^1$ satisfying

$$\left.\begin{array}{ll} \mathfrak{a}(v, u) = \langle v, f \rangle & \text{for } v \in W_{q',(1-\chi)\gamma}^1, \\[2mm] (1 - \chi)\gamma u = (1 - \chi)g & \text{on } \Gamma. \end{array}\right\} \tag{21}$$

Lastly, u is said to be a **(very weak)** L_q **solution** of (20) if $u \in L_q$ and

$$\langle \mathcal{A}^\sharp v, u \rangle = \langle v, f \rangle + \langle \partial_\mu v, (\chi - 1)g \rangle_\Gamma + \langle \gamma v, \chi g \rangle_\Gamma \tag{22}$$

for $v \in W_{q',\mathcal{B}^\sharp}^2$.

The following theorem gives a further characterization of the positivity of λ_0.

Theorem 8. *The following are equivalent:*

(i) $\lambda_0 > 0$.

(ii) *Problem (20) has for each* $(f, g) \in (L_q \times \partial W_q^2)^+$ *a unique nonnegative* W_q^2 *solution.*

(iii) *Problem (20) has for each* $(f, g) \in (W_{q,\mathcal{B}}^{-1} \times \partial W_q^1)^+$ *a unique nonnegative* W_q^1 *solution.*

(iv) *Problem (20) has for each* $(f, g) \in (W_{q,\mathcal{B}}^{-2} \times \partial W_q^0)^+$ *a unique nonnegative* L_q *solution.*

The proof of this theorem is also given in Section 16.

Remark 9. If we presuppose only the weaker hypothesis (16) then Theorem 8 remains valid if assertion (iv) is omitted. $\qquad\square$

6. The principal eigenvalue

Throughout this section we suppose that Ω is bounded. Then we can considerably improve on the results of the preceding section. For this we put

$$W_{\infty-}^2 := \bigcap_{1 < q < \infty} W_q^2.$$

We also assume that *only the weaker hypothesis* (16) *is satisfied.*

Theorem 10. *Suppose that* Ω *is bounded. Then A has a compact resolvent. Hence* $\sigma(A)$ *is discrete and each* $\lambda \in \sigma(A)$ *is an eigenvalue of finite multiplicity. Moreover,*

$$N_A(\lambda) \subset W_{\infty-}^2$$

for $\lambda \in \sigma(A)$.

This theorem, whose proof is found in Section 17, shows, in particular, that the spectrum, the eigenspaces, and the generalized eigenspaces are independent of $q \in (1, \infty)$. Thus we say that λ is an eigenvalue of $(\mathcal{A}, \mathcal{B})$ iff there exists $u \in W_{\infty-}^2 \setminus \{0\}$ satisfying

$$\mathcal{A}u = \lambda u \text{ in } \Omega, \quad \mathcal{B}u = 0 \text{ on } \Gamma. \tag{23}$$

The multiplicity of λ is, by definition, the multiplicity of λ as an eigenvalue of the L_q realization of $(\mathcal{A}, \mathcal{B})$ for some $q \in (1, \infty)$. Thanks to Theorem 10 this definition is independent of q. Thus it makes sense to say that $\sigma(A)$ is the **spectrum of** $(\mathcal{A}, \mathcal{B})$, that is, $\sigma(\mathcal{A}, \mathcal{B}) := \sigma(A)$. If $\sigma(\mathcal{A}, \mathcal{B}) \neq \varnothing$ then it follows from Theorem 6 that λ_0 is the smallest eigenvalue of $(\mathcal{A}, \mathcal{B})$, the **principal eigenvalue**, also denoted by $\lambda_0(\mathcal{A}, \mathcal{B})$, or, more precisely, by $\lambda_0(\mathcal{A}, \mathcal{B}, \Omega)$. The following theorem, whose proof is also given in Section 17, guarantees its existence.

Theorem 11. *If Ω is bounded then $\sigma(\mathcal{A}, \mathcal{B}) \neq \varnothing$, and the principal eigenvalue has a positive eigenfunction. Furthermore, $\lambda_0 > 0$ iff $(\mathcal{A}, \mathcal{B})$ is inverse positive.*

In general, λ_0 is not the only eigenvalue with a positive eigenfunction. This follows immediately by considering diagonal (that is, uncoupled) systems. Thus, in order to guarantee that λ_0 is the only eigenvalue with a positive eigenfunction we have to ascertain that the coupling is sufficiently strong. For this we need some preparation.

Let R be a ring with unit. Then $\pi \in R^{N \times N}$ is a permutation matrix iff it contains exactly one unit in every row and every column, and zeros elsewhere. The matrix $\alpha \in R^{N \times N}$ is reducible iff there exist $M \in \{1, \ldots, N-1\}$ and a permutation matrix π such that

$$\pi \alpha \pi^\top = \begin{bmatrix} \alpha_{11} & \alpha_{12} \\ 0 & \alpha_{22} \end{bmatrix} \tag{24}$$

with $\alpha_{11} \in R^{M \times M}$, and α is irreducible otherwise. It is well-known and not difficult to see that α is irreducible iff, given $r, s \in \{1, \ldots, N\}$, there exist indices $s_\rho \in \{1, \ldots, N\}$, $0 \leq \rho \leq k$, with $s_0 = s$ and $s_k = r$ such that $\alpha^{\xi\eta} \neq 0$ for (ξ, η) belonging to $\{(s_1, s_0), (s_2, s_1), \ldots, (s_k, s_{k-1})\}$.

Let R_1 and R_2 be rings with unit 1_1 and 1_2, respectively. Then $R_1 \times R_2$ is a ring, the product ring, with multiplication being defined component-wise. It has a unit, namely $(1_1, 1_2)$. Suppose that $\alpha_i \in R_i^{N \times N}$ for $i = 1, 2$. Then we denote by

$$[(\alpha_1, \alpha_2)] \in (R_1 \times R_2)^{N \times N}$$

the matrix whose entry at position (r, s) equals $(\alpha_1^{rs}, \alpha_2^{rs}) \in R_1 \times R_2$.

Note that $L_\infty(\Omega, \mathbb{R})$ and $C^{1-}(\Gamma, \mathbb{R})$ are rings with unit 1_Ω and 1_Γ, respectively, where, given any nonempty set X, we denote by 1_X the constant map $x \mapsto 1$. We also set

$$a^\triangle := \text{diag} [a^{11}, \ldots, a^{NN}], \quad b^\triangle := \text{diag} [b^{11}, \ldots, b^{NN}] \tag{25}$$

and

$$a^\bullet := a^\triangle - a, \quad b^\bullet := b^\triangle - b. \tag{26}$$

Observe that $a^\bullet$ and $b^\bullet$ are nonnegative and have zeros in the diagonals.

Suppose that $N \geq 2$. Then the pair (a, b), more precisely: $(a, \chi b \chi)$, is said to be **irreducible** iff

$$[(a^\bullet, \chi b^\bullet \chi)] \in \left(L_\infty(\Omega, \mathbb{R}) \times C^{1-}(\Gamma, \mathbb{R})\right)^{N \times N}$$

is irreducible. Thus (a, b) is irreducible iff, given any $r, s \in \{1, \ldots, N\}$, there exist $s_\rho \in \{1, \ldots, N\}$, $0 \leq \rho \leq k$, with $s_0 = s$ and $s_k = r$ such that

$$\text{either} \quad \chi^\xi b^{\xi\eta} \chi^\eta > 0 \quad \text{or} \quad a^{\xi\eta} > 0$$

for

$$(\xi, \eta) \in \{(s_1, s_0), (s_2, s_1), \ldots, (s_k, s_{k-1})\}.$$

If this is the case, we also say that $(\mathcal{A}, \mathcal{B})$ is irreducible.

A function $u \in L_{1,\text{loc}}$ is **strongly positive** if $u \in C^1(\overline{\Omega})$ and satisfies for each $r \in \{1, \ldots, N\}$ the inequalities $u^r(x) > 0$ for $x \in \Omega \cup \Gamma_1^r$ and $\partial_{\mu^r} u^r(y) < 0$ for $y \in \Gamma_0^r$ with $u^r(y) = 0$.

The next theorem is the basis for a detailed study of the principal eigenvalue. Its proof is found in Section 17.

Theorem 12. *Let Ω be bounded and suppose that either $N = 1$ or (a, b) is irreducible. Then the principal eigenvalue is simple and has a strongly positive eigenfunction. It is the only eigenvalue with a positive eigenfunction, and every other eigenvalue satisfies*

$$\text{Re}\,\lambda > \lambda_0.$$

In the scalar case this theorem is well-known and has first been proved, in the case of general boundary conditions and without a positivity restriction for b, in [3]. That proof contained a gap since it had not been asserted that the spectrum is nonempty. Motivated by this, de Pagter [33] derived a general theorem on irreducible compact positive operators on Banach lattices implying that such an operator has a strictly positive spectral radius. That theorem can be used to fill the gap (cf. the proof of [11, Theorem 2.2]). It has also been employed in papers by Sweers and coauthors (cf. [20], [71]) in the case of irreducible cooperative elliptic systems with Dirichlet boundary conditions to prove essentially Theorem 12 (under more restrictive regularity hypotheses). The proof given in Section 17 is much more elementary. It is solely based on the classical Krein-Rutman theorem and does not invoke the rather deep de Pagter result. The case of Dirichlet boundary conditions has also been considered in [56], but under the much stronger assumption that $a^{rs}(x) > 0$ for $r \neq s$ and all $x \in \Omega$, where a is supposed to be continuous.

7. The strong maximum principle

The boundary value problem $(\mathcal{A}, \mathcal{B})$ is said to satisfy the **strong maximum principle** if

$$q > n, \quad u \in W_q^2 \setminus \{0\}, \quad \mathcal{A}u \geq 0 \text{ in } \Omega, \quad \mathcal{B}u \geq 0 \text{ on } \Gamma$$

imply that u is strongly positive.

We say that u is a **strict** W_q^2 **supersolution** for $(\mathcal{A}, \mathcal{B})$ if u belongs to W_q^2 and

$$(\mathcal{A}u, \mathcal{B}u) > 0.$$

It is a **strict** W_q^1 **supersolution** for $(\mathcal{A}, \mathcal{B})$ if it satisfies (18) with a strict inequality sign either in the first place for at least one v or in the second place. Finally, u is a **strict** L_q **supersolution** for $(\mathcal{A}, \mathcal{B})$ if it satisfies (17) with a strict inequality sign for at least one v. It follows that each W_q^2 supersolution is a W_q^1 supersolution, and each W_q^1 supersolution is an L_q supersolution.

Using these concepts the following characterizations of the maximum principles are valid.

Theorem 13. *Let Ω be bounded and suppose that either $N = 1$ or (a, b) is irreducible. Then the following are equivalent:*

  (i) $\lambda_0 > 0$.

  (ii) $(\mathcal{A}, \mathcal{B})$ *satisfies the very weak maximum principle.*

  (iii) $(\mathcal{A}, \mathcal{B})$ *satisfies the weak maximum principle.*

  (iv) $(\mathcal{A}, \mathcal{B})$ *satisfies the strong maximum principle.*

  (v) $(\mathcal{A}, \mathcal{B})$ *possesses a positive strict L_q supersolution.*

Remark 14. Assume that only condition (16) is satisfied. Then Theorem 13 remains valid, provided the following modifications are implemented: assertion (ii) is omitted and 'L_q supersolution' in (v) is replaced by 'W_q^1 supersolution'. $\qquad\square$

Given the assumptions of the previous theorem, we can also improve on the solvability statements of Theorem 8. For this, $u \in L_q$ is said to be **strictly positive** if $u^r(x) > 0$ for a.a. $x \in \Omega$ and each $r \in \{1, \ldots, N\}$.

Theorem 15. *Suppose that Ω is bounded and that either $N = 1$ or (a, b) is irreducible. If $\lambda_0 > 0$ then, given any $(f, g) \in (L_q \times \partial W_q^2)^+ \setminus \{0\}$, the unique positive solution of (20) is strictly positive. If $q > n$ then it is strongly positive.*

The important new ingredient distinguishing Theorem 13 from Theorem 6 is the characterization of the validity of the maximum principle through the existence of a positive strict supersolution. In the scalar case $N = 1$ and in the framework of classical solutions, Protter and Weinberger seem to be the first to observe that the existence of a strict classical supersolution

$$w \in C^2(\Omega) \cap C^1(\Omega \cup \Gamma_1) \cap C(\overline{\Omega})$$

satisfying

$$w(x) > 0 \quad \text{for all} \quad x \in \overline{\Omega}$$

implies the existence of the maximum principle for $(\mathcal{A}, \mathcal{B})$ (cf. [63, Section II.5]). In the usual weak $\overset{\circ}{H}{}^1$ setting and with weak regularity assumptions it has also been shown in [23] that the existence of a positive H^1 supersolution characterizes the weak maximum principle for the Dirichlet problem. The fact that (i) and (iv) of Theorem 13 as well as the existence of a positive strict classical supersolution w, such that $w_r(x) > 0$ for all $x \in \overline{\Omega}$ and $1 \leq r \leq N$, are equivalent has first been observed in [56] in the case of Dirichlet problems for cooperative elliptic systems satisfying the strong irreducibility condition explained in the preceding section. In the scalar case, López-Gómez [52] could then relax the hypotheses on a classical strict supersolution w by requiring $w(y) \geq 0$ for y in Γ, keeping the assumption $w(x) > 0$ for $x \in \Omega$. This result has been extended to the case of a general boundary operator in [11, Theorem 2.4], where positive strict W_q^2 supersolutions with $q > n$ are being considered. The proof in [11] relies on [3, Theorem 6.1] which, in turn, is a consequence of the Protter-Weinberger result cited above and a construction of a strict W_q^2 supersolution w for $q > n$ satisfying $w(x) > 0$ for all $x \in \overline{\Omega}$ (see [3, Lemma 5.1]), The latter construction is somewhat involved and complicated (but see Remark 36(a)). This prompted López-Gómez [55] to give a simpler proof of Theorem 2.4 in [11] in the framework of $C^{2+\alpha}$ solutions by extending a version of the maximum principle due to Walter [76]. It should be remarked that in none of those results $(\mathcal{A}, \mathcal{B})$ is required to possess divergence form.

The fact that a strict positive supersolution in the class $W_n^2(\Omega, \mathbb{R}^N) \cap C(\overline{\Omega}, \mathbb{R}^N)$ implies the maximum principle and the existence of a unique positive eigenfunction is the main theorem in [71] for the Dirichlet problem of irreducible cooperative systems (also see [20]).

Our Theorem 13 is much more general since it applies to systems with general coupled boundary conditions and replaces W_q^2 supersolutions for $q \geq n$ by the much weaker concept of L_q supersolutions.

The importance of the results of this section is seen from the theorems in Sections 8–11. Furthermore, it should be noted that the results of Theorems 13 and 15 suffice to apply the abstract techniques of [1] to irreducible cooperative elliptic systems. By this way one obtains extensions of the existence, multiplicity, and bifurcation results contained in [1]. For example, one can extend the three solutions theorem [1, Theorem 14.2] to such systems, etc. We leave the details to interested readers.

It should also be remarked that, thanks to Theorems 12 and 15 (and their proofs), it is not difficult to extend the anti-maximum principle of Clément and Peletier [26] (also see [72]) to cooperative irreducible systems of the form $(\mathcal{A}, \mathcal{B})$. Details are also left to the readers.

8. Monotonicity of the principal eigenvalue

Throughout this section Ω *is again bounded and only the weaker hypothesis* (16) *is imposed.*

We discuss monotonicity properties of the principal eigenvalue with respect to variations of a, b, Ω, and the boundary conditions.

First we suppose that $\tilde{a} \in L_\infty(\Omega, \mathbb{R}^{N \times N})$ and $\tilde{b} \in C^{1-}(\Gamma, \mathbb{R}^{N \times N})$ are cooperative. Then we define $(\tilde{\mathcal{A}}, \tilde{\mathcal{B}})$ by replacing a and b in (8) and (9) by $\tilde{a}$ and $\tilde{b}$, respectively.

Theorem 16. *Suppose that either $N = 1$ or (a, b) and $(\tilde{a}, \tilde{b})$ are irreducible. If*

$$(a, \chi b\chi) < (\tilde{a}, \chi \tilde{b}\chi) \tag{27}$$

then

$$\lambda_0(\mathcal{A}, \mathcal{B}) < \lambda_0(\tilde{\mathcal{A}}, \tilde{\mathcal{B}}).$$

Proof. Let u_0 be a positive eigenfunction of $(\mathcal{A}, \mathcal{B})$ to the eigenvalue

$$\lambda_0 := \lambda_0(\mathcal{A}, \mathcal{B}).$$

Then

$$(\tilde{\mathcal{A}} - \lambda_0)u_0 = (\mathcal{A} - \lambda_0)u_0 + (\tilde{a} - a)u_0 = (\tilde{a} - a)u_0 \qquad \text{in } \Omega$$

and

$$\tilde{\mathcal{B}}u_0 = \mathcal{B}u_0 + \chi(\tilde{b} - b)\chi\gamma u_0 = \chi(\tilde{b} - b)\chi\gamma u_0 \qquad \text{on } \Gamma.$$

Hence it follows from (27) and the strong positivity of u_0, guaranteed by Theorem 12, that u_0 is a positive strict W_q^2 supersolution for $(\tilde{\mathcal{A}} - \lambda_0, \mathcal{B})$. Thus

$$0 < \lambda_0(\tilde{\mathcal{A}} - \lambda_0, \tilde{\mathcal{B}}) = \lambda_0(\tilde{\mathcal{A}}, \tilde{\mathcal{B}}) - \lambda_0$$

by Theorem 13 and Remark 14. □

The next theorem shows, in particular, that the principle eigenvalue decreases strictly if a Dirichlet boundary condition is replaced for at least one component of u by a Neumann boundary condition on at least one component of Γ. To make this precise we suppose that $\hat{\chi} \in C(\Gamma, \mathbb{R}_{\text{diag}}^{N \times N})$ is a boundary identification map. Then we put

$$\hat{\mathcal{B}}u := \hat{\chi}(\partial_\mu u + bu) + (1 - \hat{\chi})u.$$

Note that $\chi < \hat{\chi}$ means that $\Gamma_0^r \supset \hat{\Gamma}_0^r$ for $1 \leq r \leq N$ where at least one of the inclusions is proper.

Theorem 17. *Suppose that either $N = 1$ or $(a, \chi b\chi)$ is irreducible. Also suppose that*

$$\chi < \hat{\chi} \quad \text{and} \quad (\hat{\chi} - \chi)b^\triangle \leq 0. \tag{28}$$

Then

$$\lambda_0(\mathcal{A}, \mathcal{B}) > \lambda_0(\mathcal{A}, \hat{\mathcal{B}}).$$

Proof. First observe that the irreducibility of $(a, \chi b\chi)$ and $\chi < \hat{\chi}$ imply the one of $(a, \hat{\chi} b\hat{\chi})$. Thus, by Theorem 12, there exists a strongly positive eigenfunction $\hat{u}_0$ of $(\mathcal{A}, \hat{\mathcal{B}})$ to the eigenvalue

$$\hat{\lambda}_0 := \lambda_0(\mathcal{A}, \hat{\mathcal{B}}).$$

Thus, given $v \in (W^1_{q',(1-\chi)\gamma})^+$,

$$\mathfrak{a}(v, \widehat{u}_0) - \widehat{\lambda}_0 \langle v, \widehat{u}_0 \rangle = \widehat{\mathfrak{a}}(v, \widehat{u}_0) - \widehat{\lambda}_0 \langle v, \widehat{u}_0 \rangle + \langle \gamma v, (\chi b - \widehat{\chi} b)\gamma \widehat{u}_0 \rangle_\Gamma$$

where $\widehat{\mathfrak{a}}$ is the Dirichlet form of $(\mathcal{A}, \widehat{\mathcal{B}})$. Since $W^1_{q',(1-\chi)\gamma} \subset W^1_{q',(1-\widehat{\chi})\gamma}$, thanks to $\chi < \widehat{\chi}$, it follows that

$$\widehat{\mathfrak{a}}(v, \widehat{u}_0) - \widehat{\lambda}_0 \langle v, \widehat{u}_0 \rangle = 0, \qquad v \in (W^1_{q',(1-\chi)\gamma})^+.$$

Also observe that

$$\langle \gamma v, (\chi - \widehat{\chi})b\gamma \widehat{u}_0 \rangle_\Gamma \geq -\langle \gamma v, (\widehat{\chi} - \chi)b^\triangle \gamma u_0 \rangle_\Gamma$$

since $-b^{rs} \geq 0$ for $r \neq s$. Hence we deduce from (28) that

$$\mathfrak{a}(v, \widehat{u}_0) - \widehat{\lambda}_0 \langle v, \widehat{u}_0 \rangle \geq 0, \qquad v \in (W^1_{q',(1-\chi)\gamma})^+.$$

From $\chi < \widehat{\chi}$ and the strong positivity of $\widehat{u}_0$ we also infer that

$$(1-\chi)\gamma \widehat{u}_0 > 0 \qquad \text{on } \Gamma.$$

Thus $\widehat{u}_0$ is a strict W^1_q supersolution for $(\mathcal{A} - \widehat{\lambda}_0, \mathcal{B})$. Hence

$$0 < \lambda_0(\mathcal{A} - \widehat{\lambda}_0, \mathcal{B}) = \lambda_0(\mathcal{A}, \mathcal{B}) - \widehat{\lambda}_0$$

by Theorem 13 and Remark 14. □

Note that (28) implies that $b^{rr} \leq 0$ on each component of Γ on which a Dirichlet boundary condition is replaced by a Robin one.

For example, it follows from Theorem 17 that, given that a is irreducible if $N > 1$, the principal eigenvalues for the pure Neumann , the general Robin, and the pure Dirichlet condition satisfy

$$\lambda_0(\mathcal{A}, \partial_\mu) < \lambda_0(\mathcal{A}, \mathcal{B}) < \lambda_0(\mathcal{A}, \gamma),$$

provided $b^{rr} \leq 0$ for $1 \leq r \leq N$ with at least one strict inequality sign. Also note that, thanks to the cooperativity assumption, the second inequality in this chain is consistent with Theorem 16.

Our next theorem shows that the principle eigenvalue increases if the domain shrinks and if on each boundary component being moved inside the original boundary condition is replaced by a Dirichlet one.

Theorem 18. *Let $\Omega^\star$ be a proper C^2 subdomain of Ω with boundary $\Gamma^\star$. Denote by Σ the union of all components of $\Gamma^\star$ having a nonempty intersection with Ω, and put*

$$\Sigma' := \Gamma^\star \setminus \Sigma.$$

Define a boundary identification map $\chi^\star \in C(\Gamma^\star, \mathbb{R}^{N \times N}_{\mathrm{diag}})$ for $\Omega^\star$ by

$$\chi^\star | \Sigma := 0, \quad \chi^\star | \Sigma' := \chi | \Sigma'$$

and put

$$([a^\star_{jk}], (a^\star_1, \ldots, a^\star_n), a^\star) := ([a_{jk}], (a_1, \ldots, a_n), a) | \Omega^\star$$

and

$$\mathcal{A}^{\star} u := \mathcal{A} u, \quad \mathcal{B}^{\star} u := \chi^{\star}(\partial_{\mu^{\star}} u + b u) + (1 - \chi^{\star}) u$$

for $u \in W_q^2(\Omega^{\star}, \mathbb{R}^N)$. *Then*

$$\lambda_0(\mathcal{A}, \mathcal{B}, \Omega) < \lambda_0(\mathcal{A}^{\star}, \mathcal{B}^{\star}, \Omega^{\star})$$

provided $(a^{\star}, \chi^{\star} b \chi^{\star})$ *is irreducible if* $N > 1$.

Proof. Observe that the irreducibility of $(a^{\star}, \chi^{\star} b \chi^{\star})$ implies the one of $(a, \chi b \chi)$. Also note that the strong positivity of a positive eigenfunction u of $(\mathcal{A}, \mathcal{B})$ to the eigenvalue $\lambda_0 := \lambda_0(\mathcal{A}, \mathcal{B}, \Omega)$ implies that $u|\Sigma > 0$. Hence u is a positive strict W_q^2 supersolution for $(\mathcal{A}^{\star} - \lambda_0, \mathcal{B}^{\star}, \Omega^{\star})$. Now the assertion follows once more from Theorem 13 and Remark 14.   □

The theorems of this section, as well as their proofs, are more or less straightforward extensions and sharpenings of corresponding results established in the scalar case (i.e., $N = 1$) by López-Gómez [52] for Dirichlet and by Cano-Casanova and López-Gómez [21] for general boundary conditions (also see [37] and, for the Dirichlet problem in a weak setting, [24]). In the particular case of Theorem 16 where $\chi = 0$ and $\widetilde{\chi} = 0$, that is, for Dirichlet problems, the monotonicity of λ_0 as a function of a has also been shown in [56], given the much more restrictive assumption that $a, \widetilde{a} \in C(\overline{\Omega}, \mathbb{R}^{N \times N})$ and satisfy $\widetilde{a} \geq a$ and $\widetilde{a}^{rs}(x_0) > a^{rs}(x_0)$ for some $x_0 \in \Omega$ and all $r, s \in \{1, \ldots, N\}$.

The weak Dirichlet problem in $W_{2,\gamma}^1$ for scalar equations has attracted a lot of interest. In this case general perturbation theorems are due to Arendt and coauthors [14], [17], Stollmann [69] and, in particular, Daners [28], [29], who has perhaps the most general results. For the weak Robin boundary value problem in the scalar case we refer to [27].

9. Continuity of the principal eigenvalue

We suppose again that Ω is bounded and put

$$\mathbb{E}^1(\Omega) := C^1(\overline{\Omega}, \mathbb{R}_{\text{diag}}^{N \times N})^{n \times n} \times L_{\infty}(\Omega, \mathbb{R}_{\text{diag}}^{N \times N})^n \times L_{\infty}(\Omega, \mathbb{R}^{N \times N})$$
$$\times C^{1-}(\Gamma, \mathbb{R}^{N \times N}) \times C(\Gamma, \mathbb{R}_{\text{diag}}^{N \times N}).$$

Given

$$\alpha := ([a_{jk}], (a_1, \ldots, a_n), a, b, \chi) \in \mathbb{E}^1(\Omega),$$

we define $(\mathcal{A}(\alpha), \mathcal{B}(\alpha))$ on Ω by (8) and (9). Then we denote by $\mathcal{E}(\Omega)$ the set of all $\alpha \in \mathbb{E}^1(\Omega)$ such that χ is a boundary characterization map for Ω and $(\mathcal{A}(\alpha), \mathcal{B}(\alpha))$ is a cooperative elliptic boundary value problem on Ω such that (a, b) is irreducible if $N > 1$.

Let $\underset{\sim}{\Omega}$ be a bounded C^2 domain in $\mathbb{R}^n$ and let $\varphi : \overline{\Omega} \to \overline{\underset{\sim}{\Omega}}$ be a C^2 diffeomorphism. Then $\varphi_{\partial} := \varphi|\Gamma$ is a C^2 diffeomorphism from Γ onto $\partial \underset{\sim}{\Omega}$. Given

$$\underset{\sim}{\alpha} := ([\underset{\sim}{a}_{jk}], (\underset{\sim}{a}_1, \ldots, \underset{\sim}{a}_n), \underset{\sim}{a}, \underset{\sim}{b}, \underset{\sim}{\chi}) \in \mathbb{E}^1(\underset{\sim}{\Omega}),$$

we put

$$\varphi^* \underset{\sim}{\alpha} := \left([\varphi^* \underset{\sim}{a}_{jk}], (\varphi^* \underset{\sim}{a}_1, \ldots, \varphi^* \underset{\sim}{a}_n), \varphi^* \underset{\sim}{a}, \varphi^*_{\partial} \underset{\sim}{b}, \varphi^*_{\partial} \underset{\sim}{\chi}\right) \in \mathbb{E}^1(\Omega),$$

where, for a function $\underset{\sim}{f}$ on $\underset{\sim}{\Omega}$, the pull back of $\underset{\sim}{f}$ by φ is defined by

$$\varphi^* \underset{\sim}{f} := \underset{\sim}{f} \circ \varphi.$$

Note that $\varphi^* \underset{\sim}{\alpha} \in \mathcal{E}(\Omega)$ if $\underset{\sim}{\alpha} \in \mathcal{E}(\underset{\sim}{\Omega})$.

Let (Ω_i) be a sequence of bounded C^2 domains in $\mathbb{R}^n$. Then it is said to be C^1 converging towards Ω if there exist orientation preserving C^2 diffeomorphisms $\varphi_i : \overline{\Omega} \to \overline{\Omega}_i$ such that $\varphi_i \to \mathrm{id}_{\overline{\Omega}}$ in $C^1(\overline{\Omega}, \mathbb{R}^N)$. Such a sequence (φ_i) is said to be a representation sequence for (Ω_i).

Sequences of bounded C^2 domains being C^1 convergent towards Ω are often obtained by deformations of Ω be means of sequences of transformation groups for $\overline{\Omega}$. This is true, in particular, if the transformation group is generated by a C^2 vector field.

Example 19. Suppose that Ω is bounded and $f \in C^2(\overline{\Omega}, \mathbb{R}^n)$. For each $x \in \overline{\Omega}$ let $\varphi(\tau, x)$ be the solution at time τ of the initial value problem

$$\dot{\xi} = f(\xi), \quad \xi(0) = x. \tag{29}$$

Then there exist $\tau^- < 0 < \tau^+$ such that

$$\varphi \in C^2\left((\tau^-, \tau^+) \times \overline{\Omega}, \mathbb{R}^n\right), \tag{30}$$

and $\varphi^\tau := \varphi(\tau, \cdot)$ is for each $\tau \in (\tau^-, \tau^+)$ an orientation preserving C^2 diffeomorphism from $\overline{\Omega}$ onto $\overline{\Omega}_\tau := \varphi^\tau(\overline{\Omega})$. Thus $\Omega_\tau := \varphi^\tau(\Omega)$ is a C^2 domain in $\mathbb{R}^n$ and, given any sequence (τ_j) in (τ^-, τ^+) with $\tau_j \to 0$, the sequence (Ω_{τ_j}) is C^1 converging (in fact: C^2 converging) towards Ω, and (φ_{τ_j}) is a representation sequence for (Ω_{τ_j}).

Proof. The theory of ordinary differential equations (e.g., [6, Section 10]) implies these assertions. $\square$

Now we fix $r, s, t \in [1, \infty]$ satisfying

$$(r, s, t) = \begin{cases} (n, n/2, n-1) & \text{if } n \geq 3, \\ \in (2, \infty] \times (1, \infty] \times (1, \infty] & \text{if } n = 2, \\ (1, 1, 1) & \text{if } n = 1, \end{cases} \tag{31}$$

and put

$$\mathbb{E}_{r,s,t}(\Omega) := C(\overline{\Omega}, \mathbb{R}^{N \times N}_{\mathrm{diag}})^{n \times n} \times L_r(\Omega, \mathbb{R}^{N \times N}_{\mathrm{diag}})^n \times L_s(\Omega, \mathbb{R}^{N \times N})$$
$$\times L_t(\Gamma, \mathbb{R}^{N \times N}) \times C(\Gamma, \mathbb{R}^{N \times N}_{\mathrm{diag}}).$$

Note that $\mathbb{E}^1(\Omega)$ and $\mathbb{E}_{r,s,t}(\Omega)$ are Banach spaces satisfying $\mathbb{E}^1(\Omega) \hookrightarrow \mathbb{E}_{r,s,t}(\Omega)$.

After these preparations we can formulate the following general continuity theorem for the principle eigenvalue, whose proof is given in Section 18. Note that we allow not only all coefficients to vary but the domain as well.

Theorem 20. *Let* (31) *be satisfied. Suppose that* (Ω_i) *is a sequence of bounded* C^2 *domains* C^1 *converging towards* Ω, *and let* (φ_i) *be a representation sequence for* (Ω_i). *Also suppose that* $\alpha \in \mathcal{E}(\Omega)$ *and* $\alpha_i \in \mathcal{E}(\Omega_i)$ *such that* $(\varphi_i^* \alpha_i)$ *converges in* $\mathbb{E}_{r,s,t}(\Omega)$ *towards* α. *Then*

$$\lambda_0\big(\mathcal{A}(\alpha_i), \mathcal{B}(\alpha_i), \Omega_i\big) \to \lambda_0\big(\mathcal{A}(\alpha), \mathcal{B}(\alpha), \Omega\big) \quad as \ i \to \infty.$$

Furthermore, if u, *resp.* u_i, *is the unique positive eigenfunction of* $\big(\mathcal{A}(\alpha), \mathcal{B}(\alpha)\big)$, *resp.* $\big(\mathcal{A}(\alpha_i), \mathcal{B}(\alpha_i)\big)$, *of* W_2^1 *norm* 1 *then* $\varphi_i^* u_i \to u$ *in* W_2^1.

It should be remarked that $\varphi_i^* u_i \to u$ in W_q^1, provided r, s, and t are replaced by suitably chosen numbers ξ, η, and ζ depending on q (cf. Theorem 47).

Example 21. Suppose that Ω is bounded and $f \in C^1(\overline{\Omega}, \mathbb{R}^n)$. Let φ be the flow defined by (29) and fix $\tau^- < 0 < \tau^+$ such that (30) is true. Let V be an open neighborhood of

$$\bigcup_{\tau^- < \tau < \tau^+} \overline{\Omega}_\tau \times \{\tau\}$$

in $\mathbb{R}^n \times \mathbb{R}$ and suppose that $\big([\overline{a}_{jk}], (\overline{a}_1, \ldots, \overline{a}_n), \overline{a}, \overline{b}, \overline{\chi}\big)$ belongs to

$$C^1(V, \mathbb{R}^{N \times N}_{\mathrm{diag}})^{n \times n} \times C(V, \mathbb{R}^{N \times N}_{\mathrm{diag}})^n \times C(V, \mathbb{R}^{N \times N})$$
$$\times C(V, \mathbb{R}^{N \times N}) \times C(V, \mathbb{R}^{N \times N}_{\mathrm{diag}}).$$

For $\tau \in (\tau^-, \tau^+)$ put

$$\alpha(\tau) := \big([a_{jk}(\tau)], (a_1(\tau), \ldots, a_n(\tau)), a(\tau), b(\tau), \chi(\tau)\big),$$

where

$$a_{jk}(\tau) := \overline{a}_{jk}(\cdot, \tau)|\overline{\Omega}_\tau, \quad a_j(\tau) := \overline{a}_j(\cdot, \tau)|\overline{\Omega}_\tau, \quad a(\tau) := \overline{a}(\cdot, \tau)|\overline{\Omega}_\tau$$

and

$$b(\tau) := \overline{b}(\cdot, \tau)|\Gamma_\tau, \quad \chi(\tau) := \overline{\chi}(\cdot, \tau)|\Gamma_\tau$$

with $\Gamma_\tau := \partial\Omega_\tau = \varphi^\tau(\Gamma)$. Note that $\alpha(\tau) \in \mathbb{E}^1(\Omega_\tau)$.

Suppose that

$$a(\tau) \in \mathcal{E}(\Omega_\tau), \quad (\varphi^\tau)^* \chi(\tau) = \chi := \chi(0), \quad \tau^- < \tau < \tau^+,$$

and put

$$(\mathcal{A}_\tau, \mathcal{B}_\tau) := \big(\mathcal{A}(\alpha(\tau)), \mathcal{B}(\alpha(\tau))\big)$$

so that $(\mathcal{A}, \mathcal{B}) = (\mathcal{A}_0, \mathcal{B}_0)$. Then

$$\lambda_0(\mathcal{A}_\tau, \mathcal{B}_\tau, \Omega_\tau) \to \lambda_0(\mathcal{A}, \mathcal{B}, \Omega) \quad as \ \tau \to 0.$$

If u_τ is the unique positive eigenfunction of $(\mathcal{A}_\tau, \mathcal{B}_\tau)$ satisfying $\|u_\tau\|_{W_2^1(\Omega_\tau)} = 1$ then $(\varphi^\tau)^* u_\tau \to u_0$ in W_2^1.

Proof. It is easily verified that $(\varphi^\tau)^*\alpha(\tau) \to \alpha = \alpha(0)$ as $\tau \to 0$. Hence the assertion is a consequence of Theorem 20. □

Next we present a second continuity theorem for which we need some preparation.

We recall that the C^2 manifold $\overline{\Omega}$ is C^2 diffeomorphic to a C^∞ manifold, the latter being unique up to C^∞ diffeomorphisms (e.g., [46, Theorem II.3.4]). Thus we can assume without loss of generality that $\overline{\Omega}$ is smooth. Then Γ is a compact oriented smooth hypersurface in $\mathbb{R}^n$. Hence there exists $\rho > 0$ such that, setting

$$T_\rho := \{ y + s\nu(y) \in \mathbb{R}^n \; ; \; y \in \Gamma, \; |s| < \rho \},$$

the map

$$T_\rho \to \Gamma \times (-1, 1), \quad y + s\nu(y) \mapsto (y, s/\rho) \tag{32}$$

is a smooth diffeomorphism. In other words, T_ρ is a normal tubular neighborhood of Γ (cf. [46, Theorem IV.5.2]). Note that $\Omega \cup T_\rho$ is the open ρ neighborhood of $\overline{\Omega}$, that is,

$$\Omega \cup T_\rho = \{ x \in \mathbb{R}^n \; ; \; \mathrm{dist}(x, \overline{\Omega}) < \rho \}.$$

Assume that $\beta \in C^2(\Gamma, (-\rho, \rho))$ and put

$$\Gamma_\beta := \{ y + \beta(y)\nu(y) \in \mathbb{R}^n \; ; \; y \in \Gamma \}.$$

Then Γ_β is an oriented C^2 hypersurface in $\mathbb{R}^n$ being contained in the tubular neighborhood T_ρ of Γ. There exists a unique C^2 domain, Ω_β, in $\mathbb{R}^n$ such that it is contained in $\Omega \cup T_\rho$ and Γ_β is its boundary. We also define $\psi_\beta \in C^2(\Gamma, T_\rho)$ by

$$\psi_\beta(y) := y + \beta(y)\nu(y), \quad y \in \Gamma.$$

Suppose that

$$\beta_i \in C^2(\Gamma, (-\rho, \rho)), \quad i \in \mathbb{N}, \tag{33}$$

and put $\Omega_i := \Omega_{\beta_i}$ and $\Gamma_i := \Gamma_{\beta_i}$. Assume that

$$\alpha := \big([a_{jk}], (a_1, \ldots, a_n), a, b, \chi\big) \in \mathcal{E}(\Omega) \tag{34}$$

and

$$\alpha_i := \big([a_{jk,(i)}], (a_{1,(i)}, \ldots, a_{n,(i)}), a_{(i)}, b_{(i)}, \chi_{(i)}\big) \in \mathcal{E}(\Omega_i) \tag{35}$$

for $i \in \mathbb{N}$, where

$$a_{jk,(i)} = \overline{a}_{jk,(i)} | \overline{\Omega}$$

for some

$$\overline{a}_{jk,(i)} \in C^1(\Omega \cup T_\rho, \mathbb{R}^{N \times N}_{\mathrm{diag}}).$$

Furthermore, assume that

$$\overline{a}_{jk,(i)} \to \overline{a}_{jk} \quad \text{in } C^1(\Omega \cup T_\rho, \mathbb{R}^{N \times N}_{\mathrm{diag}}) \text{ as } i \to \infty, \tag{36}$$

where $\overline{a}_{jk} | \overline{\Omega} = a_{jk}$ for $1 \le j, k \le n$, that

$$\beta_i \to 0 \text{ in } C^1(\Gamma, \mathbb{R}) \quad \text{and} \quad \psi_i^* \chi_{(i)} = \chi, \tag{37}$$

and that

$$\int_{\Omega_i \cap \Omega} \left\{ \max_{1 \le j \le n} |a_{j,(i)} - a_j|^r + |a_{(i)} - a|^s \right\} dx \to 0 \tag{38}$$

as well as

$$\psi_i^* b_{(i)} \to b \qquad \text{in } L_t(\Gamma, \mathbb{R}^{N \times N}) \tag{39}$$

as $i \to \infty$.

Observe that conditions (36) and (38) are automatically satisfied if only the boundary of Ω is perturbed and $\mathcal{A}$ is kept fixed with its coefficients being defined on $\Omega \cup T_\rho$. Similarly, if $\bar{b} \in C^{1-}(\Omega \cup T_\rho, \mathbb{R}^{N \times N})$ and $b_{(i)} := \bar{b}|\Gamma_i$ for $i \in \mathbb{N}$ then condition (39) holds as well.

Theorem 22. *Let conditions* $(33) - (39)$ *be satisfied. Then*

$$\lambda_0(\mathcal{A}(\alpha_i), \mathcal{B}(\alpha_i), \Omega_i) \to \lambda_0(\mathcal{A}, \mathcal{B}, \Omega) \qquad \text{as } i \to \infty.$$

Furthermore, if the function u, *resp.* u_i, *is the unique positive eigenfunction of* $(\mathcal{A}, \mathcal{B})$, *resp.* $(\mathcal{A}(\alpha_i), \mathcal{B}(\alpha_i))$ *of norm* 1 *in* W_2^1, *resp.* $W_2^1(\Omega_i, \mathbb{R}^N)$, *then, given any compact subset* K *of* Ω,

$$\int_K \left\{ |u_i - u|^2 + |\nabla u_i - \nabla u|^2 \right\} dx \to 0 \tag{40}$$

as $i \to \infty$.

This theorem, which we derive in Section 18 from Theorem 20, is — except for regularity assumptions on the Dirichlet boundary — a generalization of the continuity result in [21], where the Robin boundary and the coefficients are kept fixed and the scalar case is considered only. Of course, in the case of Dirichlet boundary conditions for a scalar equation much more general results can be obtained as has been shown by Daners [28], [29] and others (see the remarks following Theorem 18). Also see [27] for the scalar Robin problem in a W_2^1 setting.

10. Minimax characterizations

Let Ω *be bounded and suppose that only the weaker hypothesis* (16) *is satisfied.*

In this section we give further characterizations of the principal eigenvalue of $(\mathcal{A}, \mathcal{B})$. Our first theorem is related to point-wise estimates.

Theorem 23. *Suppose hat either* $N = 1$ *or* (a, b) *is irreducible. Then*

$$\lambda_0(\mathcal{A}, \mathcal{B}) = \sup\{ \lambda \in \mathbb{R} ; \text{ there exist } q \in (1, \infty) \text{ and } u \in (W_q^2)^+$$
$$\text{satisfying } ((\mathcal{A} - \lambda)u, \mathcal{B}u) > 0 \}.$$

Proof. Suppose that $\lambda < \lambda_0 := \lambda_0(\mathcal{A}, \mathcal{B})$. Then

$$\lambda_0(\mathcal{A} - \lambda, \mathcal{B}) = \lambda_0 - \lambda > 0.$$

Hence, given $q \in (1, \infty)$, Theorem 8 guarantees the existence of $u \in (W_q^2)^+$ satisfying

$$((\mathcal{A} - \lambda)u, \mathcal{B}u) = (\mathbf{1}_\Omega, 0).$$

Thus, denoting by λ^* the above supremum, it follows that $\lambda_0 \leq \lambda^*$.

Suppose that $\varepsilon := (\lambda^* - \lambda_0)/2 > 0$. Then there exist $q \in (1, \infty)$ and $u \in (W_q^2)^+$ satisfying

$$((\mathcal{A} - (\lambda^* - \varepsilon))u, \mathcal{B}u) > 0.$$

Thus u is a positive strict supersolution for $(\mathcal{A} - \lambda^* + \varepsilon, \mathcal{B})$. Hence, by Theorem 13,

$$0 < \lambda_0(\mathcal{A} - \lambda^* + \varepsilon, \mathcal{B}) = \lambda_0 - \lambda^* + \varepsilon = -\varepsilon,$$

which is impossible. This proves $\lambda_0 = \lambda^*$. $\qquad\square$

Corollary 24. *Suppose that $N = 1$ and $n < q < \infty$. Then*

$$\lambda_0(\mathcal{A}, \mathcal{B}) = \sup_{u \in \mathcal{P}_\mathcal{B}} \inf_{x \in \Omega} \frac{\mathcal{A}u(x)}{u(x)},$$

where $\mathcal{P}_\mathcal{B}$ is the set of all $u \in W_q^2$ satisfying $u(x) > 0$ for $x \in \Omega$ and $\mathcal{B}u \geq 0$.

Proof. It follows from the preceding theorem by observing that Theorem 15 and the above proof show that it suffices to take the supremum with respect to all strongly positive u. $\qquad\square$

This corollary is Theorem 4.1 in [21]. A related result can be found in [20] for the Dirichlet problem for irreducible cooperative systems.

Now we prove a minimax characterization of $\lambda_0(\mathcal{A}, \mathcal{B})$ which is more in the spirit of the variational characterization of eigenvalues of self-adjoint operators.

Theorem 25. *Suppose that $N = 1$ or (a, b) is irreducible. Fix $q \in (1, \infty)$ and denote by $\mathcal{Q}$ the set of strictly positive $u \in W_q^1$ satisfying $(1 - \chi)\gamma u \geq 0$. Then*

$$\lambda_0(\mathcal{A}, \mathcal{B}) = \sup_{u \in \mathcal{Q}} \inf \frac{\mathfrak{a}(v, u)}{\langle v, u \rangle}, \tag{41}$$

the infimum being taken with respect to all nonzero $v \in (W_{q', (1-\chi)\gamma}^1)^+$.

Proof. Suppose that

$$\lambda < \lambda_0 := \lambda_0(\mathcal{A}, \mathcal{B}).$$

Then, as in the proof of Theorem 23, we see that

$$\lambda_0(\mathcal{A} - \lambda, \mathcal{B}) > 0$$

and there exists a positive strict W_q^2 supersolution u for $(\mathcal{A} - \lambda, \mathcal{B})$. Hence u is strictly positive by Theorem 15. From Green's formula we deduce that u is a positive strict W_q^1 supersolution for $(\mathcal{A} - \lambda, \mathcal{B})$. Consequently,

$$\mathfrak{a}(v, u) \geq \lambda \langle v, u \rangle, \qquad v \in (W_{q', (1-\chi)\gamma}^1)^+,$$

and $(1-\chi)\gamma u \geq 0$. Since $\langle v, u \rangle > 0$ for each $v > 0$ by the strict positivity of u, we see that

$$\lambda \leq \inf_{v \in (W^1_{q',(1-\chi)\gamma})^+ \setminus \{0\}} \frac{\mathfrak{a}(v, u)}{\langle v, u \rangle}.$$

This implies $\lambda_0 := \lambda_0(\mathcal{A}, \mathcal{B}) \leq \lambda^*$, with λ^* denoting the right-hand side of (41).

Suppose that $\varepsilon := (\lambda^* - \lambda)/2 > 0$. Then there exists $u \in \mathcal{Q}$ satisfying

$$\mathfrak{a}(v, u) > (\lambda^* - \varepsilon)\langle v, u \rangle, \qquad v \in (W^1_{q',(1-\chi)\gamma})^+ \setminus \{0\}.$$

Hence u is a positive strict W^1_q supersolution for $(\mathcal{A} - \lambda^* + \varepsilon, \mathcal{B})$. By invoking once more Theorem 13 we arrive at a contradiction as in the final part of the proof of Theorem 23. □

11. Concavity of the principal eigenvalue

Suppose again that Ω *is bounded and only assumption* (16) *is satisfied.*

In this section we give an application of the minimax characterization of Theorem 25 to the study of the behavior of $\lambda_0(\mathcal{A}, \mathcal{B})$ as a function of (a, b). For this we first note that, if $N \geq 2$, the set of all cooperative pairs (α, β) is a convex cone in the algebra

$$\left(L_\infty(\Omega, \mathbb{R}) \times C^{1-}(\Gamma, \mathbb{R})\right)^{N \times N}. \tag{42}$$

Given (a_j, b_j) in (42) for $j = 0, 1$, we put

$$(\mathcal{A}_t, \mathcal{B}_t) := (1-t)(\mathcal{A}_0, \mathcal{B}_0) + t(\mathcal{A}_1, \mathcal{B}_1), \qquad 0 \leq t \leq 1,$$

where $(\mathcal{A}_j, \mathcal{B}_j)$ are defined for $j \in \{0, 1\}$ by replacing a in (8) and b in (9) by a_j and b_j, respectively.

Theorem 26. *Suppose that* (a_0, b_0) *and* (a_1, b_1) *belong to* (42). *Also suppose that either* $N = 1$ *or* $(\mathcal{A}_t, \mathcal{B}_t)$ *is cooperative and irreducible for* $0 \leq t \leq 1$. *Then*

$$\lambda_0(\mathcal{A}_t, \mathcal{B}_t) \geq (1-t)\lambda_0(\mathcal{A}_0, \mathcal{B}_0) + t\lambda_0(\mathcal{A}_1, \mathcal{B}_1), \qquad 0 \leq t \leq 1. \tag{43}$$

Proof. Let $\mathfrak{a}_t$ be the Dirichlet form of $(\mathcal{A}_t, \mathcal{B}_t)$. Fix $q \in (1, \infty)$ and define $\mathcal{Q}$ as in the preceding theorem. Then, given $u \in \mathcal{Q}$ and $t \in (0, 1)$,

$$\mathfrak{a}_t(v, u) = (1-t)\mathfrak{a}_0(v, u) + t\mathfrak{a}_1(v, u), \qquad v \in (W^1_{q',(1-\chi)\gamma})^+,$$

implies

$$\inf_v \mathfrak{a}_t(v, u) \geq (1-t)\inf_v \mathfrak{a}_0(v, u) + t\inf_v \mathfrak{a}_1(v, u),$$

where the infima are taken with respect to

$$v \in (W^1_{q',(1-\chi)\gamma})^+ \setminus \{0\}.$$

By passing to the suprema with respect to $u \in \mathcal{Q}$ we deduce from Theorem 25 that (43) is true. □

Remark 27. Suppose that $N > 1$ and $-(a_0, b_0)$ and $-(a_1, b_1)$ are cooperative. Also suppose that either there exists a subset P of Ω of positive measure such that $a_0(x)$ and $a_1(x)$ are irreducible for all $x \in$ P, or there exists $y \in \Gamma$ with $\chi^r(y) = 1$ for $r \in \{1, \ldots, N\}$ such that $b_0(y)$ and $b_1(y)$ are irreducible. Then $(\mathcal{A}_t, \mathcal{B}_t)$ is irreducible for $0 \leq t \leq 1$. $\quad\square$

In the scalar case the concavity of the principal eigenvalue with respect to a has been shown in [21] generalizing earlier results of López-Gómez [52] and Berestycki, Nirenberg, and Varadhan [19] for the case of Dirichlet boundary conditions (also see [18], [44], and [47]). Note that our proof does not only apply to cooperative irreducible systems but is, even in the scalar case, much simpler than the earlier ones.

As shown in the scalar case in [21] and [52] (also see the references therein), Theorem 26 is an important tool for studying eigenvalue problems with weight functions. Thanks to Theorem 26 we can now consider Steklov type eigenvalue problems also where the eigenvalue parameter can occur in the boundary condition as well, that is, problems of the type

$$\mathcal{A}u = \lambda \alpha u \text{ in } \Omega, \quad \mathcal{B}u = \lambda \beta u \text{ on } \Gamma \tag{44}$$

with (α, β) belonging to (42). By means of the above theorem it can be shown that the map

$$\lambda \mapsto \lambda_0(\mathcal{A} - \lambda\alpha, \mathcal{B} - \lambda\beta) \tag{45}$$

is convex and analytic on $\mathbb{R}$ if $N = 1$, and on $\mathbb{R}^+$ if $N > 1$ and (a, b) and (α, β) are such that the latter theorem can be applied. The proofs in [21], [44], or [52] carry over without changes. Since the zeros of the function (45) correspond to eigenvalues of (44) having positive eigenfunctions, we can thus extend many of the results for the case $\beta = 0$ (and $N = 1$), known so far and cited above, to (44) (also see [35], [42] for recent developments).

We remind the reader that problems of type (44) occur naturally in the study of parameter dependent nonlinear boundary value problems of the form

$$\mathcal{A}u = \lambda f(x, u) \text{ in } \Omega, \quad \mathcal{B}u = \lambda g(x, u) \text{ on } \Gamma$$

with nonlinear boundary conditions (cf. [2] for a study of such problems in the scalar case and a regular setting).

12. Preparatory considerations

Let $E := E_0$ and E_1 be Banach spaces such that $E_1 \overset{d}{\hookrightarrow} E_0$, and suppose that E_0 is ordered and $A \in \mathcal{H}(E_1, E_0)$. Then A is said to be resolvent positive if there exists $\omega \in \mathbb{R}$ such that

$$(\omega, \infty) \subset \rho(-A) \quad \text{and} \quad (\lambda + A)^{-1} \geq 0 \text{ for } \lambda > \omega.$$

We denote by $s_+(-A)$ the infimum of all such ω.

The next two theorems are basically known and included for easy reference only.

Theorem 28.

(a) *The following are equivalent:*

    (i) *A is resolvent positive.*

    (ii) $(\lambda + A)^{-1} \geq 0$ *for* $\lambda > s(-A)$, *that is,* $s_+(-A) \leq s(-A)$.

    (iii) *The semigroup* $\mathcal{U}_A$ *is positive.*

(b) *If A is resolvent positive and* $\sigma(A) \neq \varnothing$ *then* $-s_+(-A) \in \sigma(A)$.

Proof. (a) It is known (e.g., [8, Remark II.5.1.2] or [25, Proposition 9.2]) that $s(-A)$ equals the type of $-A$, the latter being the infimum of all $\omega \in \mathbb{R}$ such that there exists $M \geq 1$ satisfying

$$\|U_A(t)\|_{\mathcal{L}(E_0)} \leq M e^{\omega t}$$

for $t \geq 0$. Hence the assertion follows from well-known results in semigroup theory (eg., [25, Proposition 7.1]).

(b) This follows from part (b) of the proof of Proposition 3.11.2 in [15] (by observing that it is valid without the standing hypotheses of those authors that the positive cone is normal and generating). $\qquad\square$

The positive cone E^+ is said to be generating if $E = E^+ - E^+$. It is normal if each order interval

$$[x, y] := \{ z \in E \, ; \, x \leq z \leq y \}$$

is bounded. Note, in particular, that E^+ is normal and generating if E is a Banach lattice.

Theorem 29. *Suppose that* E^+ *is normal and generating. Then:*

    (i) $s_+(-A) = s(-A)$;

    (ii) *If* $\lambda \in \rho(-A) \cap \mathbb{R}$ *and* $(\lambda + A)^{-1} \geq 0$ *then* $\lambda > s(-A)$.

Proof. See [15, Proposition 3.11.2] or [25, Theorem 7.4]. $\qquad\square$

The operator A is said to be inverse positive if $u \in E_1$ and $Au \geq 0$ imply $u \geq 0$.

Corollary 30. *Let* E *be a Banach lattice and suppose that* A *is resolvent positive. If* $\lambda_0(A) > 0$ *then* A *is inverse positive. Conversely, if* A *is inverse positive and* A *is surjective then* $\lambda_0(A) > 0$.

Proof. If A is inverse positive then it is injective. Thus $0 \in \rho(A)$ if, in addition, A is surjective. Now the assertion is obvious. $\qquad\square$

Let E be a Banach lattice. Then $x \in E^+$ is said to be a quasi-interior point of E^+ iff $\langle \varphi, x \rangle > 0$ for every $\varphi \in (E')^+ \setminus \{0\}$. Clearly, u is a quasi-interior point of L_q^+ iff u is strictly positive.

A positive linear operator $K \in \mathcal{L}(E)$ is called irreducible iff there exists $\mu > r(K)$, with $r(K)$ being the spectral radius of K, such that $K(\mu - K)^{-1}x$ is for each $x \in E^+ \setminus \{0\}$ a quasi-interior point of E^+. It is strongly irreducible iff Kx is for each $x \in E^+ \setminus \{0\}$ a quasi-interior point of E^+. Clearly, every strongly irreducible operator is irreducible.

The semigroup $\mathcal{U}_A$ is said to be strongly irreducible iff $U_A(t)$ has this property for each $t > 0$.

Theorem 31. *Suppose that E is a Banach lattice and A is resolvent positive. Then the following are equivalent:*

(i) *There exists $\lambda > -\lambda_0$ such that $(\lambda + A)^{-1}$ is irreducible.*

(ii) *$(\lambda + A)^{-1}$ is for each $\lambda > -\lambda_0$ strongly irreducible.*

(iii) *$\mathcal{U}_A$ is strongly irreducible.*

Proof. It follows from [65, App. 3.1] that the preceding definition of a positive irreducible bounded linear operator is equivalent to the one used in [25, Section 7.1]. Hence the assertion is implied by [25, Proposition 7.6 and Corollary 7.8]. □

Suppose that

$$E_0 \overset{d}{\hookrightarrow} E_{-1}, \quad A_{-1} \in \mathcal{H}(E_0, E_{-1}), \quad A_{-1} \supset A_0 := A. \tag{46}$$

Then we denote for $j \in \{-1, 0\}$ by $N_j(\lambda)$ the algebraic eigenspace of the eigenvalue λ of A_j.

Lemma 32. *Let (46) be satisfied. Then*

$$\sigma_p(A_{-1}) = \sigma_p(A_0).$$

Furthermore,

$$N_{-1}(\lambda) = N_0(\lambda), \quad \lambda \in \sigma_p(A_0).$$

Proof. (i) It is clear that $\sigma_p(A_0) \subset \sigma_p(A_{-1})$ and $N_0(\lambda) \subset N_{-1}(\lambda)$ for $\lambda \in \sigma_p(A_0)$.

(ii) Suppose that there exist $\lambda \in \mathbb{C}$ and $x, y \in E_0$ satisfying

$$A_{-1}x = \lambda x - y. \tag{47}$$

Fix $\omega \in \rho(-A_0) \cap \rho(-A_{-1})$. Then (47) is equivalent to

$$x = (\omega + A_{-1})^{-1}((\omega + \lambda)x - y). \tag{48}$$

From $A_{-1} \supset A_0$ it follows that

$$(\omega + A_{-1})^{-1} \supset (\omega + A_0)^{-1}.$$

Thus we deduce from $(\omega + \lambda)x - y \in E_0$ and (48), thanks to $(\omega + A_0)^{-1}(E_0) \subset E_1$, that $x \in E_1$ and, consequently, $A_0x = \lambda x - y$.

(iii) Suppose that $\lambda \in \mathbb{C}$, $m \in \mathbb{N}$, and

$$x \in \ker\big[(\lambda - A_{-1})^{m+1}\big].$$

Then there exist $x_0, \ldots, x_m \in E_0$ satisfying $x_0 = x$ and

$$A_{-1}x_k = \lambda x_k - x_{k+1}, \qquad 0 \le k \le m,$$

where $x_{m+1} := 0$. Thus we deduce from (ii) by backwards induction that $x_k \in E_1$ for $0 \le k \le m$ and that

$$x_0 \in \ker\big[(\lambda - A_0)^{m+1}\big].$$

This implies $N_{-1}(\lambda) \subset N_0(\lambda)$. $\square$

Next we prove a perturbation theorem for resolvent positive generators of analytic semi-groups.

Proposition 33. *Suppose that $\theta \in (0,1)$ and $(\cdot,\cdot)_\theta$ is an interpolation functor of exponent θ. Put $E_\theta := (E_0, E_1)_\theta$. If $B \in \mathcal{L}(E_\theta, E_0)$ then*

$$A - B \in \mathcal{H}(E_1, E_0).$$

If A is resolvent positive and $B \ge 0$ then $A - B$ is also resolvent positive.

Proof. Since $A \in \mathcal{H}(E_1, E_0)$ there exist positive constants M and ω such that

$$|\lambda|^{1-j}\,\|(\lambda + A)^{-1}\|_{\mathcal{L}(E_0, E_j)} \le M, \qquad \mathrm{Re}\,\lambda > \omega, \quad j = 0, 1.$$

Thus, by interpolation, there exists M_1 such that

$$\|(\lambda + A)^{-1}\|_{\mathcal{L}(E_0, E_\theta)} \le M_1/|\lambda|^{1-\theta}, \qquad \mathrm{Re}\,\lambda > \omega.$$

Hence we can find $\omega_1 \ge \omega$ such that

$$\|B(\lambda + A)^{-1}\|_{\mathcal{L}(E_0)} \le 1/2, \qquad \mathrm{Re}\,\lambda > \omega_1.$$

It follows that $1 - B(\lambda + A)^{-1} \in \mathcal{L}(E_0)$ has an inverse on E_0, bounded by 2, and

$$\big(1 - B(\lambda + A)^{-1}\big)^{-1} = \sum_{j=0}^\infty \big[B(\lambda + A)^{-1}\big]^j, \qquad \mathrm{Re}\,\lambda > \omega_1, \tag{49}$$

in $\mathcal{L}(E_0)$. Hence

$$[\mathrm{Re}\,\lambda > \omega_1] \subset \rho(-A + B)$$

and

$$(\lambda + A - B)^{-1} = (\lambda + A)^{-1}\big(1 - B(\lambda + A)^{-1}\big)^{-1}, \qquad \mathrm{Re}\,\lambda > \omega_1, \tag{50}$$

so that

$$\|(\lambda + A - B)^{-1}\|_{\mathcal{L}(E_0, E_j)} \le 2M_1/|\lambda|^{1-j}, \qquad \mathrm{Re}\,\lambda > \omega_1, \quad j = 0, 1.$$

This proves that $A - B \in \mathcal{H}(E_1, E_0)$. Furthermore, if $B \ge 0$ and A is resolvent positive, we deduce from (49) and (50) that $(\lambda + A - B)^{-1} \ge 0$ for $\lambda > \omega_1$. $\square$

It should be remarked that the first part of the assertion is well-known (eg., (I.2.2.2) and Theorem I.1.3.1 in [8]).

The next result shows that the set of resolvent positive operators in $\mathcal{H}(E_1, E_0)$ is closed.

Proposition 34. *Let (A_j) be a sequence in $\mathcal{H}(E_1, E_0)$ converging in $\mathcal{L}(E_1, E_0)$ towards A. If each A_j is resolvent positive then A is resolvent positive as well.*

Proof. It follows from [8, Corollary I.1.3.2] that there exist $\kappa \geq 1$ and $\omega > 0$ such that $[\operatorname{Re} z > \omega]$ belongs to $\rho(-A) \cap \rho(-A_j)$ and

$$\|(\lambda + A)^{-1}\|_{\mathcal{L}(E_0, E_1)} + |\lambda| \, \|(\lambda + A_j)^{-1}\|_{\mathcal{L}(E_0)} \leq \kappa, \qquad \operatorname{Re} \lambda > \omega, \quad j \in \mathbb{N}.$$

Hence we infer from

$$(\lambda + A_j)^{-1} - (\lambda + A)^{-1} = (\lambda + A_j)^{-1}(A - A_j)(\lambda + A)^{-1}, \qquad \operatorname{Re} \lambda > \omega,$$

that

$$\|(\lambda + A_j)^{-1} - (\lambda + A)^{-1}\|_{\mathcal{L}(E_0)} \leq |\lambda|^{-1} \kappa^2 \|A - A_j\|_{\mathcal{L}(E_1, E_0)}, \qquad \operatorname{Re} \lambda > \omega.$$

Thus, in particular,

$$(\lambda + A_j)^{-1} x \to (\lambda + A)^{-1} x$$

in E_0 for $\lambda > \omega$ and $x \in E_0$. Now the assertion is a consequence of the closedness of the positive cone. $\qquad\square$

13. The strong maximum principle for the scalar case

In this section we suppose that $N = 1$ and set $\Gamma_j := \Gamma_j^1$ for $j = 0, 1$, of course. Then we put

$$\mathcal{A}_0 u := -a_{jk}\partial_j\partial_k u + a_j\partial_j u + au,$$

where $a_{jk} = a_{kj}, a_j, a \in L_\infty$, with $[a_{jk}(x)] \in \mathbb{R}^{n \times n}$ being positive definite for a.a. $x \in \Omega$. We also put

$$\mathcal{B}_0 u := \begin{cases} u & \text{on } \Gamma_0, \\ \partial_\beta u + bu & \text{on } \Gamma_1, \end{cases}$$

where β is an outward pointing nowhere tangent C^1 vector field on Γ_1, and b is a C^{1-} function on Γ_1. Clearly, in this case the derivative ∂_β is used in the definition of strong positivity.

The following theorem is the basis for the proofs of the following sections. It slightly improves [3, Theorem 6.1]. Its importance stems from the fact that there is no sign restriction for b.

Theorem 35. *There exists $\omega_0 \in \mathbb{R}$ such that $(\mathcal{A}_0 + \omega, \mathcal{B}_0)$ satisfies for $\omega > \omega_0$ the strong maximum principle.*

Proof. If Ω is bounded, this is a reformulation of Theorem 6.1 in [3] (where the statement is incomplete since the condition $u(y) = 0$ for the validity of $\partial_\beta u(y) < 0$ is missing). Since

$$W_q^2 \hookrightarrow C_0^1(\overline{\Omega}), \qquad q > n,$$

the same proof applies if it is only supposed that Γ is compact, provided Lemma 5.1 in [3] is valid. But this follows easily from the proof of the much more general Theorem B.3 in [5]. $\qquad\square$

Remarks 36.

(a) The proof of [3, Lemma 5.1] is somewhat complicated and perhaps not too transparent. By restricting the arguments leading to Theorem B.3 in [5] to the relevant cases $k = 0$ and $k = 1$, one gets a simpler and more lucid demonstration.

(b) Suppose that $(a, b) \geq 0$. Then $\omega_0 \leq 0$. Furthermore, (A_0, B_0) satisfies the strong maximum principle unless Ω is bounded, $\Gamma = \Gamma_1$, and $(a, b) = (0, 0)$.

Proof. This is a consequence of the classical maximum principle. $\qquad\square$

14. Strong and weak solutions

We return to the case of a general $N \in \mathbb{N}^\times$ and the hypotheses of Section 2. We set $W_{q,\mathcal{B}}^0 := L_q$ and define linear operators

$$A_{k-2} \in \mathcal{L}(W_{q,\mathcal{B}}^k, W_{q,\mathcal{B}}^{k-2}), \qquad k \in \{0, 1\},$$

by

$$\langle v, A_{-1}u \rangle := \mathfrak{a}(v, u), \qquad (v, u) \in W_{q',(1-\chi)\gamma}^1 \times W_{q,(1-\chi)\gamma}^1 \tag{51}$$

and

$$\langle v, A_{-2}u \rangle := \langle A^\sharp v, u \rangle, \qquad (v, u) \in W_{q',\mathcal{B}^\sharp}^2 \times L_q, \tag{52}$$

respectively. A_{-k} is called $W_{q,\mathcal{B}}^{-k}$ **realization** of $(A, \mathcal{B})$. We also put $A_0 := A$ and denote by $N_{-j}(\lambda)$ the algebraic eigenspace of $\lambda \in \sigma_p(A_{-j})$ for $j \in \{0, 1, 2\}$.

Theorem 37. *For $j \in \{1, 2\}$*

(i) $A_{j-2} \in \mathcal{H}(W_{q,\mathcal{B}}^j, W_{q,\mathcal{B}}^{j-2})$;

(ii) $A_{-2} \supset A_{-1} \supset A_0$;

(iii) $\sigma(A_{-j}) = \sigma(A_0)$ *and* $\sigma_p(A_{-j}) = \sigma_p(A_0)$. *If* $\lambda \in \sigma_p(A_0)$ *then* $N_{-j}(\lambda) = N_0(\lambda)$.

Proof. Fix $\omega > s(-A_0)$ and let $\left[(E_\alpha, B_\alpha) \, ; \, \alpha \in \mathbb{R} \right]$ be the interpolation extrapolation scale generated by $(E_0, B_0) := (L_q, \omega + A_0)$ and the complex interpolation functors $[\cdot, \cdot]_\theta$, $0 < \theta < 1$. (We refer to [8, Chapter V] for the general interpolation extrapolation theory, and to [7, Section 6] for a summary of the main results.) Then (cf. Theorems 7.1 and 8.3 in [7] and observe that they remain valid if it is only assumed that Γ is compact) we find that

$$E_{-j/2} \doteq W_{q,\mathcal{B}}^{-j}, \quad B_{-j/2} = A_{-j},$$

where $\doteq$ means 'equal except for equivalent norms'. Hence (i), (ii), and the equality of $\sigma(A_{-j})$ and $\sigma(A_0)$ follow from (15) and the general interpolation extrapolation theory (cf. Theorems V.1.4.6 and V.2.1.3 as well as Corollary V.2.1.4 in [8]). The remaining part of (iii) is now a consequence of (i), (ii), and Lemma 32. □

Remark 38. Suppose that only the weaker assumption (16) is satisfied. Then Theorem 37 remains valid for $j = 1$ and with A_{-2} being omitted in (ii).

Proof. Of course, the interpolation extrapolation scale generated by $(L_q, \omega + A_0)$ is still well-defined. However, since in this case the dual of A is not explicitly known, the space E_{-1} cannot be identified in terms of a known space of distributions. But it is not difficult to see that

$$E_{-1/2} \doteq W_{q,\mathcal{B}}^{-1}$$

is still true. □

The next theorem concerns the solvability of the nonhomogeneous problem (20) and the parameter dependent boundary value problem

$$(\lambda + \mathcal{A})u = f \text{ in } \Omega, \quad \mathcal{B}u = g \text{ on } \Gamma. \tag{53}$$

Theorem 39.

(i) *Every strong W_q^2 solution of (20) is a weak W_q^1 solution, and each weak W_q^1 solution is a very weak L_q solution.*

(ii) *Suppose that $\lambda \in \rho(-A)$, $j \in \{0, 1, 2\}$, and $(f, g) \in W_{q,\mathcal{B}}^{j-2} \times \partial W_q^j$. Then (53) has a unique W_q^j solution.*

(iii) *If $(f, g) \in W_{q,\mathcal{B}}^{-1} \times \partial W_q^1$ then every L_q solution of (20) is a W_q^1 solution. Similarly, if $(f, g) \in L_q \times \partial W_q^2$ then every W_q^1 solution is a W_q^2 solution.*

Proof. (i) is an easy consequence of Green's formulas.

(ii) First suppose that $j = 2$. From [5, Theorem B.3] we know that there exists

$$\mathcal{R}_2 \in \mathcal{L}(\partial W_q^2, W_q^2)$$

satisfying $\mathcal{B}\mathcal{R}_2\varphi = \varphi$ for $\varphi \in \partial W_q^2$. Set $w := \mathcal{R}_2 g$. Then u is a W_q^2 solution of (53) iff $v := u - w$ satisfies

$$(\lambda + \mathcal{A})v = h \text{ in } \Omega, \quad \mathcal{B}v = 0 \text{ on } \Gamma,$$

where

$$h := f - (\lambda + \mathcal{A})w \in L_q,$$

that is, iff $(\lambda + A)v = h$ in L_q. This proves the assertion if $j = 2$.

Suppose that $j = 1$. The trace operator is a continuous retraction from the space W_q^1 onto $W_q^{1-1/q}(\Gamma)$, that is, there exists

$$\mathcal{R} \in \mathcal{L}\big(W_q^{1-1/q}(\Gamma), W_q^1\big)$$

with $\gamma \mathcal{R}\varphi = \varphi$ for $\varphi \in W_q^{1-1/q}(\Gamma)$ (e.g., [5, (B.21)–(B.23)]). Set $w := \mathcal{R}(1 - \chi)g$. Then u is a W_q^1 solution of (53) iff $v := u - w \in W_q^1$ satisfies

$$\lambda\langle\varphi, v\rangle + \mathfrak{a}(\varphi, v) = \langle\varphi, f\rangle - \lambda\langle\varphi, w\rangle - \mathfrak{a}(\varphi, w), \qquad \varphi \in W_{q',(1-\chi)\gamma}^1 \qquad (54)$$

and

$$(1 - \chi)\gamma v = 0, \qquad\qquad (55)$$

thanks to $\chi^2 = \chi$. Since $\mathfrak{a}$ is a continuous bilinear form on $W_{q',(1-\chi)\gamma}^1 \times W_q^1$, the right-hand side of (54) defines an element h in $W_{q,\mathcal{B}}^{-1}$. Thus (54) and (55) are equivalent to

$$v \in W_{q,(1-\chi)\gamma}^1, \qquad (\lambda + A_{-1})v = h.$$

Now the assertion for $j = 1$ follows from $\sigma(A_{-1}) = \sigma(A)$.

Finally, assume that $j = 0$. Note that

$$(1 - \chi)\partial_\mu \in \mathcal{L}\big(W_{q',\mathcal{B}^\sharp}^2, W_{q'}^{1-1/q'}(\Gamma_0)\big).$$

Hence

$$(\partial_\mu)'(1 - \chi) \in \mathcal{L}\big(W_q^{-1/q}(\Gamma_0), W_{q,\mathcal{B}}^{-2}\big).$$

Similarly,

$$\chi\gamma \in \mathcal{L}\big(W_{q',\mathcal{B}^\sharp}^2, W_{q'}^{2-1/q'}(\Gamma_1)\big)$$

and, consequently,

$$\gamma'\chi \in \mathcal{L}\big(W_q^{-1-1/q}(\Gamma_1), W_{q,\mathcal{B}}^{-2}\big).$$

From this and (22) we infer that u is an L_q solution of (53) iff

$$(\lambda + A_{-2})u = h,$$

where

$$h := f + (\partial_\mu)'(\chi - 1)g + \gamma'\chi g \in W_{q,\mathcal{B}}^{-2}.$$

Thus $\sigma(A_{-2}) = \sigma(A)$ implies the assertion in this case also.

(iii) Suppose that $(f, g) \in W_{q,\mathcal{B}}^{-1} \times \partial W_q^1$ and u is an L_q solution of (20). Then there exists $\omega > 0$ such that

$$\lambda_\omega := \omega + \lambda \in \rho(-A).$$

Hence (20) is equivalent to

$$(\lambda_\omega + A)u = f_\omega \text{ in } \Omega, \quad \mathcal{B}u = g \text{ on } \Gamma, \qquad (56)$$

where

$$f_\omega := f + \omega u \in W_{q,\mathcal{B}}^{-1}.$$

Thus (ii) implies that (56) has a unique W_q^1 solution v. From (i) we infer that v is an L_q solution of (56). Since it is unique, by (ii), it follows that $u = v$, that is, $u \in W_{q,\mathcal{B}}^1$. This proves the first assertion. The second one follows by similar arguments. $\qquad\square$

Remark 40. If we presuppose only condition (16) then Theorem 39 is valid with any reference to L_q solutions being omitted.

Proof. This follows from the above proof and Remark 38. □

15. Resolvent positivity

The next theorem is the basis of all the following positivity results.

Theorem 41.

(i) $W_{q,\mathcal{B}}^{-1}$ and $W_{q,\mathcal{B}}^{-2}$ are OBSs and the natural injection maps (19) are positive.

(ii) $(W_{q,\mathcal{B}}^j)^+$ is dense in $(W_{q,\mathcal{B}}^k)^+$ for $-2 \leq k < j \leq 2$.

(iii) A_{-j} are resolvent positive for $j \in \{0, 1, 2\}$.

Proof. (1) First we assume that

$$a \in L_\infty(\Omega, \mathbb{R}_{\mathrm{diag}}^{N \times N}) \quad \text{and} \quad b \in C^{1-}(\Gamma, \mathbb{R}_{\mathrm{diag}}^{N \times N}).$$

Then Theorem 35 applies to the boundary value problem $(\mathcal{A}^r, \mathcal{B}^r)$ for $1 \leq r \leq N$. Hence there exists $\omega_0 > 0$ such that

$$(\lambda + A_0)^{-1} v \geq 0, \qquad \lambda > \omega_0, \quad v \in \mathcal{D}^+, \qquad (57)$$

where $\mathcal{D} := \mathcal{D}(\Omega, \mathbb{R}^N)$ is the space of all smooth $\mathbb{R}^N$ valued functions with compact support in Ω. Since $(\lambda + A_0)^{-1} \in \mathcal{L}(L_q)$ and $\mathcal{D}^+$ is dense in L_q^+ it follows that (57) is true for all $v \in L_q^+$. Thus A_0 is resolvent positive.

The same arguments show that $A^\sharp$, the $L_{q'}$ realization of $(\mathcal{A}^\sharp, \mathcal{B}^\sharp)$, is also resolvent positive. Note that $L_{q'}^+$ is generating since $L_{q'}$ is a Banach lattice. Thus, fixing $\lambda > s(-A^\sharp)$ and setting

$$P := (\lambda + A^\sharp)^{-1} L_{q'}^+,$$

it follows from Theorem 28 that

$$P \subset (W_{q',\mathcal{B}^\sharp}^2)^+ \quad \text{and} \quad W_{q',\mathcal{B}^\sharp}^2 = P - P.$$

Hence $(W_{q',\mathcal{B}^\sharp}^2)^+$ is generating, thus total. Now (cf. the proof of Theorem 37) Theorems V.1.5.12, V.2.3.2, V.2.7.2, and Corollary V.2.7.3 in [8] imply that (i)–(iii) are true in this case.

(2) We consider the general case. First we observe that $W_{q,\mathcal{B}}^1$ and $W_{q,\mathcal{B}}^{-1}$ are independent of a and b. Hence it follows from step (1) that they are *OBSs*, the injection maps

$$W_{q,\mathcal{B}}^1 \hookrightarrow L_q \hookrightarrow W_{q,\mathcal{B}}^{-1}$$

are positive, $(W_{q,\mathcal{B}}^1)^+$ is dense in L_q^+, and the latter cone is dense in $(W_{q,\mathcal{B}}^{-1})^+$.

Recall (25) and (26). Define $(\mathcal{A}^\triangle, \mathcal{B}^\triangle)$ by replacing a and b in the definition of $(\mathcal{A}, \mathcal{B})$ by $a^\triangle$ and $b^\triangle$, respectively. Then step (1) implies that $A_{-1}^\triangle$, the $W_{q,\mathcal{B}}^{-1}$ realization of $(\mathcal{A}^\triangle, \mathcal{B}^\triangle)$, is well-defined, belongs to $\mathcal{H}(W_{q,\mathcal{B}}^1, W_{q,\mathcal{B}}^{-1})$, and is resolvent positive.

Fix $s \in (1/q, 1)$ and put

$$W_{q,\mathcal{B}}^s := W_{q,(1-\chi)\gamma}^s := \{ v \in W_q^s \ ; \ (1-\chi)\gamma v = 0 \}.$$

Then [7, Theorem 7.2] implies that

$$W_{q,\mathcal{B}}^s \doteq (W_{q,\mathcal{B}}^{-1}, W_{q,\mathcal{B}}^1)_{(1+s)/2, q}$$

where $(\cdot, \cdot)_{\theta, p}$ are the real interpolation functors for $0 < \theta < 1$ and $1 \le p \le \infty$.

By the trace theorem,

$$\gamma_s := \gamma \,|\, W_{q,\mathcal{B}}^s \in \mathcal{L}(W_{q,\mathcal{B}}^s, L_q(\Gamma_1)), \tag{58}$$

and

$$\gamma_1' \in \mathcal{L}(L_q(\Gamma_1), W_{q,\mathcal{B}}^{-1}), \tag{59}$$

where

$$\gamma_1 := \gamma \,|\, W_{q',(1-\chi)\gamma}^1 \in \mathcal{L}(W_{q',(1-\chi)\gamma}^1, L_{q'}(\Gamma_1)).$$

Consequently, setting

$$Bu := a^\bullet u + \gamma_1' \chi b^\bullet \gamma_s u, \qquad u \in W_{q,\mathcal{B}}^s,$$

it follows from $(a^\bullet, b^\bullet) \in L_\infty(\Omega, \mathbb{R}^{N \times N}) \times L_\infty(\Gamma, \mathbb{R}^{N \times N})$,

$$W_{q,\mathcal{B}}^s \hookrightarrow L_q \hookrightarrow W_{q,\mathcal{B}}^{-1}, \tag{60}$$

and (58) and (59) that $B \in \mathcal{L}(W_{q,\mathcal{B}}^s, W_{q,\mathcal{B}}^{-1})$. Furthermore, $B \ge 0$ thanks to the positivity of $(a^\bullet, b^\bullet)$, of the trace operators γ_s and γ_1, and of the injection maps (60), and thanks to the fact that L_q^+ and $(W_{q,\mathcal{B}}^{-1})^+$ are the dual cones of the positive cones of $L_{q'}$ and $W_{q',(1-\chi)\gamma}^1$, respectively. Note that

$$\langle v, (A_{-1}^\triangle - B)u \rangle = \mathfrak{a}(v, u), \qquad (v, u) \in W_{q',(1-\chi)\gamma}^1 \times W_{q,(1-\chi)\gamma}^1,$$

so that

$$A_{-1} = A_{-1}^\triangle - B.$$

Hence Proposition 33 guarantees that A_{-1} is resolvent positive.

From Theorem 37 we infer that

$$(\lambda + A_{-2})^{-1} \supset (\lambda + A_{-1})^{-1} \supset (\lambda + A_0)^{-1}, \qquad \lambda > s(-A_0).$$

Thus A_0 is resolvent positive as well. The same arguments apply to the boundary value problem $(\mathcal{A}^\sharp, \mathcal{B}^\sharp)$ and guarantee that $A_0^\sharp$ is resolvent positive. Thus we see, as in step (1), that $(W_{q,\mathcal{B}}^{-2})^+$ is a proper cone, that is, $W_{q,\mathcal{B}}^{-2}$ is an OBS. Now the remaining assertions follow by the arguments of step (1). $\qquad \square$

Remark 42. Suppose that only the weaker assumption (16) is satisfied. Then Theorem 41 remains true if all assertions involving $W_{q,\mathcal{B}}^{-2}$ and A_{-2} are omitted.

Proof. This follows from Remark 38. □

16. Proofs of the weak maximum principles

Now it is not difficult to prove the theorems presented in Sections 4 and 5.

Proof of Theorem 6   (a) Theorem 41(iii) guarantees that A is resolvent positive. Thus (3) and (4) follow from Theorems 28 and 29, since L_q is a Banach lattice.

(b) If (A, B) is inverse positive on W_q^2 then A is inverse positive. Thus, if A is surjective then $\lambda_0 > 0$ by Corollary 30.

From (52) we infer that (A, B) satisfies the very weak maximum principle iff A_{-2} is inverse positive. Suppose that $\lambda_0 > 0$. Then A is inverse positive by Corollary 30. Since $\lambda_0 = \lambda_0(-A_{-2})$ by Theorem 37(iii), it follows that

$$0 \in \rho(A_{-2}).$$

Also suppose that $u \in L_q$ and $A_{-2}u \geq 0$. Then Theorem 41(ii) guarantees the existence of a sequence (f_k) in L_q^+ converging in $W_{q,B}^{-2}$ towards $f := A_{-2}u$. Hence the sequence (u_k), where $u_k := (A_{-2})^{-1}f_k$, converges in L_q towards $u = (A_{-2})^{-1}f$. Since $A_{-2} \supset A$ by Theorem 37(ii), it follows that $(A_{-2})^{-1} \supset A^{-1}$. Thus $u_k \in W_{q,B}^2$ and

$$Au_k = f_k \geq 0.$$

Consequently, $u_k \geq 0$ by the inverse positivity of A. Hence Theorem 41(i) implies $u \geq 0$. This shows that A_{-2} is inverse positive and proves (2).

(c) Suppose that $u \in W_q^1$ satisfies assumption (18). Then we define $f \in (W_{q,B}^{-1})^+$ by

$$\langle v, f \rangle := \mathfrak{a}(v, u), \qquad v \in W_{q',(1-\chi)\gamma}^1.$$

We also set

$$g := (1 - \chi)\gamma u \in \left(W_q^{1-1/q}(\Gamma_0) \right)^+ \hookrightarrow (\partial W_q^0)^+.$$

By Greens's formula,

$$\langle A^\sharp v, u \rangle = \mathfrak{a}(v, u) + \langle (\chi - 1)\partial_\mu v, \gamma u \rangle_\Gamma$$
$$= \langle v, f \rangle + \langle \partial_\mu v, (\chi - 1)g \rangle_\Gamma \geq 0$$

for $v \in (W_{q',B^\sharp}^2)^+$. Thus $u \geq 0$ if (i) is true. Hence (i) implies (ii).

Suppose that $u \in W_q^2$ satisfies $(Au, Bu) \geq 0$. Then

$$(f, g) := (Au, Bu) \in (L_q \times \partial W_q^2)^+ \hookrightarrow (W_{q,B}^{-1} \times \partial W_q^1)^+$$

by (13). From Green's formula we infer that

$$\mathfrak{a}(v, u) = \langle v, Au \rangle + \langle \chi\gamma v, Bu \rangle_\Gamma = \langle v, f \rangle + \langle \gamma v, \chi g \rangle_\Gamma \geq 0$$

for $v \in (W_{q',(1-\chi)\gamma}^1)^+$. Thus $u \in W_q^2 \hookrightarrow W_q^1$ satisfies (18) so that $u \geq 0$ if (ii) is true. This shows that (ii) implies (iii).

Now suppose that A is surjective and $(\mathcal{A}, \mathcal{B})$ is inverse positive on W_q^2. Then A is inverse positive. Hence $\lambda_0 > 0$ by (2), so that

$$0 \in \rho(A_{-2}) = \rho(A).$$

From the second part of (b) we know that A_{-2} is inverse positive. Thus (iii) implies (i). This proves (1).   □

Proof of Remark 7(b)   Replace in the second part of step (b) of the preceding proof the very weak maximum principle by the weak one and A_{-2} by A_{-1}. Then it follows that the inverse positivity of A implies the one of A_{-1}. Hence from (51) and (18) we deduce that $\lambda_0 > 0$ implies that $(\mathcal{A}, \mathcal{B})$ is inverse positive.

Similarly, by replacing in the beginning of the last paragraph of the preceding proof A_{-2} by A_{-1}, we see that (iii) of Theorem 6 implies (ii).   □

Proof of Theorem 8.   It is an easy consequence of Theorems 6 and 39 that

$$\text{(i)} \Rightarrow \text{(iv)} \Rightarrow \text{(iii)} \Rightarrow \text{(ii)}.$$

Suppose that (ii) is true. Then it is obvious that $(\mathcal{A}, \mathcal{B})$ is inverse positive. Given $f \in L_q$, it follows from (ii) that there exists $u^{\pm} \in W_{q,\mathcal{B}}^2$ satisfying $Au^{\pm} = f^{\pm}$, where f^+ (resp. f^-) is the positive (resp. negative) part of f. Thus, setting $u := u^+ - u^- \in W_{q,\mathcal{B}}^2$, we see that $Au = f$. Hence A is surjective. Hence we infer from the second part of Theorem 6(2) that $\lambda_0 > 0$. Thus (ii) implies (i).   □

It is obvious from this proof that Theorem 8 remains valid if only the weaker assumption (16) is satisfied, provided assertion (iv) is omitted.

17. Bounded Domains

In this section we prove the theorems presented in Sections 6 and 7. We begin with a simple bootstrapping result.

Proposition 43. *Let Ω be bounded and suppose that only condition (16) is satisfied. Suppose also that $u \in W_q^2$ solves (53). If $q < p < \infty$ and $(f, g) \in L_p \times \partial W_p^2$ then $u \in W_p^2$.*

Proof. Put $r := p \wedge nq/(n - 2q)$ if $q < n/2$, and $r := p$ otherwise. Then $u \in L_r$ by Sobolev's embedding theorem. Fix $\omega := \omega_r \in \sigma(-A_{(r)})$, where $A_{(r)}$ is the L_r realization of $(\mathcal{A}, \mathcal{B})$. Then (53) is equivalent to

$$(\omega + \mathcal{A})u = f_\omega \text{ in } \Omega, \quad \mathcal{B}u = g \text{ on } \Gamma,$$

where

$$f_\omega := (\omega - \lambda)u + f \in L_r \quad \text{and} \quad g \in \partial W_r^2,$$

thanks to the boundedness of Ω (and the compactness of Γ). Hence we deduce from Theorem 39(ii) that $u \in W_r^2$. If $r < p$ we repeat this argument to arrive after finitely many steps at the assertion.   □

Proof of Theorem 10   Since Ω is bounded, embedding (14) is compact. Hence A has a compact resolvent and the assertions concerning $\sigma(A)$ follow from the general theory of linear operators with compact resolvent (e.g., [48]).

Suppose that $\lambda \in \mathbb{C}$,

$$v \in L_{\infty-} := \bigcap_{1 < p < \infty} L_p,$$

and $u \in W_q^2$ satisfies $Au = \lambda u - v$. Then Proposition 43 implies that $u \in W_{\infty-}^2$. Thus, if $\lambda \in \sigma(A) = \sigma_p(A)$ and $u \in N_A(\lambda)$, there exists $v_0, \ldots, v_m \in W_{q,\mathcal{B}}^2$ satisfying $u_0 = u$ and

$$Au_k = \lambda u_k - v_{k+1}, \qquad 0 \le k \le m,$$

where $v_{m+1} = 0$. Hence we deduce from what has just been shown, by backwards induction, that $u \in W_{\infty-}^2$. $\qquad\square$

Proposition 44. *Let Ω be bounded and suppose only that condition (16) is satisfied. Also suppose that either $N = 1$ or (a,b) is irreducible. If $q > n$ then every positive strict W_q^2 supersolution for $(\mathcal{A}, \mathcal{B})$ is strongly positive.*

Proof. Define, as in the proof of Theorem 41, $(\mathcal{A}^\triangle, \mathcal{B}^\triangle)$ by replacing (a,b) in $(\mathcal{A}, \mathcal{B})$ by $(a^\triangle, b^\triangle)$. Then Theorem 35 implies the existence of $\omega > 0$ with $\omega > -\lambda_0$ such that $(\omega + \mathcal{A}^\triangle, \mathcal{B}^\triangle)$ satisfies the strong maximum principle.

Suppose that u is a positive strict W_q^2 supersolution for $(\mathcal{A}, \mathcal{B})$. Then u belongs to $(W_q^2)^+ \setminus \{0\}$ and

$$(f, g) := (\mathcal{A}u, \mathcal{B}u) \in (L_q \times \partial W_q^2)^+ \setminus \{0\}.$$

Consequently, u satisfies

$$(\omega + \mathcal{A}^\triangle)u = (\omega + a^\bullet)u + f \text{ in } \Omega, \quad \mathcal{B}^\triangle u = \chi b^\bullet \chi \gamma u + g \text{ on } \Gamma. \tag{61}$$

Note that

$$\big((\omega + a^\bullet)u + f, \chi b^\bullet \chi \gamma u + g\big) > 0, \tag{62}$$

thanks to $u \ge 0$ and the positivity of $a^\bullet$, $b^\bullet$, and the the trace operator.

If $N = 1$ then (62) reduces to $(\omega u + f, g) > 0$. Hence (61) and the strong maximum principle imply the strong positivity of u.

Suppose that $N > 1$. Then there exists $s \in \{1, \ldots, N\}$ such that $(f^s, g^s) > 0$. Hence, by looking at equations number s of system (61), we deduce from Theorem 35 that u^s is strongly positive. Fix any $r \in \{1, \ldots, N\}$. Then the irreducibility of (a,b) guarantees the existence of $s_\rho \in \{1, \ldots, N\}$, $1 \le \rho \le k$, with $s_0 = s$ and $s_k = r$ such that

$$(a^{\xi\eta}, \chi^\xi b^{\xi\eta} \chi^\eta) > 0 \quad \text{for} \quad (\xi, \eta) \in \{(s_1, s), \ldots, (r, s_{k-1})\}.$$

Thus we infer from (62) and $u^s(x) > 0$ for all $x \in \Omega \cup \Gamma_1^s$ that component s_1 of (62) is nonzero. Thus, by looking at equations number s_1 of system (61) and invoking Theorem 35 once more, we see that u^{s_1} is strictly positive. Now we repeat these arguments with (s_1, s) replaced by (s_2, s_1), etc. to find that u^r is strongly positive. Since this is true for each index r it follows that u is strongly positive. $\qquad\square$

Proof of Theorem 12   Since the spectrum of A is independent of q we can assume that $q > n$.

Fix $\omega > -\lambda_0$. Then (23) is equivalent to

$$\mu u = (\omega + A)^{-1} u, \quad \mu := 1/(\omega + \lambda), \tag{63}$$

for $\lambda \neq -\omega$. From $\sigma(A) \subset [\operatorname{Re} \lambda \geq \lambda_0]$ it follows that $\operatorname{Re} \mu > 0$ if $\lambda \in \sigma(A)$.

Put

$$E := \{ u \in C^1(\overline{\Omega}) \; ; \; (1 - \chi)\gamma u = 0 \}.$$

Then E is an ordered Banach space whose positive cone has nonempty interior. Indeed, every strongly positive u belongs to $\operatorname{int}(E^+)$. Also set

$$T := T_\omega := (\omega + A)^{-1} | E.$$

Then the compactness of the embedding $W_{q,\mathcal{B}}^2 \hookrightarrow E$ implies that T is a compact endomorphism of E. Thus we infer from Theorem 41, Proposition 44, and $E \hookrightarrow L_q$ that T is strongly positive, that is,

$$T(E^+ \setminus \{0\}) \subset \operatorname{int}(E^+).$$

Consequently, the Krein-Rutman theorem (cf. [1, Theorem 3.2]) implies that the spectral radius $r := r_\omega$ of T is positive and a simple eigenvalue with a positive eigenvector. Moreover, it is the only eigenvalue of T with a positive eigenvector. Clearly, $u \in E$ satisfies $ru = Tu$ iff $u \in W_{q,\mathcal{B}}^2$ and u satisfies (63) with $\mu := r$. But this is equivalent to the fact that u is an eigenfunction of A to the eigenvalue

$$\lambda_\bullet := -\omega + 1/r.$$

Thus $\sigma(A) \neq \varnothing$ and $\lambda_\bullet$ is an eigenvalue of A with a positive eigenfunction $u_\bullet$. From (63) we also deduce that $u_\bullet$ is a positive strict supersolution of $(\omega + \mathcal{A}, \mathcal{B})$. Hence $u_\bullet$ is strongly positive by Proposition 44.

Suppose that there are $u_0, \ldots, u_m \in W_{q,\mathcal{B}}^2$ satisfying

$$Au_k - \lambda_\bullet u_k = u_{k+1}, \quad 0 \leq k \leq m,$$

where $u_{m+1} := 0$. Then

$$ru_k - Tu_k = rTu_{k+1}, \quad 0 \leq k \leq m.$$

Thus, if $m \geq 1$, it follows from $Tu_m = ru_m$ that

$$ru_{m-1} - Tu_{m-1} = r^2 u_m. \tag{64}$$

Since r is a simple eigenvalue of T there exists $\alpha \in \mathbb{R}$ such that $u_m = \alpha u_\bullet$. The Krein-Rutman theorem guarantees also that there exists an eigenvector φ of the dual $T' \in \mathcal{L}(E')$ to the eigenvalue r satisfying $\langle \varphi, v \rangle > 0$ for $v \in E^+ \setminus \{0\}$. By applying the functional φ to (64) it follows that

$$r^2 \alpha \langle \varphi, u_\bullet \rangle = \langle r\varphi - T'\varphi, u_{m-1} \rangle = 0.$$

Hence $\alpha = 0$. Thus we find by backwards induction that $u_k = 0$ for $1 \leq k \leq m$, which shows that

$$N_A(\lambda_\bullet) = \ker(\lambda_\bullet - A).$$

Hence the equivalence of the eigenvalue problem for A with (63) and the simplicity of r imply that $\lambda_\bullet$ is a simple eigenvalue of A.

Assume that $\lambda \in \rho(-A) \cap \mathbb{R}$ with $\lambda \neq \omega$. Note that $(\lambda + A)u = f$ for $f \in E$ is equivalent to

$$\left(\frac{1}{\omega - \lambda} - T\right)u = \frac{1}{\omega - \lambda}Tf. \tag{65}$$

Suppose that $\omega > \lambda$. Then (65) has for each $f > 0$ precisely then a positive solution if $1/(\omega - \lambda) > r$ (cf. [1, Theorem 3.2(iv)]), that is, if $\lambda > \omega - 1/r = -\lambda_\bullet$. Hence, if $(\lambda + A)^{-1} \geq 0$ and $\omega > \lambda$ then it follows that $\lambda > -\lambda_\bullet$. Clearly, given any $\lambda \in \mathbb{R}$, we can fix $\omega > \lambda$ such that the above arguments apply. (Note that $\lambda_\bullet$ is independent of ω although $T = T_\omega$ and $r = r_\omega$ depend on this choice.) This shows that $-\lambda_\bullet = s_+(-A)$. Hence Theorem 29 implies $-\lambda_\bullet = s(-A) = -\lambda_0$.

Lastly, suppose that $\lambda \in \sigma(-A) \setminus \{\lambda_0\}$ satisfies $\operatorname{Re}\lambda = \lambda_0$. Then it follows from [39, Theorem 2.4] (also see [16, Corollary C-III.2.12] or [25, Theorem 8.14]) that

$$\lambda_0 + ik\operatorname{Im}\lambda \in \sigma(A)$$

for $k \in \mathbb{Z}$. But this contradicts the fact that $\sigma(A)$ is contained in a symmetric sector around the real axis with an angle of opening less than π, as follows from $A \in \mathcal{H}(W_{q,\mathcal{B}}^2, L_q)$. Thus λ_0 is the only eigenvalue of A with $\operatorname{Re}\lambda = \lambda_0$. $\qquad\square$

Proof of Theorem 15   Thanks to Proposition 44 it suffices to show that every positive strict W_q^2 supersolution is strictly positive.

Fix $\omega > \lambda_0$ and put

$$K := (\omega + A)^{-1} \in \mathcal{L}(L_q).$$

Then $K \geq 0$. By repeated application of Proposition 43 we deduce from Proposition 44 that there exists $m \in \mathbb{N}$ such that $K^j u$ is strongly positive whenever $j > m$ and $u > 0$. Consequently, given $\mu > r(K)$,

$$K(\mu - K)^{-1}u = \sum_{j=1}^{\infty} \mu^{-j} K^j u$$

is for each $u \in (L_q^+) \setminus \{0\}$ a quasi-interior point of L_q^+.

Set $\lambda := \omega - 1/\mu$ and note that $r(K) = 1/(\omega + \lambda_0)$ implies $\lambda > -\lambda_0$. Furthermore,

$$(\lambda + A)^{-1} = \mu K(\mu - K)^{-1}.$$

Hence $(\lambda + A)^{-1}$ is strongly irreducible, thus irreducible. Now Theorem 31 implies that $(\lambda + A)^{-1}u$ is for each $\lambda > -\lambda_0$ and each $u \in (L_q^+) \setminus \{0\}$ a quasi-interior point of L_q^+, hence strictly positive.

Let u be a positive strict W_q^2 supersolution for $(\mathcal{A}, \mathcal{B})$. Set

$$(f, g) := (\mathcal{A}u, \mathcal{B}u) > 0.$$

Fix $\omega > (-\lambda_0) \vee 0$ and put $f_\omega := \omega u + f$. Then $f_\omega > 0$ and the above considerations show that
$$v := (\omega + A)^{-1} f_\omega$$
is a strictly positive element of L_q. Since $g \in (\partial W_q^2)^+$ and
$$\lambda_0(\omega + A) = \omega + \lambda_0(A) > 0,$$
it follows from Theorem 8 that there exists a unique $w \in (W_q^2)^+$ satisfying
$$(\omega + A)w = 0 \text{ in } \Omega, \qquad \mathcal{B}w = g \text{ on } \Gamma.$$
Since $u = v + w$ we see that u is strictly positive. $\square$

Proof of Theorem 13 Suppose that $\lambda_0 > 0$. Then $0 \in \rho(A)$ so that A is surjective. Hence it follows from Theorem 6(1) and (2) that $(\mathcal{A}, \mathcal{B})$ is inverse positive and that this is equivalent to (ii) and (iii). The inverse positivity of $(\mathcal{A}, \mathcal{B})$ and Proposition 44 imply that $(\mathcal{A}, \mathcal{B})$ satisfies the strong maximum principle. From this we deduce that
$$(\text{i}) \Rightarrow (\text{ii}) \Rightarrow (\text{iii}) \Rightarrow (\text{iv}).$$

(iv)$\Rightarrow$(i) Suppose that $\lambda_0 \leq 0$ and let u_0 be a positive eigenfunction of $(\mathcal{A}, \mathcal{B})$. Then
$$\mathcal{A}u_0 = \lambda_0 u_0 \leq 0 \text{ in } \Omega, \qquad \mathcal{B}u_0 = 0 \text{ on } \Gamma.$$
Hence $u_0 \in W_{\infty-}^2 \setminus \{0\}$ and the strong maximum principle imply $-u_0 > 0$, which is impossible. Thus $\lambda_0 > 0$.

(i)$\Rightarrow$(v) Every positive eigenfunction to the eigenvalue λ_0 is a positive strict W_q^2 supersolution, hence a positive strict L_q supersolution.

(v)$\Rightarrow$(i) Recall that $A^\sharp = A'$, where A is considered as an unbounded operator in L_q. Hence $\sigma(A^\sharp) = \sigma(A)$. Note that $(\mathcal{A}^\sharp, \mathcal{B}^\sharp)$ satisfies condition (7) also and the irreducibility of (a, b) implies the one of $(a^\top, b^\top)$. Thus λ_0 is also the principal eigenvalue of $(\mathcal{A}^\sharp, \mathcal{B}^\sharp)$ and it has a strongly positive eigenfunction $\varphi_0 \in W_{\infty-}^2$.

Let u be a positive strict L_q supersolution for $(\mathcal{A}, \mathcal{B})$. Fix $\omega > -\lambda_0$ and put
$$\widehat{u} := (\omega + A)^{-1} u.$$
Then $\widehat{u} \in (W_{q,\mathcal{B}}^2)^+$ by Theorem 41(iii). From $A_{-2} \supset A$ and
$$(\omega + A_{-2})^{-1} \supset (\omega + A)^{-1}$$
we deduce that
$$f := A\widehat{u} = A_{-2}(\omega + A_{-2})^{-1} u = (\omega + A_{-2})^{-1}(A_{-2}u) > 0,$$
where the last inequality sign is also a consequence of Theorem 41(iii) and Theorem 37(iii). Hence $\widehat{u}$ is a positive strict W_q^2 supersolution for $(\mathcal{A}, \mathcal{B})$, and f belongs to $L_q^+ \setminus \{0\}$. Thus the strict positivity of φ_0 implies
$$0 < \langle \varphi_0, f \rangle = \langle \varphi_0, A\widehat{u} \rangle = \langle A'\varphi_0, \widehat{u} \rangle = \lambda_0 \langle \varphi_0, \widehat{u} \rangle$$
and $\langle \varphi_0, \widehat{u} \rangle > 0$. Hence $\lambda_0 > 0$. $\square$

Proof of Remark 14   It suffices to replace in the last paragraph of the preceding proof A_{-2} by A_{-1}.   □

Proof of Theorem 11   Thanks to Theorem 12 we can assume that $N > 1$ and (a, b) is reducible. Thus we can also assume that $[(a^\bullet, \chi b^\bullet \chi)]$ has a block triangular structure of the form (24). If the first diagonal block is either one-dimensional or irreducible then we can apply Theorem 12 to the reduced system obtained by setting $u^{M+1}, \ldots, u^N$ equal to 0. This guarantees the existence of a real eigenvalue of $(\mathcal{A}, \mathcal{B})$ with a positive eigenfunction. If $M > 2$ and the first diagonal block is reducible we can repeat this argument to arrive at the existence of at least one real eigenvalue of $(\mathcal{A}, \mathcal{B})$ with a positive eigenvector. Thus $\sigma(\mathcal{A}, \mathcal{B}) \neq \varnothing$ and λ_0 is an eigenvalue of $(\mathcal{A}, \mathcal{B})$.

Fix $\omega > -\lambda_0$. Then $(\omega + A)^{-1}$ is a positive compact endomorphism of L_q, and $1/(\omega + \lambda_0)$ is its spectral radius. Hence the Krein-Rutman theorem (e.g., [1, Theorem 3.1]) guarantees that $(\omega + A)^{-1}$ has a positive eigenfunction u_0 to the eigenvalue $1/(\omega + \lambda_0)$. Thus u_0 is a positive eigenfunction of A to the eigenvalue λ_0.

Finally, if $(\mathcal{A}, \mathcal{B})$ is inverse positive then A is injective. If A is not surjective then $0 \in \sigma(A)$. Thus $\ker(A) \neq \{0\}$ since $\sigma(A) = \sigma_p(A)$. This being impossible, A is surjective and the last assertion follows from Theorem 6(2).   □

18. Domain perturbations

In this section we prove Theorems 20 and 22. For this we need some preparation.

We fix $\xi, \eta, \zeta \in [1, \infty]$ satisfying

$$\frac{1}{\xi} \leq \min\left\{\frac{1}{n}, \frac{1}{q'}\right\} \tag{66}$$

with a strict inequality sign if $q' = n$,

$$\frac{1}{\eta} \leq \min\left\{\frac{2}{n}, \frac{1}{q} + \frac{1}{n}, \frac{1}{q'} + \frac{1}{n}\right\} \tag{67}$$

with a strict inequality sign if either $q = n$ or $q' = n$, and

$$\frac{1}{\zeta} \leq \min\left\{\frac{1}{n-1}, \frac{n}{n-1}\frac{1}{q}, \frac{n}{n-1}\frac{1}{q'}\right\} \tag{68}$$

with a strict inequality sign if either $q = n$ or $q' = n$. Then we define a Banach space $\mathbb{E}_{\xi, \eta, \zeta}(\Omega)$ by

$$\mathbb{E}_{\xi, \eta, \zeta}(\Omega) := BUC(\Omega, \mathbb{R}_{\text{diag}}^{N \times N})^{n \times n} \times (L_\xi + L_\infty)(\Omega, \mathbb{R}_{\text{diag}}^{N \times N})^n$$
$$\times (L_\eta + L_\infty)(\Omega, \mathbb{R}^{N \times N}) \times L_\zeta(\Gamma, \mathbb{R}^{N \times N}) \times C(\Gamma, \mathbb{R}_{\text{diag}}^{N \times N}).$$

Given

$$\alpha := \left([a_{jk}]^{N \times N}, (a_1, \ldots, a_n), a, b, \chi\right) \in \mathbb{E}_{\xi, \eta, \zeta}(\Omega),$$

we put

$$\mathfrak{a}(\alpha)(v, u) := \langle \partial_j v, a_{jk} \partial_k u \rangle + \langle v, a_j \partial_j u + au \rangle + \langle \gamma v, \chi b \gamma u \rangle_\Gamma$$

for $(v, u) \in W_{q'}^1 \times W_q^1$.

For Banach spaces E and F we write $\mathcal{L}(E, F; \mathbb{R})$ for the Banach space of all continuous bilinear maps $E \times F \to \mathbb{R}$, endowed with its usual norm.

Given $\mathfrak{b} \in \mathcal{L}(E, F; \mathbb{R})$, let $B_\mathfrak{b}$ be the unique linear operator in $\mathcal{L}(F, E')$ satisfying

$$\langle B_\mathfrak{b} f, e \rangle_E = \mathfrak{b}(e, f), \qquad (e, f) \in E \times F.$$

It follows that the map

$$\mathcal{L}(E, F; \mathbb{R}) \to \mathcal{L}(F, E'), \quad \mathfrak{b} \mapsto B_\mathfrak{b}, \tag{69}$$

is a linear isometry.

Lemma 45. *The map*

$$\mathbb{E}_{\xi, \eta, \varsigma}(\Omega) \to \mathcal{L}(W_{q'}^1, W_q^1; \mathbb{R}), \quad \alpha \mapsto \mathfrak{a}(\alpha),$$

is well-defined, linear, and continuous.

Proof. This follows from Sobolev embeddings and the trace theorem. □

Suppose now that Ω is bounded. Let $\underset{\sim}{\Omega}$ be a bounded C^2 domain in $\mathbb{R}^n$ with boundary $\underset{\sim}{\Gamma}$ and trace operator $\underset{\sim}{\gamma}$. Also suppose that $\underset{\sim}{\chi}$ is a boundary identification map for $\underset{\sim}{\Omega}$. We set

$$X := \overline{\Omega} \quad \text{and} \quad Y := \overline{\underset{\sim}{\Omega}}$$

and denote by $(x^1, \ldots, x^n)$ and $(y^1, \ldots, y^n)$ the standard Euclidean coordinates of X and Y, respectively. Then X and Y are compact oriented n-dimensional Riemannian C^2 manifolds with boundary and the standard Euclidean metric

$$(\cdot | \cdot)_X := dx^j \otimes dx^j \quad \text{and} \quad (\cdot | \cdot)_Y := dy^j \otimes dy^j,$$

respectively.

Also suppose that $\varphi : X \to Y$ is an orientation preserving C^2 diffeomorphism satisfying $\varphi_\partial^* \underset{\sim}{\chi} = \chi$. Then

$$\varphi^* \in \mathcal{L}\mathrm{is}\big(W_{p,(1-\underset{\sim}{\chi})\underset{\sim}{\gamma}}^j(\underset{\sim}{\Omega}, \mathbb{R}^N), W_{p,(1-\chi)\gamma}^j\big), \quad j \in \{-1, 0, 1\}, \quad p \in (1, \infty), \tag{70}$$

with inverse

$$\varphi_* := (\varphi^{-1})^*.$$

Indeed, this is easily verified if $j \in \{0, 1\}$, and follows by duality if $j = -1$.

Put

$$(\cdot|\cdot)_M := \varphi^*(dy^j \otimes dy^j) = g_{jk}\, dx^j \otimes dx^k, \tag{71}$$

where

$$g_{jk} := \left(\frac{\partial \varphi}{\partial x^j} \bigg| \frac{\partial \varphi}{\partial x^k}\right)_X, \qquad 1 \le j, k \le n.$$

Then $M := (X, (\cdot|\cdot)_M)$ is an oriented n-dimensional Riemannian C^2 manifold with boundary, and φ is an orientation preserving C^2 diffeomorphism from M onto Y. The volume element, ω_M, of M is given by

$$\omega_M = \varphi^*\omega_Y = \sqrt{g}\, dx^1 \wedge \cdots \wedge dx^n \tag{72}$$

with

$$g := g_\varphi := \det[g_{jk}].$$

Since M and Y are compact, there exists a constant $\kappa \ge 1$ such that $\kappa^{-1} \le \sqrt{g} \le \kappa$. Hence

$$L_p(M, \mathbb{R}^N) := L_p(M, \mathrm{vol}_M, \mathbb{R}^N) \doteq L_p, \qquad 1 \le p \le \infty, \tag{73}$$

where vol_M is the Lebesgue volume measure on M induced by ω_M.

Moreover, φ_∂ is an orientation preserving C^2 diffeomorphism from $\partial X = \Gamma$ onto $\partial Y = \tilde{\Gamma}$, and ∂M is an $(n-1)$-dimensional Riemannian C^2 manifold, oriented by means of the outer unit normal, and with the volume element

$$\omega_{\partial M} = \varphi_\partial^* \omega_{\partial Y}.$$

There exists a unique $J \in C^1(\Gamma, \mathbb{R})$ satisfying

$$\omega_{\partial M} = J \omega_{\partial X},$$

the Jacobian of φ_∂. Consequently,

$$\mathrm{vol}_{\partial M} = J\, d\sigma. \tag{74}$$

By means of local coordinates and the compactness of Γ one finds a constant $\kappa_1 \ge 1$ such that $\kappa_1^{-1} \le J \le \kappa_1$. Hence

$$L_p(\partial M, \mathbb{R}^N) := L_p(\partial M, \mathrm{vol}_{\partial M}, \mathbb{R}^N) \doteq L_p(\Gamma), \qquad 1 \le p \le \infty.$$

Given a Riemannian manifold R, we denote by $\mathcal{V}_p^j(R)$ the Banach space of all W_p^j vector fields on R for $1 \le p \le \infty$ and $j = 0, 1$, and grad_R is the gradient operator on R. We identify $\vec{w} \in \mathcal{V}_p^j(Y)$ with

$$(w^1, \ldots, w^n) \in W_p^j(\Omega, \mathbb{R}^n)$$

by setting

$$\vec{w} = w^j \partial/\partial y^j.$$

Hence $\varphi_* \vec{v}$, the push forward of $\vec{v} \in \mathcal{V}^j(M)$, is given by

$$\varphi_* \vec{v} = \varphi_*(\partial\varphi)(\vec{v} \circ \varphi^{-1}), \tag{75}$$

where $\partial\varphi \in C^1(\overline{\Omega}, \mathbb{R}^{n \times n})$ is the derivative of φ. It follows that

$$\varphi_* \in \mathcal{L}\mathrm{is}(\mathcal{V}_p^j(M), \mathcal{V}_p^j(Y)), \qquad 1 \le p \le \infty, \quad j = 0, 1,$$

and that its inverse is given by

$$\varphi^* := (\varphi^{-1})_* = \big(\vec{w} \mapsto \varphi^*(\partial \varphi^{-1})(\vec{w} \circ \varphi)\big).$$

Since

$$\varphi^*(\varphi^{-1}) = \varphi^{-1} \circ \varphi = \mathrm{id}_\Omega,$$

the chain rule implies

$$\varphi^*(\partial \varphi^{-1})\partial \varphi = 1.$$

Consequently,

$$\varphi^* \vec{w} = (\partial \varphi)^{-1}(\vec{w} \circ \varphi), \qquad w \in \mathcal{V}_p^j(Y), \quad 1 \le p \le \infty, \quad j = 0, 1. \tag{76}$$

Also note that

$$[g_{jk}] = (\partial \varphi)^\top \partial \varphi.$$

Hence

$$[g^{jk}] := [g_{jk}]^{-1} = (\partial \varphi)^{-1}\big((\partial \varphi)^{-1}\big)^\top,$$

where we identify $\mathbb{R}^{n \times n}$ and $\mathcal{L}(\mathbb{R}^n)$ by means of the standard basis. Thus, using the representation of grad_M in the x-coordinates, it follows that

$$\mathrm{grad}_M f = g^{jk} \frac{\partial f}{\partial x^k} \frac{\partial}{\partial x^j} = (\partial \varphi)^{-1}\big((\partial \varphi)^{-1}\big)^\top \mathrm{grad} f, \qquad f \in \mathcal{V}_p^1(\Omega), \tag{77}$$

for $1 \le p \le \infty$, where we write $\mathrm{grad} := \mathrm{grad}_X$ and, later, $(\cdot \,|\, \cdot) := (\cdot \,|\, \cdot)_X$.

Observe that

$$(\vec{v} \,|\, \vec{w})_M = \varphi^*(\varphi_* \vec{v} \,|\, \varphi_* \vec{w})_Y, \qquad (\vec{v}, \vec{w}) \in \mathcal{V}_{p'}^j(M) \times \mathcal{V}_p^j(M), \quad j = 0, 1, \tag{78}$$

for $1 \le p \le \infty$. Thus, given $f \in W_p^1(\Omega, \mathbb{R})$ and $\vec{v} \in \mathcal{V}_{p'}^0(\Omega)$, the definition of the gradient implies

$$\begin{aligned}
\big(\mathrm{grad}_M(\varphi^* f) \,|\, \vec{v}\big)_M &= d(\varphi^* f)\vec{v} = \varphi^*(df)\vec{v} = \varphi^*\big(df(\varphi_* \vec{v})\big) \\
&= \varphi^*(\mathrm{grad}_Y f \,|\, \varphi_* \vec{v})_Y = (\varphi^* \mathrm{grad}_Y f \,|\, \vec{v})_M,
\end{aligned}$$

using standard properties of the pull back and push forward operators. (We refer to [10] for the theory of differential forms and vector fields as well as the elementary Riemannian geometry used in this section). Hence

$$\mathrm{grad}_M = \varphi^* \circ \mathrm{grad}_Y \circ \varphi_* \in \mathcal{L}\big(\mathcal{V}_p^1(M), \mathcal{V}_p^0(M)\big), \qquad 1 \le p \le \infty. \tag{79}$$

Now suppose that

$$\underset{\sim}{\alpha} := \big([\underset{\sim}{a}_{jk}], (\underset{\sim}{a}_1, \dots, \underset{\sim}{a}_n), \underset{\sim}{a}, \underset{\sim}{b}, \underset{\sim}{\chi}\big) \in \mathcal{E}(\underset{\sim}{\Omega})$$

such that

$$\varphi_\partial^* \underset{\sim}{\chi} = \chi,$$

where χ is a fixed boundary identification map for Ω. Put

$$\underset{\sim}{\mathfrak{a}} := \mathfrak{a}(\underset{\sim}{\alpha})$$

and observe that

$$\underset{\sim}{\mathfrak{a}} = \sum_{r=1}^{N} \underset{\sim}{\mathfrak{a}}^{r},$$

with

$$\underset{\sim}{\mathfrak{a}}^{r}(\underset{\sim}{v}, \underset{\sim}{u}) = \int_{\Omega} \left\{ (\text{grad}_Y \underset{\sim}{v}^{r} \,|\, \underset{\sim}{a}^{r} \text{grad}_Y \underset{\sim}{u}^{r})_Y + \underset{\sim}{v}^{r} \Big[(\underset{\sim}{\vec{a}}^{r} \,|\, \text{grad}_Y \underset{\sim}{u}^{r})_Y + \sum_{s=1}^{N} \underset{\sim}{a}^{rs} \underset{\sim}{u}^{s} \Big] \right\} dy$$

$$+ \int_{\Gamma} \gamma \underset{\sim}{v}^{r} \underset{\sim}{\chi}^{r} \sum_{s=1}^{N} \underset{\sim}{b}^{rs} \gamma \underset{\sim}{v}^{s} \omega_{\Gamma}$$

for $1 \le r \le N$ and

$$(\underset{\sim}{v}, \underset{\sim}{u}) \in W_{q'}^1(Y, \mathbb{R}^N) \times W_q^1(Y, \mathbb{R}^N),$$

where

$$\boldsymbol{a}^r := [\underset{\sim}{a}_{jk}^r], \quad \vec{\underset{\sim}{a}}^r := (\underset{\sim}{a}_1^r, \ldots, \underset{\sim}{a}_n^r), \quad 1 \le r \le N.$$

Note that (73), (78), and (79), imply

$$\left(\text{grad}_Y (\varphi_* v^r) \,\big|\, \boldsymbol{a}^r \text{grad}_Y (\varphi_* u^r) \right)_Y = \varphi_* \left(\text{grad}_M v^r \,\big|\, \varphi^* \big(\underset{\sim}{\boldsymbol{a}}^r \varphi_* (\text{grad}_M u^r) \big) \right)_M \quad (80)$$

for $(v, u) \in W_{q'}^1 \times W_q^1$. From (75)–(77) we deduce that

$$\varphi^* \big(\underset{\sim}{\boldsymbol{a}}^r \varphi_* (\text{grad}_M u^r) \big) = (\partial \varphi)^{-1} \big(\varphi^* \underset{\sim}{\boldsymbol{a}}^r \big((\partial \varphi)^{-1} \big)^{\top} \text{grad}\, u^r.$$

By inserting this in (80) and recalling (71) and the representation (77) of grad_M in the x-coordinates we arrive at

$$\left(\text{grad}_Y (\varphi_* v^r) \,\big|\, \boldsymbol{a}^r \text{grad}_Y (\varphi_* u^r) \right)_Y = \varphi_* \left(\text{grad}\, v^r \,\big|\, (\partial \varphi)^{-1} (\varphi^* \underset{\sim}{\boldsymbol{a}}^r) \big((\partial \varphi)^{-1} \big)^{\top} \text{grad}\, u^r \right).$$

Similarly,

$$\big(\vec{\underset{\sim}{a}}^r \,\big|\, \text{grad}_Y (\varphi_* u^r) \big)_Y = \varphi_* \big((\partial \varphi)^{-1} (\vec{\underset{\sim}{a}}^r \circ \varphi) \,\big|\, \text{grad}\, u^r \big). \quad (81)$$

Now, using (72), (74), (80), (81), and the (global) transformation theorem (e.g., [10, Theorem XII.2.3]), we see that

$$\underset{\sim}{\mathfrak{a}}^r (\varphi_* v, \varphi_* u) = \int_{\Omega} \left\{ \left(\text{grad}\, v^r \,\big|\, (\partial \varphi)^{-1} (\varphi^* \underset{\sim}{\boldsymbol{a}}^r) \big((\partial \varphi)^{-1} \big)^{\top} \text{grad}\, u^r \right) \right.$$

$$+ v^r \Big[\big((\partial \varphi)^{-1} (\vec{\underset{\sim}{a}}^r \circ \varphi) \,\big|\, \text{grad}\, u^r \big) + \sum_{s=1}^{N} (\varphi^* \underset{\sim}{a}^{rs})\, u^s \Big] \right\} \sqrt{g}\, dx \quad (82)$$

$$+ \int_{\Gamma} \gamma v^r \chi^r \sum_{s=1}^{N} (\varphi_{\partial}^* \underset{\sim}{b}^{rs}) \gamma u^s J\, d\sigma$$

for $1 \le r \le N$ and $(v, u) \in W_{q'}^1 \times W_q^1$.

Set

$$\underset{\sim}{a}_{jk,\varphi}^r := \sqrt{g} \Big((\partial \varphi)^{-1} (\varphi^* \underset{\sim}{\boldsymbol{a}}^r) \big((\partial \varphi)^{-1} \big)^{\top} \Big)_{jk} \quad (83)$$

and

$$\underset{\sim}{a}_{j,\varphi}^{r} := \sqrt{g} \, (\partial\varphi)^{-1} (\underset{\sim}{a}_{j}^{r} \circ \varphi) \tag{84}$$

for $1 \le j, k \le n$, as well as

$$\underset{\sim}{a}_{\varphi} := \sqrt{g} \, \varphi^{*} \underset{\sim}{a}, \quad \underset{\sim}{b}_{\varphi} := J \varphi_{\partial}^{*} \underset{\sim}{b}. \tag{85}$$

Then

$$\alpha_{\varphi} := \left([\, \underset{\sim}{a}_{jk,\varphi}], (\underset{\sim}{a}_{1,\varphi}, \dots, \underset{\sim}{a}_{n,\varphi}), \underset{\sim}{a}_{\varphi}, \underset{\sim}{b}_{\varphi}, \chi \right) \in \mathcal{E}(\Omega). \tag{86}$$

Also set

$$\varphi^{*} \underset{\sim}{\mathfrak{a}}(u, v) := \underset{\sim}{\mathfrak{a}}(\varphi_{*} u, \varphi_{*} v), \qquad (v, u) \in W_{q'}^{1} \times W_{q}^{1}.$$

Then it follows from (82) that

$$\varphi^{*} \underset{\sim}{\mathfrak{a}} = \mathfrak{a}(\alpha_{\varphi}), \tag{87}$$

that is, $\varphi^{*} \underset{\sim}{\mathfrak{a}}$ is the Dirichlet form of $\left(\mathcal{A}(\alpha_{\varphi}), \mathcal{B}(\alpha_{\varphi}) \right)$.

Given $(v, u) \in L_{q'} \times L_{q}$, we deduce from (72) and the transformation theorem that

$$\begin{aligned}
\langle \varphi_{*} v, \varphi_{*} u \rangle_{L_{q}(Y, \mathbb{R}^{N})} &= \int_{Y} \varphi_{*} v \cdot \varphi_{*} w \, \omega_{Y} \\
&= \int_{Y} \varphi_{*} (v \cdot w \sqrt{g} \, dx^{1} \wedge \cdots \wedge dx^{n}) \\
&= \int_{\Omega} v \cdot w \sqrt{g} \, dx = \langle v, \sqrt{g} w \rangle.
\end{aligned}$$

Thus, thanks to (70) and a density argument,

$$\langle \varphi_{*} v, \varphi_{*} w \rangle_{W_{q,(1-\underset{\sim}{\chi})\underset{\sim}{\gamma}}^{-1}(\Omega, \mathbb{R}^{N})} = \langle v, \sqrt{g} \, w \rangle_{W_{q,(1-\chi)\gamma}^{-1}}, \qquad (v, w) \in W_{q',(1-\chi)\gamma}^{1} \times W_{q,(1-\chi)\gamma}^{1}.$$

Hence, denoting by $A_{-1}(\underset{\sim}{\alpha})$ the $W_{q,(1-\chi)\gamma}^{-1}(\underset{\sim}{\Omega}, \underset{\sim}{\mathbb{R}}^{N})$ realization of $\left(\mathcal{A}(\alpha), \mathcal{B}(\alpha) \right)$, etc., we obtain from (87) and $\underset{\sim}{\mathfrak{a}} = \mathfrak{a}(\underset{\sim}{\alpha})$ that

$$\begin{aligned}
\langle v, A_{-1}(\underset{\sim}{\alpha}\varphi) u \rangle_{W_{q,(1-\chi)\gamma}^{-1}} &= \mathfrak{a}(\underset{\sim}{\alpha}\varphi)(v, u) = \varphi^{*} \mathfrak{a}(\underset{\sim}{\alpha})(v, u) = \mathfrak{a}(\underset{\sim}{\alpha})(\varphi_{*} v, \varphi_{*} u) \\
&= \langle \varphi_{*} v, A_{-1}(\underset{\sim}{\alpha}) \varphi_{*} u \rangle_{W_{q,(1-\underset{\sim}{\chi})\underset{\sim}{\gamma}}^{-1}(\Omega, \mathbb{R}^{N})} \\
&= \langle \varphi_{*} v, \varphi_{*} \big(\varphi^{*} \circ A_{-1}(\underset{\sim}{\alpha}) \varphi_{*} u \big) \rangle_{W_{q,(1-\underset{\sim}{\chi})\underset{\sim}{\gamma}}^{-1}} \\
&= \langle v, \sqrt{g} \, \big(\varphi^{*} \circ A_{-1}(\underset{\sim}{\alpha}) \circ \varphi_{*} \big) u \rangle_{W_{q,(1-\chi)\gamma}^{-1}}
\end{aligned}$$

for

$$(v, u) \in W_{q',(1-\chi)\gamma}^{1} \times W_{q,(1-\chi)\gamma}^{1}.$$

This shows that

$$\varphi^{*} \circ A_{-1}(\underset{\sim}{\alpha}) \circ \varphi_{*} = \frac{1}{\sqrt{g}} A_{-1}(\underset{\sim}{\alpha}\varphi). \tag{88}$$

Next we give an easy proof of a perturbation theorem for simple eigenvalues.

Theorem 46. *Let E_0 and E_1 be Banach spaces such that $E_1 \hookrightarrow E_0$. Suppose that X is a topological space and*

$$B(\cdot) \in C\big(X, \mathcal{L}(E_1, E_0)\big).$$

Also suppose that $x_0 \in X$ and β_0 is an isolated simple eigenvalue of $B_0 := B(x_0)$. Then, given a corresponding eigenvector e_0 of norm 1 in E_1, there exist an open neighborhood $U \times V \times W$ of (x_0, β_0, e_0) in $X \times \mathbb{C} \times E_1$ and a map $\big(\beta(\cdot), e(\cdot)\big) \in C(U, V \times W)$ such that

$$\big(\beta(x_0), e(x_0)\big) = (\beta_0, e_0), \quad \|e(x)\|_{E_1} = 1,$$

and

$$B(x)e(x) = \beta(x)e(x)$$

for $x \in U$. Furthermore, $\beta(x)$, resp. $e(x)$, is, for $x \in U$, the only eigenvalue, resp. eigenvector, of $B(x)$ in V, resp. W.

Proof. Since β_0 is an isolated simple eigenvalue of B_0, it follows that B_0, considered as a linear operator in E_0, has a nonempty resolvent set. Consequently, B_0 is closed in E_0. Thus spectral theory implies that

$$E_0 = \mathbb{C}_{e_0} \oplus \operatorname{im}(\beta_0 - B_0). \tag{89}$$

Furthermore, B_0', the dual of B_0 (in E_0), has β_0 as a simple eigenvalue as well, and there exists a unique eigenvector e_0' of B_0' to the eigenvalue β_0 satisfying $\langle e_0', e_0 \rangle = 1$. In addition,

$$\operatorname{im}(\beta_0 - B_0) = \big\{ e \in E_0 \, ; \, \langle e_0', e \rangle = 0 \big\}$$

(e.g., [48, Sections III.6.5 and III.6.6]).

Now we define a C^1 map

$$f \, : \, \mathcal{L}(E_1, E_0) \times (\mathbb{C} \times E_1) \to \mathbb{C} \times E_0$$

by

$$f\big(B, (\beta, e)\big) := \big(\langle e_0', e \rangle - 1, Be - \beta e\big).$$

Note that

$$f\big(B_0, (\beta_0, e_0)\big) = 0$$

and

$$\partial_2 f\big(B_0, (\beta_0, e_0)\big)(\beta, e) = \big(\langle e_0', e \rangle, Be - \beta_0 e - \beta e_0\big), \qquad (\beta, e) \in \mathbb{C} \times E_1.$$

Hence $\partial_2 f\big(B_0, (\beta_0, e_0)\big)(\beta, e) = 0$ implies $\langle e_0', e \rangle = 0$, that is,

$$e \in \operatorname{im}(\beta_0 - B_0) \quad \text{and} \quad Be - \beta_0 e = \beta e_0.$$

Thus we deduce from (89) that

$$(\beta, e) = 0,$$

which shows that the linear operator $\partial_2 f\big(B_0, (\beta_0, e_0)\big)$ is injective. From (89) we also infer that this map is surjective. Hence

$$\partial_2 f\big(B_0, (\beta_0, e_0)\big) \in \mathcal{L}\mathrm{is}(\mathbb{C} \times E_1, \mathbb{C} \times E_0)$$

by the open mapping theorem. Now the implicit function theorem guarantees the existence of an open neighborhood $B \times V \times W$ of (B_0, β_0, e_0) in $\mathcal{L}(E_1, E_0) \times \mathbb{C} \times E_1$ and a map $h \in C^1(B, V \times W)$ such that

$$\big(B, (\beta, e)\big) \in B \times V \times W \text{ and } f\big(B, (\beta, e)\big) = 0 \quad \text{iff} \quad B \in B \text{ and } (\beta, e) = h(B).$$

Since $B(\cdot) \in C\big(X, \mathcal{L}(E_1, E_0)\big)$ there exists an open neighborhood U of x_0 in X such that $B(U) \subset B$. Thus the assertion follows by setting

$$\big(\beta(x), e_1(x)\big) := h\big(B(x)\big), \quad e(x) := e_1(x)/\|e_1(x)\|_{E_1}$$

for $x \in U$. $\square$

This proof is a modification and an extension to the infinite dimensional case of the one of [6, Proposition (26.24)]. An alternative proof of Theorem 46 can also be derived from the general perturbation results in [48, Section IV.3.4].

After these preparations we are ready to derive the theorems stated in Section 9. In fact, instead of Theorem 20 we prove the following more precise version.

Theorem 47. *Let the hypotheses of Theorem 20 be satisfied with the exception that it is assumed that*

$$\varphi_i^* \alpha_i \to \alpha \qquad \text{in } \mathbb{E}_{\xi, \eta, \zeta}(\Omega). \tag{90}$$

Then the assertions of Theorem 20 are valid, except that $\varphi_i^* u_i \to u$ in W_q^1.

Proof. It follows from $\varphi_i \to \mathrm{id}_\Omega$ in $C^1(\overline{\Omega}, \mathbb{R}^n)$ that $\partial \varphi_i \to 1$ in $C(\overline{\Omega}, \mathbb{R}^{n \times n})$ and $\sqrt{g_{\varphi_i}} \to 1$ in $C(\overline{\Omega}, \mathbb{R})$ as well as $J_{\varphi_i} \to 1$ in $C(\Gamma, \mathbb{R})$. Hence we deduce from (90) and (83)–(86) that

$$\alpha_{i, \varphi_i} \to \alpha \qquad \text{in } \mathbb{E}_{\xi, \eta, \zeta}(\Omega). \tag{91}$$

This implies

$$\varphi_{i, \partial}^* \chi_i = \chi \tag{92}$$

for all sufficiently large i. Thus we can assume that (92) is true for all $i \in \mathbb{N}$. Now we infer from (91), (92), and Lemma 45 that

$$\mathfrak{a}(\alpha_{i, \varphi_i}) \to \mathfrak{a}(\alpha) \qquad \text{in } \mathcal{L}(W_{q', (1-\chi)\gamma}^1, W_{q, (1-\chi)\gamma}^1; \mathbb{R}).$$

Consequently, (69) and (88) imply, setting $(\mathcal{A}, \mathcal{B}) := \big(\mathcal{A}(\alpha), \mathcal{B}(\alpha)\big)$,

$$C_i := \varphi_i^* \circ A_{-1}(\alpha_i) \circ \varphi_{i, *} \to A_{-1} \qquad \text{in } \mathcal{L}(W_{q, \mathcal{B}}^1, W_{q, \mathcal{B}}^{-1}). \tag{93}$$

Since, by (70),

$$\varphi_i^* \in \mathcal{L}is\big(W_{q, \mathcal{B}(\alpha_i)}^j(\Omega_i, \mathbb{R}^N), W_{q, \mathcal{B}}^j\big), \qquad j = \pm 1,$$

with inverse $\varphi_{i, *}$, we see that

$$\sigma(C_i) = \sigma\big(A_{-1}(\alpha_i)\big)$$

and

$$\ker(\lambda - C_i) = \varphi_i^*\big(\ker\big(\lambda - A_{-1}(\alpha_i)\big)\big), \qquad \lambda \in \mathbb{C},$$

for $i \in \mathbb{N}$.

Set

$$X := \{(\mathrm{id}_{\overline{\Omega}}, \alpha)\} \cup \{(\varphi_i, \alpha_i) ; i \in \mathbb{N}\}.$$

Endow X with the topology induced by $C^1(\overline{\Omega}, \mathbb{R}^n) \times \mathbb{E}_{\xi, \eta, \zeta}(\Omega)$. Put

$$E_0 := W^{-1}_{q, \mathcal{B}} \quad \text{and} \quad E_1 := W^1_{q, \mathcal{B}},$$

and define $B(x) \in \mathcal{L}(E_1, E_0)$ by

$$B(x) := \begin{cases} A_{-1} & \text{if } x = x_0 := (\mathrm{id}_\Omega, \alpha), \\ C_i & \text{if } x \neq x_0. \end{cases}$$

Then (93) implies $B(\cdot) \in C(X, \mathcal{L}(E_1, E_0))$. Furthermore, Theorem 12 guarantees that $\beta_0 := \lambda_0(\mathcal{A}, \mathcal{B})$ is a simple isolated eigenvalue of $B(x_0)$ and that $e_0 := u$ is a corresponding eigenvector. Thus Theorem 46 guarantees the existence of an eigenvalue β_i of C_i and corresponding eigenvector e_i such that $\beta_i \to \beta$ and $e_i \to e_0$ in $W^1_{q, \mathcal{B}}$ with $\|e_i\|_{W^1_q} = 1$. From the upper semicontinuity of the spectrum (e.g., [48, Theorem IV.3.1]) it easily follows that

$$\beta_i = \lambda_0(\mathcal{A}(\alpha_i), \mathcal{B}(\alpha_i), \Omega_i)$$

for all sufficiently large i. Finally, $u \geq 0$ implies $e_i = \varphi_i^* u_i$, where u_i is a positive eigenfunction of $(\mathcal{A}(\alpha_i), \mathcal{B}(\alpha_i))$, since $\lambda_0(\mathcal{A}(\alpha_i), \mathcal{B}(\alpha_i), \Omega_i)$ is also a simple eigenvalue of $(\mathcal{A}(\alpha_i), \mathcal{B}(\alpha_i))$ by Theorem 12. $\qquad\square$

Clearly, (r, s, t), defined in (31), is an admissible choice for (ξ, η, ζ) if $q = 2$. Hence Theorem 20 is a particular case of Theorem 47.

In order to prove Theorem 22 we prepare the following technical lemma, where we use the notations introduced in Section 9.

Lemma 48. *Suppose that $\beta \in C^2(\Gamma, (-\rho, \rho))$ and $\|\beta\|_\infty \leq \rho/3$. Then there exists an orientation preserving C^2 diffeomorphism $\varphi : \overline{\Omega} \to \overline{\Omega}_\beta$ satisfying $\varphi|\Gamma = \psi_\beta$ and $\varphi(x) = x$ for $x \in \Omega$ with $\mathrm{dist}(x, \Gamma) \geq 2 \|\beta\|_\infty$.*

Proof. Denote by $\widetilde{\beta} \in C^2(\Gamma, (-1, 1))$ the image of β under the C^2 diffeomorphism h of (32) and let $\widetilde{\Gamma}$ be the graph of $\widetilde{\beta}$. Then $\widetilde{\Gamma}$ is an oriented $(n-1)$-dimensional C^2 manifold lying in $\Gamma \times [-\widetilde{\delta}, \widetilde{\delta}]$, where $\widetilde{\delta} := \|\widetilde{\beta}\|_\infty < 1/3$. Fix $\sigma \in C^2(\mathbb{R}, [0, 1])$ satisfying $\sigma(s) = 1$ for $|s| \leq \widetilde{\delta}$ and $\sigma(s) = 0$ for $|s| \geq 2\widetilde{\delta}$, and being strictly decreasing on $(\widetilde{\delta}, 2\widetilde{\delta})$ and even. Then

$$\widetilde{\psi} : \Gamma \times (-1, 1) \to \Gamma \times (-1, 1), \quad (y, t) \mapsto (y, t - \sigma(t)\widetilde{\beta}(y)) \tag{94}$$

is an orientation preserving C^2 diffeomorphism onto $\Gamma \times (-1, 1)$. It satisfies

$$\widetilde{\psi}(y, t) = (y, t) \text{ for } y \in \Gamma \text{ and } |t| \geq 2\widetilde{\delta} \quad \text{and} \quad \widetilde{\psi}(\widetilde{\Gamma}) = \Gamma \times \{0\}.$$

Let $\widetilde{\varphi}$ be the restriction of $\widetilde{\psi}^{-1}$ to

$$\{(y, t) \in \Gamma \times (-1, 1) ; t \geq 0\}$$

and set
$$\widehat{\varphi} := h^{-1} \circ \widetilde{\varphi} \circ h.$$
Then $\widehat{\varphi}$ is a C^2 diffeomorphism from $\overline{\Omega} \cap T_\rho$ onto $\overline{\Omega}_\beta \cap T_\rho$ satisfying $\widehat{\varphi}(x) = x$ for $x \in \overline{\Omega} \cap T_\rho$ with $\mathrm{dist}(x, \Gamma) \geq 2 \|\beta\|_\infty$. Define $\varphi : \overline{\Omega} \to \overline{\Omega}_\beta$ by setting $\varphi(x) := \widehat{\varphi}(x)$ for $x \in \overline{\Omega} \cap T_\rho$, and $\varphi(x) := x$ for $x \in \Omega \setminus T_\rho$. Then φ is an orientation preserving C^2 diffeomorphism possessing the stated properties. $\square$

Now we can prove a more precise version of Theorem 22 in which W_2^1 convergence of the eigenfunction is replaced by W_q^1 convergence.

Theorem 49. *Let Ω be bounded and suppose that conditions (33)–(39) be satisfied, except that r, s, and t are replaced by ξ, η, and ζ, respectively. Then the assertion of Theorem 22 are valid, provided the exponents 2 in (40) are replaced by q.*

Proof. We can assume that $\|\beta_i\|_\infty \leq \rho/3$ for all $i \in \mathbb{N}$. For each i fix an orientation preserving C^2 diffeomorphism $\varphi_1 : \overline{\Omega} \to \overline{\Omega}_{\beta_i}$ such that $\varphi_i(x) = x$ for $x \in \Omega$ with $\mathrm{dist}(x, \Gamma) \geq 2 \|\beta_i\|_\infty$. Lemma 48 guarantees that this is possible, and it follows from (94) and $\beta_i \to 0$ in $C^1(\Gamma, \mathbb{R})$ that $\varphi_i \to \mathrm{id}_{\overline{\Omega}}$ in $C^1(\overline{\Omega}, \mathbb{R}^n)$ as $i \to \infty$. It is easily verified that $\varphi_i^* \alpha_i \to \alpha$ in $\mathbb{E}_{\xi,\eta,\zeta}(\Omega)$. Hence Theorem 47 implies the assertion. $\square$

Similarly as above, it is clear that Theorem 49 contains Theorem 22 as a particular case.

19. Elliptic comparison theorems

In this section we study the weak maximum principle under optimal regularity assumptions for the coefficients of $\mathcal{A}$ and $\mathcal{B}$. More precisely, we assume that

$$
\left.
\begin{aligned}
&\bullet \quad \xi, \eta, \text{ and } \zeta \text{ satisfy (66), (67), and (68), respectively;} \\
&\bullet \quad a_{jk} = a_{kj} \in BUC(\Omega, \mathbb{R}_{\mathrm{diag}}^{N \times N}), \ 1 \leq j, k \leq n; \\
&\bullet \quad \boldsymbol{a}^r(x) \in \mathbb{R}^{n \times n} \text{ is positive definite for } 1 \leq r \leq N, \\
&\qquad \text{uniformly with respect to } x \in \Omega; \\
&\bullet \quad a_j \in (L_\xi + L_\infty)(\Omega, \mathbb{R}_{\mathrm{diag}}^{N \times N}), \ 1 \leq j \leq n; \\
&\bullet \quad a \in (L_\eta + L_\infty)(\Omega, \mathbb{R}^{N \times N}), \ b \in L_\zeta(\Gamma, \mathbb{R}^{N \times N}); \\
&\bullet \quad -a \text{ and } -b \text{ are cooperative;} \\
&\bullet \quad \chi \text{ is a boundary characterization map for } \Omega.
\end{aligned}
\right\} \tag{95}
$$

Then we define $(\mathcal{A}, \mathcal{B})$ and its Dirichlet form as before. Note that
$$W_{q,\mathcal{B}}^1 := W_{q,(1-\chi)\gamma}^1 \quad \text{and} \quad W_{q,\mathcal{B}}^{-1} := (W_{q,\mathcal{B}}^1)'$$
are still well-defined.

It follows from Lemma 45 that

$$\mathfrak{a} \in \mathcal{L}(W^1_{q',(1-\chi)\gamma}, W^1_{q,(1-\chi)\gamma}; \mathbb{R}).$$

Hence there exits a unique

$$A_{-1} \in \mathcal{L}(W^1_{q,\mathcal{B}}, W^{-1}_{q,\mathcal{B}})$$

satisfying

$$\langle v, A_{-1}u \rangle = \mathfrak{a}(v, u), \qquad (v, u) \in W^1_{q',(1-\chi)\gamma} \times W^1_{q,(1-\chi)\gamma},$$

the $W^{-1}_{q,\mathcal{B}}$ **realization** of $(\mathcal{A}, \mathcal{B})$.

Observe that in the present situation $\mathcal{A}(W^2_q) \not\subset L_q$, in general, so that the L_q realization of $(\mathcal{A}, \mathcal{B})$ cannot be defined as in the earlier sections and is not useful for our purposes.

Theorem 50. *Let* (95) *be satisfied. Then*

$$A_{-1} \in \mathcal{H}(W^1_{q,\mathcal{B}}, W^{-1}_{q,\mathcal{B}})$$

and is resolvent positive.

Proof. In [9] it is shown (by an amplification of the proof of [4, Theorem 2.1], where Ω is supposed to be bounded and more regularity is required for the lower order coefficients) that $A_{-1} \in \mathcal{H}(W^1_{q,\mathcal{B}}, W^{-1}_{q,\mathcal{B}})$.

Choose a sequence $\alpha_k \in \mathbb{E}^1(\Omega)$ converging in $\mathbb{E}_{\xi,\eta,\varsigma}(\Omega)$ towards

$$\alpha := \big([a_{jk}], (a_1, \ldots, a_n), a, b, \chi\big).$$

Since $\mathcal{D}$ is dense in L_p for $1 \le p < \infty$, it is clear that such a sequence exists. Then Lemma 45 and (69) imply that

$$A_{-1}(\alpha_k) \to A_{-1} \quad \text{in } \mathcal{L}(W^1_{q,\mathcal{B}}, W^{-1}_{q,\mathcal{B}}).$$

By Theorems 37 and 41 we know that

$$A_{-1}(\alpha_k) \in \mathcal{H}(W^1_{q,\mathcal{B}}, W^{-1}_{q,\mathcal{B}})$$

and is resolvent positive. Thus Proposition 34 implies that A_{-1} is also resolvent positive. $\square$

Clearly, definition (18) of the weak maximum principle is valid in the present situation also, as is the definition (21) of a (weak) W^1_q solution.

Corollary 51. *Let assumption* (95) *be satisfied and suppose that*

$$\lambda_0(A_{-1}) = -s(-A_{-1}) > 0. \tag{96}$$

Then

(i) $(\mathcal{A}, \mathcal{B})$ *satisfies the weak maximum principle in* W^1_q;

(ii) *The boundary value problem*

$$\mathcal{A}u = f \text{ in } \Omega, \quad \mathcal{B}u = g \text{ on } \Gamma$$

has for each $(f, g) \in W_{q,\mathcal{B}}^{-1} \times \partial W_q^1$ *a unique* W_q^1 *solution* u, *and*

$$u \geq 0 \quad \text{if } (f, g) \geq 0.$$

Proof. (i) follows from Theorems 28 and 50.

(ii) Assumption (96) guarantees that $0 \in \rho(A_{-1})$. Thus the proof of case $j = 1$ of Theorem 39(ii) applies to give the unique solvability. The last part of the assertion is a consequence of (i). □

Remarks 52.

(a) Note that Theorem 28(b) guarantees that $\lambda_0(-A_{-1})$ belongs to $\sigma(A_{-1})$ if the latter set is not empty. However, we do not know whether this is true, in general, even if the domain is bounded (in which case A_{-1} has a compact resolvent) and even if $N = 1$.

(b) In the weak setting studied above it is natural to consider operators of the form

$$\tilde{\mathcal{A}}u := -\partial_j(a_{jk}\partial_k u + \tilde{a}_j u) + a_j \partial_j + au,$$

where $\tilde{a}_j : \Omega \to \mathbb{R}_{\text{diag}}^{N \times N}$ satisfy appropriate regularity assumptions. It is not difficult to determine these optimal conditions and to show that Theorem 50 and its corollary hold in this case also. Note that the corresponding boundary operator is now — formally — given by

$$\tilde{\mathcal{B}}u := \chi(\partial_\mu u + \nu^j \cdot \tilde{a}_j u + bu) + (1 - \delta)u.$$

We leave the details to the interested reader. □

The scalar case ($N = 1$ has been studied by many authors (see [23], [24], [38], [49], [61], [68], [73]–[75], and the references therein). However, in all those papers only the case $q = 2$ is considered. In that situation one can, of course, weaken the regularity conditions on Γ considerably, and it suffices to assume that the a_{jk} are only bounded and measurable (in fact, Trudinger [73]–[75] considers even the case of nonuniformly elliptic equations). It is well-known that this is no longer true if $q \neq 2$. We do not know of any work dealing with weak maximum principles in a W_q^1 setting, except for [7, Theorem 8.7], where the resolvent positivity of A_{-1} is proved if $N = 1$ and the lower order coefficients satisfy stronger regularity assumptions.

Now we can easily derive a comparison theorem for semilinear elliptic boundary value problems. For this we recall that, given a σ-finite measure space (X, m) and Banach spaces E and F, a function $f : X \times E \to F$ is said to be a Carathéodory function if

$$f(x, \cdot) : E \to F$$

is continuous for m-a.a. $x \in X$, and

$$f(\cdot, \xi) : X \to F$$

is m-measurable for each $\xi \in E$. We denote by $\mathrm{Car}(X \times E, F)$ the set of all such functions.

We assume that

$$
\left.
\begin{array}{l}
\bullet \quad f \in \mathrm{Car}(\Omega \times \mathbb{R}^N, \mathbb{R}^N),\ g_1 \in \mathrm{Car}(\Gamma \times \mathbb{R}^N, \mathbb{R}^N); \\[4pt]
\bullet \quad r, s \in (1, \infty)\ with \\[4pt]
\qquad r \leq (n+q)/(n-q)\ and\ s \leq n/(n-q)\ if\ q < n; \\[4pt]
\bullet \quad \alpha_0 \in L_{r_0}(\Omega, \mathbb{R}^+)\ and\ \beta_0 \in L_{s_0}(\Gamma, \mathbb{R}^+),\ where\ r_0, s_0 \in [1, \infty) \\[4pt]
\qquad satisfy\ r_0 \geq nq/(n+q)\ and\ s_0 \geq q(n-1)/n; \\[4pt]
\bullet \quad |f(\cdot, \xi)| \leq \alpha_0 + \alpha\, |\xi|^r,\ |g_1(\cdot, \xi)| \leq \beta_0 + \beta\, |\xi|^s,\ \xi \in \mathbb{R}^N, \\[4pt]
\qquad where\ \alpha\ and\ \beta\ are\ positive\ constants; \\[4pt]
\bullet \quad g_0 \in W_q^{1-1/q}(\Gamma, \mathbb{R}^N).
\end{array}
\right\} \tag{97}
$$

We also set

$$
g(\cdot, \xi) := (1 - \chi)g_0 + \chi g_1(\cdot, \xi), \qquad \xi \in \mathbb{R}^N,
$$

as well as

$$
F(u) := f(\cdot, u(\cdot)), \quad G(u) := g(\cdot, \gamma u(\cdot)).
$$

Then Sobolev embeddings, the trace theorem, and an obvious duality argument imply that

$$
\big(F(u), G(u)\big) \in W_{q,\mathcal{B}}^{-1} \times \partial W_q^1, \qquad u \in W_q^1.
$$

A function $\widehat{u}$ is said to be a W_q^1 **supersolution** of the nonlinear boundary value problem

$$
\mathcal{A}u = f(x, u)\ in\ \Omega, \quad \mathcal{B}u = g(x, u)\ on\ \Gamma \tag{98}
$$

if $\widehat{u} \in W_q^1$ and

$$
\mathcal{A}\widehat{u} \geq f(x, \widehat{u})\ in\ \Omega, \quad \mathcal{B}\widehat{u} \geq g(x, \widehat{u})\ on\ \Gamma
$$

in the weak sense, that is,

$$
\left.
\begin{array}{ll}
\mathfrak{a}(v, \widehat{u}) \geq \big\langle v, F(\widehat{u}) \big\rangle + \big\langle \gamma v, G(\widehat{u}) \big\rangle_\Gamma & for\ v \in (W_{q', (1-\chi)\gamma}^1)^+, \\[6pt]
(1 - \chi)\gamma \widehat{u} \geq (1 - \chi)g_0 & on\ \Gamma.
\end{array}
\right\} \tag{99}
$$

If both inequalities in (99) are reversed then $\widehat{u}$ is said to be a W_q^1 **subsolution** of (98).

Theorem 53. *Let* (95) *and* (97) *be satisfied and suppose that*

$$
\lambda_0(A_{-1}) > 0.
$$

If $\overline{u}$ *is a* W_q^1 *subsolution and* $\widehat{u}$ *is a* W_q^1 *supersolution of* (98) *such that*

$$
\big(F(\overline{u}), G(\overline{u})\big) \leq \big(F(\widehat{u}), G(\widehat{u})\big)
$$

then

$$
\overline{u} \leq \widehat{u}.
$$

Proof. The assumptions imply that $\widehat{u} - \overline{u} \in W_q^1$ satisfies

$$\mathfrak{a}(v, \widehat{u} - \overline{u}) \geq 0 \quad \text{for } v \in (W_{q',(1-\chi)\gamma}^1)^+,$$

$$(1 - \chi)\gamma(\widehat{u} - \overline{u}) \geq 0 \quad \text{on } \Gamma.$$

Hence the assertion follows from Corollary 51. □

It should be clear to the reader that in order to guarantee the validity of the very weak maximum principle the regularity assumption on a in (7) can be weakened also. We refrain from giving details.

References

1. Amann, H., Fixed point equations and nonlinear eigenvalue problems in ordered Banach spaces, *SIAM Rev.* **18** (1976), 620–709.
2. Amann, H., Nonlinear elliptic equations with nonlinear boundary conditions, in *New Developments in Differential Equations (Proc. 2nd Scheveningen Conf., Scheveningen, 1975)*, pages 43–63, North–Holland Math. Studies 21, North-Holland, Amsterdam, 1976.
3. Amann, H., Dual semigroups and second order linear elliptic boundary value problems, *Israel J. Math.* **45** (1983), 225–254.
4. Amann, H., Dynamic theory of quasilinear parabolic systems, III. Global existence, *Math. Z.* **202** (1989), 219–250.
5. Amann, H., Dynamic theory of quasilinear parabolic equations, II. Reaction-diffusion systems, *Diff. Int. Eqns.* **3** (1990), 13–75.
6. Amann, H., *Ordinary Differential Equations*, de Gruyter Studies in Mathematics 13, Walter de Gruyter & Co., Berlin, 1990.
7. Amann, H., Nonhomogeneous linear and quasilinear elliptic and parabolic boundary value problems, in *Function Spaces, Differential Operators and Nonlinear Analysis (Friedrichroda, 1992)*, pages 9–126, Teubner-Texte Math. 133, Teubner, Stuttgart, 1993.
8. Amann, H., *Linear and Quasilinear Parabolic Problems, Vol. I*, Monographs in Mathematics 89. Birkhäuser Boston Inc., Boston, MA, 1995.
9. Amann, H., Maximal regularity and weak solutions of linear parabolic equations, to appear.
10. Amann, H., and Escher, J., *Analysis, Vol. III*, Grundstudium Mathematik, Birkhäuser Verlag, Basel, 2001.
11. Amann, H., and López-Gómez, J., A priori bounds and multiple solutions for superlinear indefinite elliptic problems, *J. Diff. Eqns.* **146** (1998), 336–374.
12. Amann, H., and Quittner, P., Elliptic boundary value problems involving measures: existence, regularity, and multiplicity, *Adv. Diff. Eqns.* **3** (1998), 753–813.
13. Arendt, W., Gaussian estimates and interpolation of the spectrum in L^p, *Diff. Int. Eqns.* **7** (1994), 1153–1168.

14. Arendt, W., and Batty, Ch.J.K., Principal eigenvalues and perturbation, in *Operator Theory in Function Spaces and Banach Lattices*, pages 39–55, Oper. Theory Adv. Appl. 75, Birkhäuser, Basel, 1995.
15. Arendt, W., Batty, Ch.J.K., Hieber, M., and Neubrander, F., *Vector-valued Laplace Transforms and Cauchy Problems*, Monographs in Mathematics 96, Birkhäuser Verlag, Basel, 2001.
16. Arendt, W., Grabosch, A., Greiner, G., Groh, U., Lotz, H.P., Moustakas, U., Nagel, R., Neubrander, F., and Schlotterbeck, F., *One-parameter Semigroups of Positive Operators*, Lecture Notes in Mathematics 1184, Springer-Verlag, Berlin, 1986.
17. Arendt, W., and Monniaux, S., Domain perturbation for the first eigenvalue of the Dirichlet Schrödinger operator, in *Partial Differential Operators and Mathematical Physics (Holzhau, 1994)*, pages 1–19, Oper. Theory Adv. Appl. 78, Birkhäuser, Basel, 1995.
18. Beltramo, A., and Hess, P., On the principal eigenvalue of a periodic-parabolic operator, *Comm. Part. Diff. Eqns.* **9** (1984), 919–941.
19. Berestycki, H., Nirenberg, L., and Varadhan, S.R.S., The principal eigenvalue and maximum principle for second-order elliptic operators in general domains, *Comm. Pure Appl. Math.* **47** (1994), 47–92.
20. Birindelli, I., Mitidieri, E., and Sweers, G., Existence of the principal eigenvalue for cooperative elliptic systems in a general domain, *Differ. Uravn.* **35** (1999), 325–333.
21. Cano-Casanova, S., and López-Gómez, J., Properties of the principal eigenvalues of a general class of non-classical mixed boundary value problems, *J. Diff. Eqns.* **178** (2002), 123–211.
22. Cantrell, R.S., and Schmitt, K., On the eigenvalue problem for coupled elliptic systems, *SIAM J. Math. Anal.* **17** (1986), 850–862.
23. Chicco, M., Principio di massimo generalizzato e valutazione del primo autovalore per problemi ellittici del secondo ordine di tipo variazionale, *Ann. Mat. Pura Appl. (4)* **87** (1970), 1–9.
24. Chicco, M., Some properties of the first eigenvalue and the first eigenfunction of linear second order elliptic partial differential equations in divergence form, *Boll. Un. Mat. Ital.* **5** (1972), 245–254.
25. Clément, Ph., Heijmans, H.J.A.M., Angenent, S., van Duijn, C.J., and de Pagter, B., *One-Parameter Semigroups*, CWI Monographs 5, North-Holland Publishing Co., Amsterdam, 1987.
26. Clément, Ph., and Peletier, L.A., An anti-maximum principle for second-order elliptic operators, *J. Diff. Eqns.* **34** (1979), 218–229.
27. Dancer, E.N., and Daners, D., Domain perturbation for elliptic equations subject to Robin boundary conditions, *J. Diff. Eqns.* **138** (1997), 86–132.
28. Daners, D., Existence and perturbation of principal eigenvalues for a periodic-parabolic problem, in *Proceedings of the Conference on Nonlinear Differential Equations (Coral Gables, FL, 1999)*, pages 51–67 (electronic), Electron. J. Differ. Equ. Conf. 5, San Marcos, TX, 2000.
29. Daners, D., Dirichlet problems on varying domains, *J. Diff. Eqns.* **188** (2003), 591–624.
30. Dautray, R., and Lions, J.L., *Mathematical Analysis and Numerical Methods for Science and Technology, Vol. 1*, Springer-Verlag, Berlin, 1990.

31. Davies, E.B., L^p spectral independence for certain uniformly elliptic operators, in *Partial differential Equations and Mathematical Physics (Copenhagen, 1995; Lund, 1995)*, pages 122–125, Progr. Nonlinear Differential Equations Appl. 21, Birkhäuser Boston, Boston, MA, 1996.

32. de Figueiredo, D.G., and Mitidieri, E., Maximum principles for cooperative elliptic systems, *C. R. Acad. Sci. Paris Sér. I Math.* **310** (1990), 49–52.

33. de Pagter, B., Irreducible compact operators, *Math. Z.* **192** (1986), 149–153.

34. Denk, R., Hieber, M., and Prüss, J., *R-boundedness, Fourier Multipliers and Problems of Elliptic and Parabolic Type*, Memoirs Amer. Math. Soc. 166, Providence, R.I., 2003.

35. Fleckinger, J., Hernández, J., and de Thélin, F., Existence of multiple principal eigenvalues for some indefinite eigenvalue problems, *Boll. Unione Mat. Ital.* **7** (2004), 159–188.

36. Fleckinger, J., Hernández, J., and de Thélin, F., A maximum principle for linear cooperative elliptic systems, in *Differential Equations with Applications to Mathematical Physics*, pages 79–86, Math. Sci. Engrg. 192, Academic Press, Boston, MA, 1993.

37. Fraile, J.M., Koch Medina, P., López-Gómez, J., and Merino, S., Elliptic eigenvalue problems and unbounded continua of positive solutions of a semilinear elliptic equation, *J. Diff. Eqns.* **127** (1996), 295–319.

38. Gilbarg, D., and Trudinger, N.S., *Elliptic Partial Differential Equations of Second Order*, Classics in Mathematics, Springer-Verlag, Berlin, 2001.

39. Greiner, G., Zur Perron-Frobenius-Theorie stark stetiger Halbgruppen, *Math. Z.* **177** (1981), 401–423.

40. Hernández, J., Some existence and stability results for solutions of reaction-diffusion systems with nonlinear boundary conditions, in *Nonlinear Differential Equations (Proc. Internat. Conf., Trento, 1980)*, pages 161–173, Academic Press, New York, 1981.

41. Hernández, J., Positive solutions of reaction-diffusion systems with nonlinear boundary conditions and the fixed point index, in *Nonlinear Phenomena in Mathematical Sciences (Arlington, Tex., 1980)*, pages 525–535, Academic Press, New York, 1982.

42. Hernández, J., Mancebo, F.J., and Vega, J.M., On the linearization of some singular, nonlinear elliptic problems and applications, *Analyse Non Linéaire, Ann. Inst. H. Poincaré* **19** (2002), 777–813.

43. Hess, P., On the eigenvalue problem for weakly coupled elliptic systems, *Arch. Rat. Mech. Anal.* **81** (1983), 151–159.

44. Hess, P., *Periodic-Parabolic Boundary Value Problems and Positivity*, Pitman Research Notes in Mathematics Series 247, Longman Scientific & Technical, Harlow, 1991.

45. Hieber, M., and Schrohe, E., L^p spectral independence of elliptic operators via commutator estimates, *Positivity* **3** (1999), 259–272.

46. Hirsch, M.W., *Differential Topology*, Graduate Texts in Mathematics 33, Springer-Verlag, New York, 1976.

47. Kato, T., Superconvexity of the spectral radius, and convexity of the spectral bound and the type, *Math. Z.* **180** (1982), 265–273.

48. Kato, T., *Perturbation Theory for Linear Operators*, Classics in Mathematics, Springer-Verlag, Berlin, 1995.

49. Kinderlehrer, D., and Stampacchia, G., *An Introduction to Variational Inequalities and their Applications*, Classics in Applied Mathematics 31, Society for Industrial and Applied Mathematics (SIAM), Philadelphia, PA, 2000.

50. Kunstmann, P.Ch., Heat kernel estimates and L^p spectral independence of elliptic operators, *Bull. London Math. Soc.* **31** (1999), 345–353.

51. López-Gómez, J., Nonlinear eigenvalues and global bifurcation application to the search of positive solutions for general Lotka-Volterra reaction diffusion systems with two species, *Diff. Int. Eqns.* **7** (1994), 1427–1452.

52. López-Gómez, J., The maximum principle and the existence of principal eigenvalues for some linear weighted boundary value problems, *J. Diff. Eqns.* **127** (1996), 263–294.

53. López-Gómez, J., Large solutions, metasolutions, and asymptotic behaviour of the regular positive solutions of sublinear parabolic problems, in *Proceedings of the Conference on Nonlinear Differential Equations (Coral Gables, FL, 1999)*, pages 135–171 (electronic), Electron. J. Differ. Equ. Conf. 5, San Marcos, TX, 2000.

54. López-Gómez, J., *Spectral Theory and Nonlinear Functional Analysis*, Chapman & Hall/CRC Research Notes in Mathematics 426, Chapman & Hall/CRC, Boca Raton, FL, 2001.

55. López-Gómez, J., Classifying smooth supersolutions in a general class of elliptic boundary value problems, *Adv. in Diff. Eqns.* **8** (2003), 1025–1042.

56. López-Gómez, J., and Molina-Meyer, M., The maximum principle for cooperative weakly coupled elliptic systems and some applications, *Diff. Int. Eqns.* **7** (1994), 383–398.

57. López-Gómez, J., and Pardo, R., Multiparameter nonlinear eigenvalue problems: positive solutions to elliptic Lotka-Volterra systems, *Appl. Anal.* **31** (1988), 103–127.

58. López-Gómez, J., and Sabina de Lis, J., Coexistence states and global attractivity for some convective diffusive competing species models, *Trans. Amer. Math. Soc.* **347** (1995), 3797–3833.

59. Mitidieri, E., and Sweers, G., Existence of a maximal solution for quasimonotone elliptic systems, *Diff. Int. Eqns.* **7** (1994), 1495–1510.

60. Mitidieri, E., and Sweers, G., Weakly coupled elliptic systems and positivity, *Math. Nachr.* **173** (1995), 259–286.

61. Nečas, J., *Les Méthodes Directes en Théorie des Équations Elliptiques*, Masson et Cie, Éditeurs, Paris, 1967.

62. Pao, C.V., *Nonlinear Parabolic and Elliptic Equations*, Plenum Press, New York, 1992.

63. Protter, M.H., and Weinberger, H.F., *Maximum Principles in Differential Equations*, Springer-Verlag, New York, 1984.

64. Sattinger, D.H., Monotone methods in nonlinear elliptic and parabolic boundary value problems, *Indiana Univ. Math. J.* **21** (1971), 979–1000.

65. Schaefer, H.H., *Topological Vector Spaces*, Graduate Texts in Mathematics 3, Springer-Verlag, New York, 1971.

66. Schmitt, K., Boundary value problems for quasilinear second-order elliptic equations, *Nonlinear Anal.* **2** (1978), 263–309.

67. Smoller, J., *Shock Waves and Reaction-Diffusion Equations*, Grundlehren der Mathematischen Wissenschaften 258, Springer-Verlag, New York, 1983.

68. Stampacchia, G., Le problème de Dirichlet pour les équations elliptiques du second ordre à coefficients discontinus, *Ann. Inst. Fourier (Grenoble)* **15** (1965), 189–258.

69. Stollmann, P., A convergence theorem for Dirichlet forms with applications to boundary value problems with varying domains, *Math. Z.* **219** (1995), 275–287.

70. Sturm, K.Th., On the L^p-spectrum of uniformly elliptic operators on Riemannian manifolds, *J. Funct. Anal.* **118** (1993), 442–453.

71. Sweers, G., Strong positivity in $C(\overline{\Omega})$ for elliptic systems. *Math. Z.* **209** (1992), 251–271.

72. Takáč, P., An abstract form of maximum and anti-maximum principles of Hopf's type, *J. Math. Anal. Appl.* **201** (1996), 339–364.

73. Trudinger, N.S., Maximum principles for linear, non-uniformly elliptic operators with measurable coefficients, *Math. Z.* **156** (1977), 291–301.

74. Trudinger, N.S., On the positivity of weak supersolutions of nonuniformly elliptic equations, *Bull. Austral. Math. Soc.* **19** (1978), 321–324.

75. Trudinger, N.S., On the first eigenvalue of nonuniformly elliptic boundary value problems, *Math. Z.* **174** (1980), 227–232.

76. Walter, W., A theorem on elliptic differential inequalities with an application to gradient bounds, *Math. Z.* **200** (1989), 293–299.

77. Walter, W., The minimum principle for elliptic systems, *Appl. Anal.* **47** (1992), 1–6.

Ten Mathematical Essays on Approximation in Analysis and Topology
J. Ferrera, J. López-Gómez, F. R. Ruiz del Portal, Editors
© 2005 Elsevier B.V. All rights reserved

On Some Approximation Problems in Topology

A. N. Dranishnikov

Department of Mathematics, University of Florida,
P.O. Box 118105, 358 Little Hall, Gainesville, FL 32611-8105, USA

Abstract

There are two basic types of approximation problems in topology:

1) *to approximate a given topological space by better spaces;*

2) *to approximate a given map by better maps.*

Here we shall discuss examples of both types. We consider two approximation problems which appeared in different areas of topology as keystone problems. One of them is the area of intensive research formed around the Novikov Higher Signature conjecture. The other deals with the transversality of mappings of compacta to euclidean space. In both cases the problems are partially solved and in each case any further progress on them could lead to a major breakthrough in the area.

Key words: Coarse category, anti-Čech approximation, expanders, coarse embeddings, cohomological dimension

1. Anti-Čech Approximation

Problem 1. *Which finitely presented groups Γ admit spherical anti-Čech approximations?*

We note if a group Γ admits either such an approximation, then several famous Novikov type conjectures hold true for manifolds M with the fundamental group

$$\pi_1(M) = \Gamma.$$

Below we give corresponding definitions and outline the proof of this statement.

1.1. *Large scale topology (Large vs Small)*

The classical topology is a small scale science in a sense that it deals with the continuity which is defined in the small scale terms. To see that we restrict ourself to the case of metric spaces. We recall the classical $\epsilon - \delta$ definition of a uniformly continuous map $f : X \to Y$ between metric spaces (X, d_X) and (Y, d_Y):

$$\forall \epsilon > 0 \; \exists \delta > 0 \; : \; d_X(x, x') < \delta \; \Rightarrow \; d_Y(f(x), f(x')) < \epsilon.$$

In this definition we assume that ϵ and δ are small numbers. Here is a large scale analog of the uniform continuity:

$$\forall \delta < \infty \; \exists \epsilon < \infty \; : \; d_X(x, x') < \delta \; \Rightarrow \; d_Y(f(x), f(x')) < \epsilon.$$

Here we assume that ϵ and δ are large.

Perhaps for some readers this analogy does not look very analogous because of the difference in the order of quantifiers: We use $\forall \epsilon \exists \delta$ in the first case and $\forall \delta \exists \epsilon$ in the second. To make the analogy visual we reformulate these definitions as follows.

Small scale (classic): *A map $f : X \to Y$ is uniformly continuous if there exists a function $\rho : \mathbb{R}_+ \to \mathbb{R}_+$ with $\lim_{t \to 0} \rho(t) = 0$ such that*

$$d_Y(f(x), f(x')) < \rho(d_X(x, x')) \quad \textit{for all} \;\; x, x' \in X.$$

Large scale: *A map $f : X \to Y$ is uniformly continuous if there exists a function $\rho : \mathbb{R}_+ \to \mathbb{R}_+$ with $\lim_{t \to \infty} \rho(t) = \infty$ such that*

$$d_Y(f(x), f(x')) < \rho(d_X(x, x')) \quad \textit{for all} \;\; x, x' \in X.$$

By analogy we can say that the large scale topology must be a discipline about the large scale continuity. To define it in the most general setting one should introduce new structure generalizing metric in the large scale direction similarly as the topology structure does it in the small scale aspect. The idea of large scale continuity is presented in the above definition of large scale uniformly continuous maps between metric spaces. For the purpose of our paper it suffices to have only this notion.

Some seeds of the large scale topology can be found in [22], [33], [34], [7], [32], [40], [36] and others.

1.2. *Coarse category*

The coarse category was defined in [33]. Let $\mathcal{M}$ be a category of metric spaces which morphisms are the large scale uniformly continuous maps. Following [33] we will call such maps shortly 'uniform'. Then we define the quotient category by means of the following equivalence relation on the set of morphisms between X and Y for each pair of objects X

and Y: Two maps $f, g : X \to Y$ are equivalent (we will call them *coarsely equivalent* in that case) if there is a constant D such that

$$d_Y(f(x), g(x)) < D \quad \text{for all} \quad x \in X.$$

This equivalence leads to the notion of coarsely equivalent metric spaces: X and Y are *coarsely equivalent* if there are uniform maps $f : X \to Y$ and $g : Y \to X$ such that $g \circ f \sim 1_X$ and $f \circ g \sim 1_Y$. For example, the integers $\mathbb{Z}$ and the reals $\mathbb{R}$ are coarsely equivalent being considered as metric spaces with the metric

$$d(x, y) = |x - y|.$$

We call a metric space X ϵ-*discrete* if $d_X(x, x') \geq \epsilon$ for all $x, x' \in X$, $x \neq x'$. We call it *discrete* if it is ϵ-discrete for some ϵ.

Proposition 1. *Every metric space X is coarsely equivalent to a discrete metric space.*

Proof. By transfinite induction one can construct a 1-discrete subset $S \subset X$ with the property $d_X(x, S) \leq 1$ for all $x \in X$. The inclusion $S \subset X$ is a coarse equivalence whose inverse is any map $g : X \to S$ with the property $d(x, g(x)) \leq d(x, S) + 1$. $\quad\square$

We will consider only proper metric spaces of bounded geometry. A metric space X is called *proper* if every closed ball $B_r(x)$ in X is compact. In the definition of the coarse category the morphisms are assumed to be metrically proper. We recall that a map $f : X \to Y$ is called *metrically proper* if the preimage $f^{-1}(B)$ is bounded for every bounded set $B \subset X$. Note that the set of all metrically proper maps between metric spaces X and Y is invariant under the coarse equivalence. The ϵ-*capacity* $c_\epsilon(W)$ of a subset $W \subset X$ of a metric space X is the maximal cardinality of ϵ-discrete set in W. A metric space X has *bounded geometry* if there are $\epsilon > 0$ and a function $c : \mathbb{R}_+ \to \mathbb{R}_+$ such that

$$c_\epsilon(B_r(x)) \leq c(r) \quad \text{for all} \quad x \in X.$$

In this article we will consider geodesic metric spaces. We recall that a metric space (X, d_X) is called *geodesic* if for every pair of its points x and y there is an isometric embedding of the interval $[0, d(x, y)]$ into X with the end points x and y.

Finitely generated groups give us one of the main sources of examples of metric spaces of bounded geometry. Let $S = S^{-1}$ be a finite symmetric set of generators of a group Γ. Then the word metric d_S is defined as

$$d_S(x, y) = \|x^{-1}y\|_S,$$

where the S-norm $\|a\|_S$ of an element $a \in \Gamma$ is the shortest length of presentation of a in the alphabet S. We note that if S' is another finite symmetric generating set of Γ, then the metric spaces (Γ, d_S) and $(\Gamma, d_{S'})$ are coarsely equivalent. Also we note that a finitely generated group Γ with a word metric is coarsely equivalent to its Cayley graph $C_S(\Gamma)$ which is a geodesic metric space. We recall that the vertices of $C_S(\Gamma)$ are the elements of Γ and $a, b \in \Gamma$ are joined by an edge if and only if $d_S(a, b) = 1$. Every connected graph carries a natural geodesic metric with respect to which every edge is isometric to the unit interval $[0, 1]$.

Every discrete metric space (X, d) can be isometrically embedded into a geodesic metric space $(GX, \bar{d})$ constructed as follows:

(1) GX is a complete graph on X;

(2) every edge $[x, x']$ in GX is isometric to the interval $[0, d(x, x')]$;

(3) for every two points $x, y \in GX$ the distance $\bar{d}(x, y)$ is the length of the shortest path joining x and y.

In view of Proposition 1 every metric space is coarsely embeddable in a geodesic metric space.

A metric space (X, d) is called *discrete geodesic* if there is a number $c > 0$ such that for every pair of points $x, y \in X$ there is a chain $z_0, z_1, \ldots, z_n$ with

$$d(z_i, z_{i+1}) \leq c, \quad z_0 = x, \quad z_n = y, \quad \sum_{i=1}^{n} d(z_i, z_{i+1}) = d(x, y).$$

A typical example of a discrete geodesic metric space is a finitely generated group with the word metric.

1.3. Čech and anti-Čech approximations

There is an analogy between the category of compact metric spaces and the category of proper metric spaces with bounded geometry which is based on the existence of approximations by finite dimensional polyhedra in both cases. In the first case it is called Čech approximation, in the second case, anti-Čech approximation. We recall that a *Čech approximation* of a compact metric space X is a sequence of finite covers $\{\mathcal{U}_n\}$ such that, $\mathcal{U}_{n+1}$ is a refinement of $\mathcal{U}_n$ for all n, and

$$\lim_{n \to \infty} D(\mathcal{U}_n) = 0,$$

where

$$D(\mathcal{U}_n) = \max\{\operatorname{diam} U \mid U \in \mathcal{U}_n\}.$$

Definition 2. (**[33]**) An anti-Čech approximation of a metric space Y is a sequence of uniformly bounded locally finite covers $\mathcal{U}_n$ such that $\mathcal{U}_n$ is a refinement of $\mathcal{U}_{n+1}$, and

$$\lim_{n \to \infty} L(\mathcal{U}_n) = \infty,$$

where $L(\mathcal{U}_n)$ is the Lebesgue number of $\mathcal{U}_n$.

We call a cover $\mathcal{U}$ *locally finite* if every element $U \in \mathcal{U}$ has nonempty intersection with finitely many other elements. The Lebesgue number is defined by the formula:

$$L(\mathcal{U}) = \inf_{y \in Y} \sup_{U \in \mathcal{U}} d(y, Y \setminus U).$$

We recall that the *nerve* of a cover $\mathcal{U}$ is the simplicial complex with vertices labeled by the elements of $\mathcal{U}$, and vertices $U_0, \dots U_n$ form an n-simplex $[U_0, \dots, U_n]$ if and only if $U_0 \cap \cdots \cap U_n \neq \emptyset$. A refinement of covers $\mathcal{V} \leq \mathcal{U}$ of a space X defines a simplicial map

$$\phi : N(\mathcal{V}) \to N(\mathcal{U})$$

with the property $V \subset \phi(V)$ for all $V \in \mathcal{V}$. Thus, a Čech approximation leads to an inverse sequence of polyhedra $\{N(\mathcal{U}_n), \phi_n^{n+1}\}$ and an anti-Čech approximation defines a direct system $\{N(\mathcal{U}_n), \phi_{n+1}^n\}$. These systems are sufficient to define homology and cohomology in both cases. In the case of compactum X we have the Čech cohomology

$$\check{H}^i(X) = \lim_{\to}\{H^i(N(\mathcal{U}_n)), (\phi_n^{n+1})^*\}$$

and the Steenrod homology

$$0 \to \lim_{\leftarrow}{}^1\{H_{i+1}(N(\mathcal{U}_n)), (\phi_n^{n+1})_*\} \to H_i(X) \to \lim_{\leftarrow}\{H_i(N(\mathcal{U}_n)), (\phi_n^{n+1})_*\} \to 0.$$

Similarly in the case of a metric space Y there are coarse (anti-Čech) homology:

$$HX_i(Y) = \lim_{\to}\{H_i^{lf}(N(\mathcal{U}_n)), (\phi_{n+1}^n)^*\}$$

and Roe's coarse cohomology

$$0 \to \lim_{\leftarrow}{}^1\{H_c^{i_1}(N(\mathcal{U}_n)), (\phi_{n+1}^n)^*\} \to HX^i(Y) \to \lim_{\leftarrow}\{H_c^i(N(\mathcal{U}_n)), (\phi_{n+1}^n)^*\} \to 0.$$

In the case of anti-Čech approximation we use the simplicial homology and cohomology of the second type, namely, the simplicial homology with infinite locally finite chains and the simplicial cohomology with compact supports.

Proposition 3. ([28]) *Every proper metric space of bounded geometry Y admits an anti-Čech approximation by covers with finite dimensional nerves.*

Proof. In view of Proposition 1 we may assume that Y is 1-discrete. It suffices to construct a uniformly bounded cover of Y with the Lebesgue number greater than given number L and of bounded multiplicity. The cover by L-balls $\{B_L(x)\}_{x \in Y}$ has all above properties. The multiplicity of this cover at given point $y \in Y$ is bounded by the cardinality of the ball $B_L(y)$ which is uniformly bounded due to the bounded geometry property. $\qquad\square$

A well-known result, arising to Alexandroff, states that a compact metric space X has the covering dimension $\dim X \leq k$ if and only if X admits a Čech approximation with at most k-dimensional nerves. This fact together with the micro-macro analogy leads to Gromov's definition of a large scale dimension, called the *asymptotic dimension* [22]: $\operatorname{asdim} X \leq k$ if and only if X admits an anti-Čech approximation with at most k-dimensional nerves.

The following proposition easy follows from definition.

Proposition 4. *Let GX be a geodesic extension of a metric space X constructed in Section 2. Then*

$$\operatorname{asdim} GX = \max\{\operatorname{asdim} X, 1\}.$$

A partition of unity $\{f_\alpha\}$ subordinated to a cover $\mathcal{U} = \{U_\alpha\}$ of a space X defines a *projection to the nerve*

$$p : X \to N(\mathcal{U})$$

by the formula

$$p(x) = \sum_\alpha f_\alpha(x)e_\alpha.$$

Here $\{e_\alpha\}$ is a basis in a vector space V indexed by the same set as the cover.

The first part of the following lemma is well known.

Lemma 5.

1. *Let $\{\mathcal{U}_n\}$ be a Čech approximation of a compactum X. Then there are a sequence $\{n_k\}$, mappings*
$$q_k^{k+1} : N(\mathcal{U}_{n_{k+1}}) \to N(\mathcal{U}_{n_k}),$$
and projections to the nerve $p_k : X \to N(\mathcal{U}_{n_k})$ such that
$$p_k = q_k^{k+1} \circ p_{k+1}$$
and ϕ_k^{k+1} is a simplicial approximation of q_k^{k+1} for all k.

2. *Let $\{\mathcal{U}_n\}$ be an anti-Čech approximation of a metric space X. Then there are a sequence $\{n_k\}$, mappings*
$$q_{k+1}^k : N(\mathcal{U}_{n_k}) \to N(\mathcal{U}_{n_{k+1}}),$$
and projections to the nerve $p_k : X \to N(\mathcal{U}_{n_k})$ such that
$$p_{k+1} = q_{k+1}^k \circ p_k$$
and ϕ_{k+1}^k is a simplicial approximation of q_{k+1}^k for all k.

In the first case the nerves form an inverse sequence whose limit is X. In the second case the nerves form a direct system with an obscure meaning of the limit space.

1.4. *Geometry of nerves*

Let $\mathcal{U} = \{U_i\}_{i \in A}$ be a cover of X and let V be a vector space with a basis $\{e_i\}_{\alpha \in A}$. Then the nerve $N = N(\mathcal{U})$ can be realized in V with vertices at e_i. Assume that V is a Banach space. Then every simplex Δ in N is given the induced metric. We define a metric on N as the largest metric d such that the restriction of d to every simplex Δ defines the induced metric. In other words the distance $d(x,y)$ is the supremum of the lengths of piece-wise linear paths with every leg lying in a simplex of N. Note that this is a geodesic metric on N.

In the case when $V = l_2$ is a Hilbert space we call this metric on N *euclidean*. The unit sphere in V can be given two natural metrics: the metric restricted from V, and the geodesic metric induced by the metric on V. The radial projection of a simplex

$$\Delta = \{(x_i) \mid \sum x_i = 1, x_i \geq 0, \}$$

to the unit sphere brings these metrics to Δ. In similar fashion these metrics on simplicies define a global metrics on N which will be called a *V-spherical* metric and a *geodesic V-spherical* metric. Thus, a V-spherical metric on N is the largest metric d with the property that d restricted to any simplex Δ gives a spherical metric restricted from V. An l_2-spherical metric will be called *spherical*. We don't consider the geodesic spherical metric separately, since it is 2-Lipschitz equivalent to the spherical metric.

Definition 6. Let $f : X \to Y$ be a large scale uniform map between metric spaces. The coarse dilatation of f is the following number (or ∞):

$$\mathrm{dil}(f) = \lim_{K \to \infty} \sup_{d_X(x,x') \geq K} \frac{d_Y(f(x), f(x'))}{d_X(x, x')}.$$

We note that if X is a geodesic metric space, then the dilatation of f is always finite. Also we note that

$$\mathrm{dil}(f) = \mathrm{dil}(g)$$

for coarsely equivalent maps f and g.

It easy to check that a projection of a geodesic metric space to the nerve $p : X \to N(\mathcal{U})$ is a coarse isomorphism for any V-metric or V-spherical metric on $N(\mathcal{U})$.

Let V be a normed vector space. An anti-Čech approximation $\{\mathcal{U}_n\}$ of a metric space X is called a *V-approximation (spherical V-approximation)* if

$$\lim_{n \to \infty} \mathrm{dil}(p_n) = 0,$$

where

$$p_n : X \to N(\mathcal{U}_n)$$

is a projection to the nerve of the cover $\mathcal{U}_n$ supplied with a V-metric (spherical V-metric). If $V = l_2$, an anti-Čech V-approximation is called *euclidean*, and an anti-Čech spherical V-approximation is called *spherical*.

The definition of the anti-Čech V-approximation allows to assume that the size of simplicies in the nerves $N(\mathcal{U}_n)$ tends to infinity and all projections p_n are 1-Lipschitz.

The Banach space of bounded sequences $x = \{x_n\}$ with the norm $\|x\| = \sup |x_n|$ will be denoted as l_∞.

Proposition 7. *Every metric space X admits an l_∞-spherical anti-Čech approximation.*

Proof. Since every metric space is coarsely embeddable into a geodesic metric space, it suffices to prove the proposition in the case when X is geodesic. Let $\mathcal{U}$ be an open uniformly bounded locally finite cover of X of with the Lebesgue number $L(\mathcal{U}) > 2\lambda$. We may assume that

$$\max_{x \in X} d(x, X \setminus U) > 2\lambda \quad \text{for all} \quad U \in \mathcal{U}.$$

Otherwise we can drop off such U and still have a cover with $L(\mathcal{U}) > 2\lambda$. We denote

$$\phi(x) = d(x, X \setminus U).$$

Let $p : X \to N(\mathcal{U})$ be the projection to the nerve defined by means of the following partition of unity

$$f_U(x) = \frac{\phi_U(x)}{\sum_{V \in \mathcal{U}} \phi_V(x)}.$$

Since

$$\|\phi'(x) - \phi'(y)\|_\infty = \max_U |d(x, X \setminus U) - d(y, X \setminus U)| \leq d(x, y)$$

for all $x, y \in X$, the map $\phi' : X \to l_\infty$ defined by the family $\{\phi_U\}_{U \in \mathcal{U}}$ is 1-Lipschitz. Note that

$$\phi'(X) \subset l_\infty \setminus B_{2\lambda}(0).$$

The following inequalities show that the radial projection π of $H \setminus B_{2\lambda}(0)$ onto the unit sphere $S_1(0)$ is $\frac{1}{\lambda}$-Lipschitz in every Banach space H:

$$\|\pi(x) - \pi(y)\| = \|\frac{x}{\|x\|} - \frac{y}{\|y\|}\| \leq \frac{1}{\|x\|}\|x - y\| + \|y\|\|\frac{1}{\|x\|} - \frac{1}{\|y\|}\|$$
$$\leq \frac{1}{2\lambda}\|x - y\| + \frac{1}{\|x\|}|\|x\| - \|y\|| \leq \frac{1}{\lambda}\|x - y\|.$$

Then the composition

$$p = \pi \circ \phi' : X \to S_1(0)$$

is $\frac{1}{\lambda}$-Lipschitz. The image $p(X)$ lies in the n-skeleton of the spherical infinite dimensional simplex

$$\Delta_0^\infty = S_1(0) \cap l_\infty^+.$$

In other words, the nerve $N(\mathcal{U})$ of the cover $\mathcal{U}$ has the right geometry and it is contained as a subcomplex in the n-skeleton of Δ_0^∞.

Thus, the projection $p : X \to N(\mathcal{U})$ is $\frac{1}{\lambda}$-Lipschitz when the spherical realization of $N(\mathcal{U})$ is supplied with the metric restricted from l_∞. Therefore, the restriction

$$p_\Delta = p|_{p^{-1}(\Delta)} : p^{-1}(\Delta) \to \Delta$$

is $\frac{1}{\lambda}$-Lipschitz for every simplex $\Delta \subset N(\mathcal{U})$. For every two points $x, y \in X$ a geodesic J joining them can be divided in smaller intervals J_k such that $J_k \subset p^{-1}(\Delta_k)$ for some Δ_k. Then

$$d(p(x), p(y)) \leq \sum_k d(p(x_k), p(x_{k+1})) \leq \sum_k \frac{1}{\lambda} d_X(x_k, x_{k+1}) = \frac{1}{\lambda} d(x, y),$$

where x_k and x_{k+1} are the endpoints of the geodesic interval J_k. $\square$

Corollary 8. *Every metric space X with $\mathrm{asdim}X < \infty$ admits a euclidean anti-Čech approximation.*

Proof. In view of Proposition 4 we may assume that X is a geodesic metric space. Let asdim$X = n$. Then we may assume that $\dim N(\mathcal{U}) = n$. There is a number c_n such that for every n-dimensional complex K the identity map of K, denoted by $h : K \to K$, where the domain is supplied with the spherical l_∞-metric and the target with the l_2-metric has the property $\mathrm{dil}(h) < c_k$. Let $p_n : X \to N(\mathcal{U}_n)$ denote the projection to the nerve supplied with the l_2-metric and let p'_n denote the same projection for the spherical l_∞-metric. Then

$$\mathrm{dil}(p_n) \le c_k \, \mathrm{dil}(p'_n) \xrightarrow[n\to\infty]{} 0,$$

which concludes the proof. $\qquad\square$

Proposition 9. *If a metric space admits an anti-Čech l_1-approximation, then it admits l_p-approximation for all $p \ge 1$.*

Proof. Clearly, we have that $|a - b|^{p-1} \le 1$ for all $1 \ge a, b \ge 0$. Hence

$$\|x - y\|_p^p = \sum_i |x_i - x_j)|^p = \sum_i |x_i - y_i||x_i - y_i|^{p-1} \le \sum_i |x_i - y_i| = \|x - y\|_1.$$

Then an l_1-approximation $\{p_n\}$ automatically determines l_p-approximations for all $p \ge 1$. $\qquad\square$

1.5. Property A

By l_p, $p \ge 1$, we denote the Banach space of sequences $\{x_n\}$ with the norm

$$\|x\|_p = (\sum_{n=1}^\infty |x_n|^p)^{\frac{1}{p}}.$$

Also $l_p(Z)$ will denote the corresponding Banach space with a basis indexed by Z. Thus, $l_p = l_p(\mathbb{N})$.

Definition 10. Let $p \in \mathbb{R}_+ \cup \{\infty\}$. A discrete metric space X has Property A_p if there is a sequence of maps $a^n : X \to l_p(X)$ satisfying $\|a_z^n\|_p = 1$ and $a_z^n(y) \ge 0$ for all $y, z \in X$, such that

(1) there is a function $R = R(n)$ such that

$$\mathrm{supp}(a_z^n) \subset B_{R(n)}(z) \quad \text{for all } z \in X;$$

(2) for every $K > 0$,

$$\lim_{n\to\infty} \sup_{d(z,w)<K} \|a_z^n - a_w^n\|_p = 0.$$

Clearly, the property A_p is invariant under coarse equivalences. Thus, in view of Proposition 1 it can be applied to any metric space. We leave to the reader to show that Property A_1 is equivalent to existence of an anti-Čech l_1-approximation (see Lemma 5).

Proposition 11. *Every discrete metric space has Property A_∞.*

Proof. We define
$$a_z^n(x) = \max\{1 - \frac{1}{n}d(x,z)), 0\}.$$
Then $\operatorname{supp}(a_z^n) \subset B_n(z)$. We note that for $z, w \in X$ with $d(z,w) \leq K$ we have
$$\|a_z^n - a_w^n\|_\infty = \max\left\{\sup|(1 - \frac{1}{n}d(x,z)) - (1 - \frac{1}{n}d(w,z))|,\right.$$
$$\left.\sup\{1 - \frac{1}{n}d(x,z)|d(x,w) > n\}, \sup\{1 - \frac{1}{n}d(x,w))|d(x,z) > n\}\right\}.$$
The first supremum here can be estimated as
$$\frac{1}{n}d(z,w) \leq K/n.$$
Since
$$d(x,y) \geq d(x,w) - d(x,z) \geq n - K$$
in the second case, we have
$$\sup\{1 - \frac{1}{n}d(x,z)|d(x,w) > n\} \leq 1 - (n - K)/n = K/n.$$
A similar estimate holds in the third case. Thus,
$$\|a_z^n - a_w^n\|_\infty \leq K/n$$
and the condition (2) is satisfied. $\qquad\qquad\qquad\qquad\qquad\qquad\qquad\qquad\qquad\square$

Proposition 12. *For every $p \geq 1$ Property A_1 is equivalent to Property A_p.*

Proof. First we show that for every $p \geq 1$ Property A_1 implies A_p. Assume that X has Property A_1. Let $a^n : X \to l_1(X)$ be a sequence of functions satisfying the conditions (1)-(2) from the definition of A_1. Then
$$\sum_{y \in X} a_z^n(y) = 1$$
and $a_z^n(y) \geq 0$ for all $z, y \in X$. We define
$$b_z^n(y) = a_z^n(y)^{\frac{1}{p}}.$$
Then $\|b_z^n\|_p = 1$. The condition (1) is satisfied automatically. We check the condition (2). In view of the obvious inequality
$$t^{p/m} + (1 - t)^{p/m} \leq (t + (1 - t))^p = 1$$
for $p \geq 1, t \in [0,1]$, we have that $|a - b|^p \leq |a^p - b^p|$ for all $a, b \geq 0$. Hence
$$\|b_z^n - b_w^n\|_p^p = \sum_{x \in X} |b_z^n(x) - b_w^n(x)|^p \leq \sum_{x \in X} |b_z^n(x)^p - b_w^n(x)^p| = \|a_z^n - a_w^n\|_1.$$
This implies the condition (2).

Then we show that A_p, $p \geq 1$, implies A_m for $m \geq p$. Assume that X has Property A_p. Let $a^n : X \to l_p(X)$ be a sequence of functions satisfying the conditions (1)-(2) from the definition of A_1. Then

$$\sum_{y \in X} (a_z^n(y))^p = 1$$

and $a_z^n(y) \geq 0$ for all $z, y \in X$. We define

$$b_z^n(y) = a_z^n(y)^{\frac{p}{m}}.$$

Then $\|b_z^n\|_m = 1$. The condition (1) is satisfied automatically. We check the condition (2). In view of the inequality $t^{p/m} + (1-t)^{p/m} \geq 1$ for $p \leq m$, $t \in [0,1]$, we have that $|a - b|^{p/m} \geq |a^{p/m} - b^{p/m}|$ for all $a, b \geq 0$. Hence

$$\|b_z^n - b_w^n\|_m^m = \sum_{x \in X} |b_z^n(x) - b_w^n(x)|^m = \sum_{x \in X} |(a_z^n(x))^{p/m} - (a_w^n(x))^{p/m}|^m$$

$$\leq \sum_{x \in X} |a_z^n(x) - a_w^n(x)|^p = \|a_z^n - a_w^n\|_p.$$

This implies the condition (2).

Now assume that X has Property A_p, $p \in \mathbb{N}$ and show that X has Property A_1. Let $b^n : X \to l_p(X)$ be a corresponding sequence of functions. We define $a^n = (b^n)^p$ and check that

$$\|a_z^n - a_w^n\|_1 = \sum_{x \in X} |(b_z^n(x))^p - (b_w^n(x))^p|$$

$$= \sum_{x \in X} |b_z^n(x) - b_w^n(x)| \times |\sum_{i=0}^{p-1} (b_z^n(x))^i (b_w^n(x))^{p-1-i}|$$

By the Hölder inequality we have

$$\|a_z^n - a_w^n\|_1 \leq \|b_z^n - b_w^n\|_p \|\sum_{i=0}^{p-1} (b_z^n)^i (b_w^n)^{p-1-i}\|_q,$$

where $\frac{1}{p} + \frac{1}{q} = 1$. Since $pq = p + q$, we have

$$(b_z^n(x))^{iq} (b_w^n(x))^{(p-1-i)q} = (b_z^n(x)/b_w^n(x))^{iq} (b_w^n(x))^p.$$

Therefore,

$$(b_z^n(x))^{iq} (b_w^n(x))^{(p-1-i)q} \leq (b_w^n(x))^p,$$

provided $b_z^n(x) \leq b_w^n(x)$. Symmetrically we obtain that

$$(b_z^n(x))^{iq} (b_w^n(x))^{(p-1-i)q} \leq (b_z^n(x))^p,$$

provided $b_z^n(x) \geq b_w^n(x)$.

Thus,

$$(b_z^n(x))^{iq} (b_w^n(x))^{(p-1-i)q} \leq (b_z^n(x))^p + (b_w^n(x))^p.$$

Hence

$$\|(b_z^n)^i (b_w^n)^{p-1-i}\|_q = \sum_x (b_z^n(x))^{iq} (b_w^n(x))^{(p-1-i)q}$$

$$\leq \sum_x (b_z^n(x))^p + (b_w^n(x))^p = 2.$$

Therefore,

$$\|a_z^n - a_w^n\|_1 \leq 2p \|b_z^n - b_w^n\|_p,$$

and, consequently, the condition (2) holds for a^n. $\square$

The equality $A_1 = A_2$ was proven in [40]. It was shown in [29] that Property A_1 coincides with Yu's Property A [43] for metric spaces X of bounded geometry. Also it was proven there that Property A ($= A_1 = A_2$) for finitely generated groups Γ with word metrics is equivalent to the topological amenability of the natural action of Γ on the Stone-Čech compactification $\beta\Gamma$.

Lemma 13. *For a discrete geodesic metric space X of bounded geometry the following conditions are equivalent:*

(1) *X admits an l_p-spherical anti-Čech approximation;*

(2) *X has Property A_p.*

Proof. Let $\{\mathcal{U}_n, p_n\}$ be an l_p-spherical anti-Čech approximation of X. Let $K_n = N(\mathcal{U}_n)$.

We assume that n is fixed. Let $\{e^i\}$ be the basis of l_p that corresponds to the cover $\mathcal{U}_n$, each e^i corresponds to an element $U_i \in \mathcal{U}_n$. Then

$$K_n^{(0)} \subset \{e^i\}.$$

We consider an injective map $s_n : K^{(0)} \to X$ such that $s_n(e^i) \in U_i$ for all i. Let $p_n = (\alpha_n^1, \alpha_n^2, \dots)$. We define $a_z^n(y) = \alpha_n^i(z)$, where $y = s_n(e^i)$. If $y \notin s_n(K^{(0)})$ we set $a_y^n = 0$. Since

$$\|p_n(z)\|_p = \left(\sum_i \alpha_n^i(z)^p\right)^{\frac{1}{p}} = 1,$$

we have

$$\|a_z^n\|_p = \left(\sum_{x \in X} a_z^n(x)^p\right)^{\frac{1}{p}} = \left(\sum_i \alpha_n^i(z)^p\right)^{\frac{1}{p}} = 1.$$

The condition (1) of the Property A_p follows from the facts that

$$\mathrm{supp}(a_z^n) \subset \bigcup_{U \in \mathcal{U}_n, z \in U} U$$

and the cover $\mathcal{U}_n$ is uniformly bounded. Note that

$$\|a_z^n - a_w^n\|_p = \left(\sum_{x \in X} |a_z^n(x) - a_w^n(x)|^p\right)^{\frac{1}{p}}$$

$$= \|p_n(z) - p_n(w)\|_p \leq \mathrm{dil}(p_n) d(z, w).$$

Since $\text{dil}(p_n)$ tends to zero, we have for every $K > 0$,

$$\lim_{n \to \infty} \sup_{d(z,w) < K} \|a_z^n - a_w^n\|_p = 0.$$

The condition (2) is checked.

Assume that X has Property A_p and let $a^n : X \to l_p(X)$ be the corresponding functions. We define $p_n : X \to l_p$ as $p_n = a^n$. Since $\|a_z^n\|_p = 1$ for all $z \in X$, the image $p_n(X)$ lies in the l_p-sphere. For every $z \in X$ the support of a_z^n spans a simplex Δ_z in $S_1(0)$. Since the support of a_z^n is uniformly bounded and X has bounded geometry, there is a number k such that all simplices Δ_z are at most k-dimensional. They form a simplicial complex K_n. We supply K_n with l_p-spherical metric as in Section 4. Now we define a cover

$$\mathcal{U}_n = \{p_n^{-1}(\text{St}(v, K_n)) \mid v \in K^{(0)}\}.$$

Then $N(\mathcal{U}_n) = K_n$ and $p_n : X \to K_n$ is a projection to the nerve. To show that $\{\mathcal{U}_n, p_n\}$ is an l_p-spherical anti-Čech approximation we have to check that

$$\text{dil}(p_n) \to 0.$$

We may assume that $d(x, y) \geq 1$ for all points in X. Let c be a constant from the definition of discrete geodesic metric space applied to X. Then a coarse dilatation of a map $f : X \to Y$ is bounded from above by

$$\sup_{d_X(x,x') \leq d} d_Y(f(x), f(x')).$$

In our case

$$\text{dil}(p_n) \leq \sup_{d_X(z,z') \leq 1} d_{K_n}(p_n(z), p_n(z')).$$

The Property A_p implies that

$$\|p_n(z) - p_n(z')\|_p \leq \epsilon(n),$$

where $\lim \epsilon(n) = 0$. For large enough n we have

$$\text{supp}(a_z^n) \cap \text{supp}(a_{z'}^n) \neq \varnothing.$$

Let Δ_1, Δ_2, and Δ be simplices in $l_p(X)$ spanned by

$$\text{supp}(a_z^n), \quad \text{supp}(a_{z'}^n), \quad \text{and} \quad \text{supp}(a_z^n) \cap \text{supp}(a_{z'}^n).$$

respectively. Clearly, for $x \in \Delta_1$, $d(x, \Delta_2) = d(x, \Delta)$. Hence,

$$\epsilon(n) > \|p_n(z) - p_n(z')\|_p \geq d(p_n(z), \Delta), d(p_n(z'), \Delta).$$

Let

$$d(p_n(z), \Delta) = \|p_n(z) - w\|_p$$

and

$$d(p_n(z'), \Delta) = \|p_n(z') - w'\|_p,$$

where $w, w' \in \Delta$. By the triangle inequality we have $\|w - w'\|_p \leq 3\epsilon(n)$. Then

$$d_{K_n}(p_n(z), p_n(z')) \leq \|p_n(z) - w\|_p + \|w - w'\|_p + \|w' - p_n(z')\|_p \leq 5\epsilon(n).$$

Thus,

$$\mathrm{dil}(p_n) \le 5\epsilon(n) \to 0.$$

$\square$

Proposition 14. *A metric space X with Property A admits a euclidean anti-Čech approximation.*

Proof. The equality

$$\|x - y\|_2^2 \le \sum |x_i - y_i||x_i + y_i|$$

implies that the identity map of the standard n-simplex in l_1 with image considered as in l_2 is $\sqrt{2}$-Lipschitz for all n. By Lemma 13, X has an l_1-spherical anti-Čech approximation. We note that l_1-spherical approximation is the same thing as l_1-approximation. The latter determines a euclidean approximation. $\square$

1.6. Coarse approach to the Novikov conjecture

Let Γ be a finitely presented group and let $B\Gamma = K(\Gamma, 1)$ be its classifying space. Let M be a closed manifold with the fundamental group $\pi_1(M) = \Gamma$ and let $\alpha : M \to B\Gamma$ be a map that induces an isomorphism of the fundamental groups. A rational cohomology class $x \in H^*(B\Gamma; \mathbb{Q})$ defines an element $\alpha^*(x) \in H^*(M; \mathbb{Q})$. A *higher signature* of M is by the definition the following number:

$$\mathrm{sign}_x(M, \alpha) = \langle L(M) \cap \alpha^*(x), [M] \rangle \in \mathbb{Q},$$

where $L(M)$ is the Hirzebruch class, a certain formal power series in the rational Pontryagin classes. The classical Novikov conjecture states that the higher signatures of a manifold are homotopy invariant [21]. We say that the Novikov conjecture holds for a group Γ if for every $x \in H^*(B\Gamma; \mathbb{Q})$, for every closed oriented manifold M and every map $\alpha : M \to B\Gamma$ there is the equality

$$\mathrm{sign}_x(M, \alpha) = \mathrm{sign}_x(N, \alpha \circ h)$$

for every orientation preserving homotopy equivalence $h : N \to M$. It is known that the Novikov conjecture holds true for all manifolds with the fundamental group Γ if and only if the rational assembly map from the surgery exact sequence

$$H_*(B\Gamma; \mathbb{L}) \otimes \mathbb{Q} \to L_*(\mathbb{Z}[\Gamma]) \otimes \mathbb{Q}$$

is a monomorphism [21].

The Novikov conjecture follows from the so-called *strong Novikov conjecture* which states that the analytic assembly map

$$A : K_*(B\Gamma) \to K_*(C_r^*(\Gamma))$$

is injective [21]. Here $C_r^*(\Gamma)$ is a reduced group C^*-algebra. The strong Novikov conjecture for torsion free Γ is a partial case of the Baum-Connes conjecture which states that A

is an isomorphism. The Baum-Connes conjecture can be restated in terms of the universal cover $E\Gamma$ of $B\Gamma$ as

$$A : K_*^{\Gamma}(E\Gamma) :\to K_*(C_r^*(\Gamma))$$

is an isomorphism, where K_*^{Γ} is the equivariant K-theory with 'locally finite chains'. We note here that $E\Gamma$ is a coarse object with a metric induced from $B\Gamma$. The Baum-Connes conjecture has a coarse analog which can be formulated for all metric spaces X:

$$A : KX_*(X) \to K_*(C^*(X))$$

is an isomorphism, where KX_* is Roe's (homology) K-theory (see §1.3.) and $C^*(X)$ is Roe's algebra [34]. It is known that the coarse Baum-Connes conjecture for a group Γ as a metric space with a word metric implies the Novikov conjecture [34]. The monomorphism version of the coarse Baum-Connes conjecture is often called the coarse Novikov conjecture. If A is an equivariant split monomorphism for $X = \Gamma$, then the original Novikov conjecture for Γ holds true. G. Yu proved the following remarkable theorem [43] (see also [27], [36]).

Theorem 15. (Yu) *If a finitely presented group Γ supplied with the word metric admits a coarse embedding in Hilbert space, then the coarse Baum-Connes and, hence, the Novikov conjecture hold for Γ.*

1.7. Coarse embeddings

A map between metric spaces $f : X \to Y$ is a *coarse embedding* if it is a coarse equivalence onto the image $f(X)$, i.e. there is a coarse morphism $g : f(X) \to Y$ such that the maps $g \circ f$ and $f \circ g$ are coarsely equivalent to 1_X and $1_{f(X)}$ respectively. The following proposition is obvious.

Proposition 16. *A map between metric spaces $f : X \to Y$ is a coarse embedding if and only if there exist two monotone tending to infinity functions $\rho_1, \rho_2 : \mathbb{R}_+ \to \mathbb{R}_+$ such that*

$$\rho_1(d_X(x, x')) \le d_Y(f(x), f(x')) \le \rho_2(d_X(x, x')) \quad \text{for all} \ \ x, \ x' \in X.$$

We note that embeddings in coarse category, i.e. coarse embeddings very often are called in the literature *uniform embeddings*.

Lemma 17. *A discrete metric space X with Property A_p admits a coarse embedding in l_p.*

Proof. Let a^n be the sequence of maps from the definition of Property A_p. By switching to a subsequence we may assume that

$$\sup_{d(z,w)<n} \|a_z^n - a_w^n\|_p^p < 1/2^n.$$

Let $z_0 \in X$ be a base point. We define a map $f : X \to l_p(X \times \mathbb{N})$ by the formula

$$f(z)(x, n) = a_z^n(x) - a_{z_0}^n(x).$$

We shall show that f is a coarse embedding. We may assume that the function R in the definition of Property A_p is strictly monotone. Let $S = R^{-1}$ be the inverse function. We define

$$\rho_1(t) = (2S(t/2) - 2)^{1/p} \quad \text{and} \quad \rho_2(t) = (2t + 1)^{1/p}.$$

According to Proposition 11 it suffices to check two inequalities. We have

$$\|f(z) - f(w)\|_p^p = \sum_{n=1}^{\infty} \|a_z^n - a_w^n\|_p^p \leq \sum_{n=1}^{[d(z,w)]} \|a_z^n - a_w^n\|_p^p + \sum_{[d(z,w)]+1}^{\infty} \|a_z^n - a_w^n\|_p^p$$

$$\leq \sum_{n=1}^{[d(z,w)]} \|a_z^n - a_w^n\|_p^p + \sum_{[d(z,w)]+1}^{\infty} 1/2^n$$

$$\leq 2d(z,w) + 1 = \rho_2^p(d(z,w)),$$

and

$$\|f(z) - f(w)\|_p^p = \sum_{n=1}^{\infty} \|a_z^n - a_w^n\|_p^p \geq \sum_{n=1}^{[S(d(z,w)/2)]} \|a_z^n - a_w^n\|_p^p$$

$$= 2[S(d(z,w)/2] \geq \rho_1^p(d(z,w)).$$

Here we used the fact that the inequality $n \leq [S(d(z,w)/2)]$ implies the inequality

$$R(n) \leq d(z,w)/2$$

which implies that

$$\operatorname{supp}(a_z^n) \cap \operatorname{supp}(a_w^n) = \varnothing.$$

The latter implies that

$$\|a_z^n - a_w^n\|_p^p = \|a_z^n\|_p^p + \|a_w^n\|_p^p = 2,$$

which concludes the proof. $\qquad\square$

Now Lemmas 13 and 17 together with Yu's theorem, Theorem 15, imply the main result of §1.

Theorem 18. *If a finitely presented group Γ admits a spherical anti-Čech approximation, then the Novikov conjecture holds for Γ.*

We conclude this section by a conjecture which appeared from my conversation with G. Yu.

Conjecture 19. *A discrete metric space X has Property A if and only if X is coarsely embeddable in l_1.*

Lemma 17 proves the conjecture in the 'only if' direction. According to Lemma 2 and the fact that the standard simplex in l_1 lies on the unit sphere, to settle the opposite direction it suffices to show that a subspace $X \subset l_1$ admits an anti-Čech l_1-approximation. The conjecture is also supported by the theorem in [17] which states that a discrete metric space of bounded geometry coarsely embeddable in l_1 admits a coarse embedding in l_2.

The l_2 analog of the conjecture is also of interest. It should be stated as follows:

X admits a euclidean anti-Čech approximation if and only if it is coarsely embeddable in l_2.

This conjecture is open in both directions. In [8] it was shown that proper spaces X with euclidean anti-Čech approximations admit coarse quasi-embeddings in l_2. The latter means that given a monotone function $f : \mathbb{R}_+ \to \mathbb{R}_+$ tending to infinity one can find an 1-Lipschitz map $\phi : X \to l_2$ such that for every $R > 0$ there is $S > 0$ such that

$$\mathrm{diam}\phi^{-1}(B_R(x)) \leq f(\|x\|)$$

whenever $\|x\| > S$.

It seems likely that the Novikov conjecture holds for the groups which admit euclidean anti-Čech approximations.

1.8. *Expanders*

Let X be a finite graph, we denote by V the set of vertices and by E the set of edges in X. We will identify the graph X with its set of vertices V. Every graph is a metric space with respect to the natural metric where every edge has the length one. For a subset $A \subset X$ we define the boundary

$$\partial A = \{x \in X \mid dist(x, A) = 1\}.$$

Let $|A|$ denote the cardinality of A.

Definition 20. (Lubotzky [30]) An *expander* with a conductance number c and the degree d is an infinite sequence of finite graphs $\{X_n\}$ with the degree d such that $|X_n|$ tends to infinity and for every $A \subset X_n$ with $|A| \leq |X_n|/2$ there is the inequality $|\partial A| \geq c|A|$.

Let X be a finite graph, we denote by P all nonordered pairs of distinct points in X. For every nonconstant map $f : X \to l_p$ to the Banach space we introduce the number

$$D_f^p = \frac{\frac{1}{|P|} \sum_{\{x,y\}\in P} \|f(x) - f(y)\|_p^p}{\frac{1}{|E|} \sum_{\{x,y\}\in E} \|f(x) - f(y)\|_p^p}.$$

If X is a graph with the degree d and with $|X| = n$, then

$$|P| = n(n-1)/2 \quad \text{and} \quad |E| = dn/2.$$

The following lemma can be derived from [31, Proposition 3]. For $p = 2$ it also can be obtain from the equality

$$\lambda_1(X) = \inf \left\{ \frac{\|df\|^2}{\|f\|^2} \;\middle|\; \sum f(x) = 0 \right\}$$

for the first positive eigenvalue of the Laplacian on X and Cheeger's inequality (see [30, Proposition 4.2.3]).

Lemma 21. *Let $\{X_n\}$ be an expander. Then there is a constant c_0 such that*

$$D^p_{f_n} \le c_0$$

for all n for all possible maps $f_n : X_n \to l_p$ to the Banach space l_p, $p \ge 1$.

Corollary 22. *For every sequence of 1-Lipschitz maps $f_n : X_n \to l_p$ there is the inequality*

$$\frac{1}{|P_n|} \sum_{P_n} \|f_n(x) - f_n(y)\|_p^p \le c_0$$

for every n.

Proof. In the case of 1-Lipschitz map we have

$$\frac{1}{|E|} \sum_{\{x,y\} \in E} \|f(x) - f(y)\|_p^p \le 1.$$

Then the required inequality follows.   □

Corollary 23. *Assume that for a sequence of 1-Lipschitz maps $f_n : X_n \to l_2$ we have*

$$\sum_{x \in X_n} f_n(x) = 0 \quad \text{for every} \quad n.$$

Then

$$\frac{1}{|X_n|} \sum_{x \in X_n} \|f_n(x)\|_2^2 \le c_0/2 \quad \text{for all} \quad n.$$

Proof. According Corollary 22 we have

$$n(n-1)c_0/2 = |P_n|c_0 \ge \sum_{P_n} \|f_n(x) - f_n(y)\|_2^2$$

$$= \sum_{P_n} \|f_n(x) - f_n(y)\|_2^2 + \left(\sum_{X_n} f(x_n)\right)^2$$

$$= \sum_{P_n} (f_n(x) - f_n(y))^2 + \sum_{X_n} f_n(x)^2 + \sum_{P_n} 2(f_n(x), f_n(y))$$

$$= |X_n| \sum_{X_n} f_n(x)^2 = n \sum_{X_n} \|f_n(x)\|_2^2.$$

Then the required inequality follows.   □

We say that a metric space X contains an expander $\{X_n\}$ if there is a sequence of isometric embeddings $X_n \to X$.

The following fact was discovered by M. Gromov for $p = 2$.

Theorem 24. *Suppose that a metric space X can be coarsely embedded in l_p. Then X does not contain an expander.*

Proof. Assume that X contains an expander $\{X_n\}$ of degree d and let c_0 be the constant from Lemma 21. We assume that $|X_n| = n$ where n runs over a subset of $\mathbb{N}$. Let $f : X \to l_p$ be a coarse embedding. Then there are monotone functions $\rho_1, \rho_2 \to \infty$ such that

$$\rho_1(d(x,y)) \leq \|f(x) - f(y)\|_p \leq \rho_2(d(x,y)).$$

We may assume that $\rho_2(1) \leq 1$. We take r such that $\rho_1(r) > (2c_0)^p$. For every point $v \in X$ and every n we have an estimate

$$|B_k(v) \cap X_n| \leq 1 + d + d^2 + \cdots + d^k \leq 2d^k.$$

Let $k = \log_d(n/4)$. We denote by $P_{n,k}$ the set of pairs $\{x,y\}$ in X_n with $d(x,y) \geq k$. Then

$$|P_{n,k}| \geq \frac{1}{2} \sum_{v \in X_n} (n - |B_k(v) \cap X_n|) \leq n^2/4.$$

Therefore we have

$$
\begin{aligned}
c_0 &\geq \frac{1}{|P_n|} \sum_{P_n} \|f(x) - f(y)\|_p^p \geq \frac{1}{|P_n|} \sum_{P_{n,k}} \|f(x) - f(y)\|_p^p \\
&\geq \frac{n^2}{8|P_n|} \min_{P_{n,k}} \|f(x) - f(y)\|_p^p = \frac{n^2}{4n(n-1)} \|f(\tilde{x}) - f(\tilde{y})\|_p^p \\
&\geq \frac{1}{4} \|f(\tilde{x}) - f(\tilde{y})\|_p^p = \frac{1}{4} \|g(\phi(\tilde{x})) - g(\phi(\tilde{y}))\|_p^p \\
&\geq \frac{1}{4} \rho^p(d_N(\phi(\tilde{x}), \phi(\tilde{y}))) \geq \frac{1}{4} \rho^2(r) > c_0.
\end{aligned}
$$

This contradiction completes the proof. $\qquad\square$

Theorem 24 together with the following result of Gromov (announced in [25]) restricts the anti-Čech approximation approach to the Novikov conjecture.

Theorem 25. (Gromov) *There exists a group Γ that contains an expander and has a finite complex $K(\Gamma, 1)$.*

It was shown in [8] that a metric space that admits a euclidean anti-Čech approximation cannot contain an expander.

1.9. *Polynomial dimension growth*

By Proposition 3, every discrete metric space of bounded geometry X admits a uniformly bounded cover $\mathcal{U}$ with finite multiplicity $m(\mathcal{U})$ and with the Lebesgue number $L(\mathcal{U})$ greater than given $L > 0$. We define a function

$$d(L) = \min_{L(\mathcal{U}) > L} m(\mathcal{U}) - 1.$$

Clearly, $d(L)$ is monotone. The limit

$$\lim_{L \to \infty} d(L)$$

equals the Gromov asymptotic dimension $\operatorname{asdim} X$ defined in §3. We will refer to the function $d(L)$ as to the asymptotic dimension of X as well.

Theorem 26. *Suppose that the asymptotic dimension $d(L)$ of finitely presented group Γ has a polynomial growth. Then the Novikov conjecture holds for Γ.*

Proof. Since $d(L)$ has a polynomial growth, there is $p > 1$ such that

$$\lim_{L \to \infty} d(L)/L^p = 0.$$

In view of Proposition 12, Lemma 17 and Yu's theorem, Theorem 15, it suffices to show that Γ has Property A_p.

Given $L > 0$ take an open uniformly bounded covering $\mathcal{U} = \{U_i\}_{i \in J}$ of Γ with the Lebesgue number $> L$ and with the multiplicity $\leq d(L) + 1$. For every $i \in J$ we fix $y_i \in U_i$. This gives us an embedding $l_p(J) \subset l_p(\Gamma)$. Denote

$$\phi_i(x) = d(x, \Gamma \setminus U_i).$$

Then for fixed $x \in \Gamma$ the family $\{\phi_i(x)\}$ defines a nonzero element $\phi_x \in l_p(J)$. We define a map

$$a^L : \Gamma \to l_p(J) \subset l_p(\Gamma)$$

as

$$a_z^L = \phi_z / \|\phi_z\|_p \quad \text{for every} \quad z \in \Gamma.$$

Assume that

$$\operatorname{diam} U < R \quad \text{for all} \quad U \in \mathcal{U}.$$

Then

$$\operatorname{supp}(a_z^L) = \{y_i \mid \phi_i(z) \neq 0\} = \{y_i \mid z \in U_i \in \mathcal{U}\} \subset B_R(z)$$

and the condition (1) from the definition of Property A is satisfied. To verify the condition (2) we take $w, z \in \Gamma$ with $d(z, w) \leq K$ and consider the triangle inequality

$$\|a_z^L - a_w^L\|_p \leq \|a_z^L - \frac{\phi_w}{\|\phi_z\|_p}\|_p + \|\frac{\phi_w}{\|\phi_z\|_p} - a_w^L\|_p.$$

Since

$$|d(z, \Gamma \setminus U_j) - d(w, \Gamma \setminus U_j)| < d(z, w),$$

$\|\phi_z\|_p \geq L$, and $m(\mathcal{U}) \leq 2d(L)$, we obtain the following estimate

$$\|a_z^L - \frac{\phi_w}{\|\phi_z\|_p}\|_p = \frac{1}{\|\phi_z\|_p}\|\phi_z - \phi_w\|_p$$

$$= \frac{1}{\|\phi_z\|_p}(\Sigma_J |d(z, \Gamma \setminus U_j) - d(w, \Gamma \setminus U_j)|^p)^{1/p}$$

$$\leq \frac{1}{\|\phi_z\|_p}(4d(L)d(z, w)^p)^{1/p}$$

$$\leq 4K \frac{d(L)^{1/p}}{L}.$$

We note that

$$\|\frac{\phi_w}{\|\phi_z\|_p} - a_w^L\|_p = (|\frac{1}{\|\phi_z\|_p} - \frac{1}{\|\phi_w\|_p}|)\|\phi_w\|_p = \frac{|\|\phi_z\|_p - \|\phi_w\|_p|}{\|\phi_z\|_p}$$

$$\leq \frac{\|\phi_z - \phi_w\|_p}{\|\phi_z\|_p} \leq 4K \frac{d(L)^{1/p}}{L}$$

by the above estimate. Thus

$$\|a_z^L - a_w^L\|_p \leq 8K \frac{d(L)^{1/p}}{L} \to 0,$$

which completes the proof. □

Theorem 26 was proved by G. Yu for bounded $d(L)$ in [42] and extended for sublinear $d(L)$ in [7].

1.10. *Nonpositively curved manifolds*

The following theorem was proved by Gromov [26] and Higson. We present here a modification of their proof designed for an extension to the case of non-Riemannian manifolds.

Theorem 27. (Gromov-Higson) *A contractible Riemannian manifold with nonpositive sectional curvature does not contain an expander.*

Proof. Let M be a contractible Riemannian manifold with nonpositive sectional curvature and let dim $M = m$. Assume the contrary: there is an expander $\{X_n\}$, $X_n \subset M$. The curvature condition implies that the exponential map

$$\exp_x : T_x = \mathbb{R}^m \to M$$

is a diffeomorphism with the 1-Lipschitz inverse map

$$\log_x : M \to \mathbb{R}^m.$$

We show that for any n there is a point $y_n \in M$ such that

$$\sum_{x \in X_n} \log_{y_n}(x) = 0.$$

Assume that

$$w_y = \sum_{x \in X_n} \log_y(x) \neq 0 \quad \text{for all} \quad y \in M.$$

The vector w_y defines a point s_y on the visual sphere at infinity $S(\infty)$, namely the geodesic ray in M issued from y in the direction of w_y gives us s_y. This defines a continuous map $f : M \rightarrow S(\infty)$ which naturally extends to the map $\bar{f} : \bar{M} \rightarrow S(\infty)$, where

$$\bar{M} = M \cup S(\infty).$$

It is easy to see that this map has no fixed point. This yields a contradiction.

Corollary 23 of Lemma 21 implies that there is a number $r \in \mathbb{N}$ such that

$$\| \log_{y_n} x \| \leq r$$

for more that a half points x in X_n for all n. It means that

$$|(\log_{y_n})^{-1}(B_r(0)) \cap X_n| > n/2.$$

Now, note that

$$(\log_{y_n})^{-1}(B_r(0)) = \exp_{y_n}(B_r(0)) = B_r(y_n)$$

and, therefore, there exists $v_n \in X_n$ such that

$$|B_{2r}(v_n) \cap X_n| > n/2.$$

On the other hand

$$|B_{2r}(v_n)| \leq k^{2r},$$

where k is the valency of the expander. This contradicts with fact that n tends to infinity and, consequently, concludes the proof. $\qquad\square$

As it was shown in [14] the Gromov-Higson theorem can be extended to universal covers of spherical manifolds with Lipschitz cohomology.

2. Mapping Intersection Problem

Problem 2. *Can every two maps $f : X \rightarrow \mathbb{R}^n$ and $g : Y \rightarrow \mathbb{R}^n$ of compacta with*

$$\dim(X \times Y) < n$$

be approximated arbitrarily close with maps whose images are disjoint ?

This problem is called the Mapping Approximation Problem. It is still open in the case when $\dim X = n-2$ or $\dim Y = n-2$. The solution of it in the codim $\neq 2$ case appeared in [10] (see also [13] and [20]) and it was preceded by a sequence of papers [37], [38], [12], [18], [19] and others. The difficulties there are due to the fact that $\dim X + \dim Y$ could be much greater than $\dim(X \times Y)$. Below we discuss the reduction of Problem 2 to other approximation problems and give a sketch of the proof in the codim $\neq 2$ case.

2.1. *Cohomological dimension*

The cohomological dimension of a compact metric space X with respect to a coefficient group G is defined as follows

$$\dim_G X = \max\{n \mid \check{H}^n(X, A; G) \neq 0, A \subset_{Cl} X\}.$$

Clearly, $\dim_G X \leq \dim X$. The definition can be restated in terms of cohomology groups with compact supports as follows:

$$\dim_G X = \max\{n \mid H_c^n(U; G) \neq 0, U \subset_{Op} X\}.$$

The following lemma is useful for detecting the cohomological dimension. It is [15, Proposition 1.12].

Lemma 28. *Suppose that a compactum X with $\dim_G X < \infty$ has an open basis $\mathcal{B}$ with the property*

$$H_c^k(U; G) = 0 \ \text{for all} \ k > n \ \text{and} \ U \in \mathcal{B}.$$

Then

$$\dim_G X \leq n.$$

The equivalent definition is that $\dim_G X \leq n$ if and only if the following extension problem

$$A \xrightarrow{f} K(G, n)$$
$$i \downarrow$$
$$X$$

always has a solution, where $i : A \to X$ is the inclusion of a closed subset and $K(G, n)$ is an Eilenberg-MacLane space, i.e., a CW complex that has trivial homotopy groups in all dimensions $\neq n$ and the n-th homotopy group equal to G. We say in this case that $K(G, n)$ is an absolute extensor for X and we denote this condition as $K(G, n) \in AE(X)$. We recall that the covering dimension can be defined in similar terms: $\dim X \leq n$ if and only if $S^n \in AE(X)$, i.e. the following extension problem

$$A \xrightarrow{f} S^n$$
$$i \downarrow$$
$$X$$

always has a solution. P.S. Alexandroff proved the following theorem (see [1] and [41]).

Theorem 29. (Alexandroff) *For a compact metric space*

$$\dim X = \dim_{\mathbb{Z}} X,$$

provided $\dim X < \infty$.

Given a prime number p we denote
$$\mathbb{Z}_{p^k} = \mathbb{Z}/p^k\mathbb{Z}, \qquad \mathbb{Z}_{p^\infty} = \varinjlim \mathbb{Z}_{p^k}$$
and
$$\mathbb{Z}_{(p)} = \left\{ \frac{m}{n} \mid (n, p) = 1 \right\} \in \mathbb{Q}.$$
Also, we set
$$\sigma_p = \{\mathbb{Z}_{(p)}, \mathbb{Z}_p, \mathbb{Z}_{p^\infty}\}.$$
The family
$$\sigma = \mathbb{Q} \cup_p \sigma_p$$
is called the Bockstein family of abelian groups. We combine the results of Bockstein [2], [3] and Boltyanskij [4] into the following result.

Theorem 30. *For finite dimensional metric compacta X and Y the following conditions are equivalent:*

(1) $\dim_G X = \dim_G Y$ *for all abelian groups G;*

(2) $\dim_G X = \dim_G Y$ *for $G \in \sigma$;*

(3) $\dim(X \times Z) = \dim(Y \times Z)$ *for every compactum Z.*

If the above conditions hold for X and Y we say that X and Y have the same *dimension type* denoted as
$$DIM(X) = DIM(Y).$$
Clearly, a dimension type of a compactum X is defined by a function $\beta_X : \sigma \to \mathbb{N}$,
$$\beta_X(G) = \dim_G X.$$
Not all functions $\beta : \sigma \to \mathbb{N}$ correspond to dimension types. It turns out that valid functions are those for which
$$\beta(\sigma_p) = \beta(\mathbb{Q}) \quad \text{or} \quad \beta(\mathbb{Z}_{(p)}) = \max\{\beta(\mathbb{Q}), \beta(\mathbb{Z}_{p^\infty}) + 1\}$$
and
$$\beta(\mathbb{Z}_p) \geq \beta(\mathbb{Z}_{p^\infty}) \geq \beta(\mathbb{Z}_p) - 1$$
for every prime p. Such functions are called *Bockstein functions*. More generally we consider all not only positive bounded functions $\beta : \sigma \to \mathbb{Z}$ satisfying the above conditions and call them Bockstein functions. It is easy to verify that the maximum of any family of Bockstein functions is a Bockstein function. Also it is clear that every constant function is a Bockstein function. For every compact space X we denote by β_X the function defined by the formula
$$\beta_X(G) = \dim_G X.$$
It is an exercise to show that β_X is indeed a Bockstein function. Note that for a Bockstein function β_X representing the dimension type $DIM(X)$ of a finite dimensional compactum X there is the equality
$$\max\{\beta_X(G) \mid G \in \sigma\} = \dim X.$$
The following realization theorem goes back to [9], [10].

Theorem 31. *For every Bockstein function $\beta \geq 1$ with $\max \beta = n$ there is a compact set $Y_\beta \subset \mathbb{R}^{n+2}$ such that*

$$\dim_G Y_\beta = \beta(G) \quad \text{for all} \quad G \in \sigma,$$

i.e., $\beta = \beta_{Y_\beta}$.

It is easy to check that the sum of a Bockstein function and a constant is a Bockstein function. Here we define the product $\wedge$ on Bockstein functions. First for positive Bockstein functions α and β we define the product

$$\alpha \wedge \beta = \beta_{(Y_\alpha \times Y_\beta)}.$$

In view of Theorem 30 it is well-defined operation. In general case we define $\alpha \wedge \beta$ as

$$\alpha \wedge \beta = (\alpha + n) \wedge (\beta + m) - n - m$$

for large enough natural numbers n and m. Clearly, $\alpha \wedge n = \alpha + n$ for every $n \in \mathbb{Z}$.

We note that the set of Bockstein functions inherits a partial order from the linear order on $\mathbb{Z}$. For every Bockstein function β we define the *Bockstein dual*

$$\beta^* = \max\{\alpha \mid \alpha \wedge \beta \leq 0\}.$$

It is not difficult to show the equality

$$(\beta^*)^* = \beta$$

for all Bockstein functions β and the equivalence

$$\alpha \leq \beta \Leftrightarrow \alpha^* \geq \beta^*$$

for all α and β. We note that $n^* = -n$ for $n \in \mathbb{Z}$ [19], [13], [15], [35].

2.2. *Extending maps to CW complexes*

Let M be a CW complex. We say that M is an absolute extensor for a space X and write $M \in AE(X)$ if for every closed subset $A \subset X$ every extension problem

$$A \xrightarrow{f} M$$

$$i \downarrow$$

$$X$$

has a solution.

We recall that a complex M is called *simple (k-simple)* if the action of the fundamental group $\pi_1(M)$ on all homotopy groups $\pi_i(M)$ ($i \leq k$) is trivial. This amounts to say that every map of the wedge $f : S^1 \vee S^i \to M$ has an extension $\bar{f} : S^1 \times S^i \to M$ for all k. Here we identify $S^1 \vee S^i$ with

$$S^1 \times \{x_0\} \cup \{y_0\} \times S^k \subset S^1 \times S^i, \quad x_0 \in S^i, \quad y_0 \in S^1.$$

We note that a 1-simple space is exactly a space with the abelian fundamental group.

The following theorem first appeared in [11], [12] for a simply connected M. The case of simple M is contained in [5].

Theorem 32. *Let M be a simple CW complex and let X be a finite-dimensional compactum. Then the following are equivalent:*

(1) $M \in AE(X)$;

(2) $\dim_{H_k(M)} X \leq k$ for all k;

(3) $\dim_{\pi_k(M)} X \leq k$ for all k.

We need the following version of this theorem.

Corollary 33. *Let M be a k-simple CW complex and let X be a $k+1$-dimensional compactum. Then the following are equivalent:*

(1) $M \in AE(X)$;

(2) $\dim_{H_i(M)} X \leq i$ for all i;

(3) $\dim_{\pi_i(M)} X \leq i$ for all $i \leq k$.

Proof. The implications (1) $\Rightarrow$ (2) and (2) $\Rightarrow$ (3) follow from Theorem 32. We show the implication (3) $\Rightarrow$ (1). We attach to M cells of dimension $\geq k+2$ to kill all homotopy groups in dimensions $\geq k+1$. We obtain a complex

$$M' = M \cup_\alpha B_\alpha^{m_\alpha}$$

with $\pi_i(M') = 0$ for $i \geq k+1$ and with $\pi_i(M') = \pi_i(M)$ for $i \leq k$. Then by condition (3) and Theorem 32 we obtain that $M' \in AE(X)$. Since $\dim X \leq k+1$, every map $f : X \to M'$ can be swept to M. This implies that $M \in AE(X)$. $\square$

Remark 34. The implication (1) $\Rightarrow$ (2) holds without any restriction on M, [11].

2.3. Negligibility Criterion

Definition 35. Given a compactum X, a compactum $Y \subset \mathbb{R}^n$ is said to be X-**negligible** if every map : $X \to \mathbb{R}^n$ is a limit of maps $f_i : X \to \mathbb{R}^n \setminus Y$.

Let $C(X, \mathbb{R}^n)$ be the space of all continuous mappings of a compactum X to $\mathbb{R}^n$ with the C^0-topology. We note that a compactum $Y \subset \mathbb{R}^n$ is X-negligible if and only if the subset $C(X, \mathbb{R}^n \setminus Y)$ is dense (and open) in $C(X, \mathbb{R}^n)$.

Proposition 36. *For a compactum $Y \subset \mathbb{R}^n$ the following are equivalent*

(1) Y is X-negligible;

(2) $(U \setminus Y) \in AE(X)$ for every open set $U \subset \mathbb{R}^n$ homeomorphic to $\mathbb{R}^n$.

Proof. (1) $\Rightarrow$ (2). Suppose Y is X-negligible, and let $\phi : A \to U \setminus Y$ be a continuous map of a closed subset $A \subset X$. Since A is compact, there is ϵ_0 such that the ϵ_0-neighborhood N of $\phi(A)$ in $\mathbb{R}^n$ lies in $U \setminus Y$. Then every ϵ_0-close to ϕ map $\psi : A \to N$ is homotopic to ϕ in N by means of obvious homotopy

$$H(x,t) = (1-t)\phi(x) + t\psi(x).$$

Since $U \in AR$, there is an extension $\bar{\phi} : X \to U$ of the map ϕ. Let ϵ' be the distance from $\bar{\phi}(X)$ to ∂U. we take $\epsilon = \min\{\epsilon_0, \epsilon'\}$. Since Y is X-negligible, there is an ϵ-approximation $f : X \to \mathbb{R}^n \setminus Y$ of $\bar{\phi}$. Note that $f(X) \subset U \setminus Y$ and the restriction $f|_A$ is ϵ_0-close to ϕ. Therefore $f|_A$ is homotopic to ϕ. Since $U \setminus Y$ is an ANR, by the Homotopy Extension Theorem there is an extension $\phi' : X \to U \setminus Y$ of the map ϕ.

(2) $\Rightarrow$ (1). Now suppose a map $f : X \to \mathbb{R}^n$ and an $\epsilon > 0$ are given. We consider an $\epsilon/4$-approximation $g : X \to K$ with the image a subpolyhedron of $\mathbb{R}^n$ with respect to a triangulation τ of $\mathbb{R}^n$ with mesh $< \epsilon/4$. For every simplex $\sigma \subset K$ we denote by U_σ the interior of the star of σ in the second barycentric subdivision of τ. Note that if $\sigma_0 = \sigma_1 \cap \sigma_2$, then $U_{\sigma_0} = U_{\sigma_1} \cap U_{\sigma_2}$. Furthermore, it is easily seen that the space U_σ is homeomorphic to an open topological ball. By induction on i we define a map

$$g_i : g^{-1}(K^{(i)}) \to \mathbb{R}^n \setminus Y,$$

compatible with g_{i-1} and such that $g_i(f^{-1}(\sigma)) \subset U_\sigma$. For $i = 0$ it can be defined by choosing a point $x_\sigma \in U_\sigma \setminus Y$ for every $\sigma \in K^{(0)}$. $\square$

We call a basis $\mathcal{B}$ of topology on $\mathbb{R}^n$ *special* if $U \cap V \in \mathcal{B}$ for every $U, V \in \mathcal{B}$ and all elements $U \in \mathcal{B}$ are homeomorphic to $\mathbb{R}^n$. We note that in Proposition 36 the condition (2) can be replaced by anyone of the following conditions

(3) *There exists a special basis $\mathcal{B}$ such that $(U \setminus Y) \in AE(X)$ for every $U \in \mathcal{B}$.*

(4) *For every special basis $\mathcal{B}$, $(U \setminus Y) \in AE(X)$ for every $U \in \mathcal{B}$.*

Definition 37. A compactum $Y \subset \mathbb{R}^n$ is called *k-simple* if there is a special basis $\mathcal{B}$ in $\mathbb{R}^n$ such that $U \setminus Y$ is k-simple.

Criterion 38. (of Negligibility) *Suppose a compactum $Y \subset \mathbb{R}^n$ is $(n-3)$-simple. Then Y is X-negligible if and only if*

$$\dim(X \times Y) < n.$$

Proof. If $\dim X = n$, then these conditions are vacuous. If $\dim X = n - 1$, then it's not difficult to show that both conditions are equivalent to the equality $\dim Y = 0$.

We assume that $\dim X \leq n-2$. Let $\mathcal{B}$ be a special basis of open n-balls such that $U \setminus Y$ is $(n-3)$-simple. Then $U \setminus Y$ satisfies the conditions of Corollary 33. By an extended version of Proposition 36 we obtain that a compactum Y is X-negligible if and only if

$(U \setminus Y) \in AE(X)$ for all $U \in \mathcal{B}$. By Corollary 33 we obtain that $(U \setminus Y) \in AE(X)$ for all $U \in \mathcal{B}$ if and only if

$$\dim_{H_k(U \setminus Y)} X \leq k$$

for all $U \in \mathcal{B}$ and for all k. By the Alexander duality the latter condition is equivalent to

$$\dim_{H_c^{n-k-1}(Y \cap U)} X \leq k \quad \forall k, \ \forall U \in \mathcal{B}.$$

By the definition of cohomological dimension this is equivalent to the condition

$$H_c^{k+1}(V; H^{n-k-1}(Y \cap U)) = 0 \quad \forall k, \ \forall U \in \mathcal{B}, \ \forall V \subset_{Op} X.$$

By the Künneth formula the latter is equivalent to

$$H_c^n(V \times (U \cap Y); \mathbb{Z}) = 0 \quad \forall U \in \mathcal{B}, V \subset_{Op} X.$$

By Lemma 28 this condition is equivalent to the inequality $\dim_{\mathbb{Z}}(X \times Y) < n$. In view of the Alexandroff theorem it is equivalent to the inequality $\dim(X \times Y) < n$. $\quad\square$

Remark 39. It is easy to check that the Remark 34 implies that the 'only if' part of the Negligibility Criterion holds for any compactum Y.

2.4. *Reduction to other approximation problems*

We consider the following.

Conjecture 40. (of Dimension Approximation) *Let X a compactum with*

$$\dim X \leq n - 2.$$

Then the set of maps $f : X \to \mathbb{R}^n$ with

$$DIM(f(X)) = DIM(X)$$

is dense G_δ in $C(X, \mathbb{R}^n)$.

A theorem of Bockstein (cf. [15]) implies that this conjecture follows from its weak form:

Conjecture 41. (weak form) *For every group $G \in \sigma$ the set*

$$\{f : X \to \mathbb{R}^n) \mid \dim_G f(X) = \dim_G X\}$$

is dense G_δ in $C(X, \mathbb{R}^n)$.

Conjecture 42. (of Simple Approximation) *Let X a compactum with*

$$\dim X \leq n - 2.$$

Then the set of maps $f : X \to \mathbb{R}^n$ with $(n - 3)$-simple image $f(X)$ is dense G_δ in $C(X, \mathbb{R}^n)$.

Assertion 43. *The Dimension Approximation Conjecture together with the Simple Approximation Conjecture give an affirmative answer to Problem 2.*

Proof. Let $\dim(X \times Y) < n$. To exclude the degenerating cases we assume that

$$\max\{\dim X, \dim Y\} \leq n - 2.$$

By the above conjectures we can take an arbitrary close approximation $g' : Y \to \mathbb{R}^n$ of g with

$$DIM(g'(Y)) = DIM(Y)$$

and $(n-3)$-simple image $g'(Y)$. By Theorem 30

$$\dim(X \times g'(Y)) = \dim(X \times Y) < n.$$

By the Negligibility Criterion $g'(Y)$ is X-negligible. Hence there exists an approximation $f' : X \to \mathbb{R}^n$ of f with $f'(X) \cap g'(Y) = \varnothing$. $\qquad\square$

Proposition 44. *Let $X \subset \mathbb{R}^n$ be an arbitrary compact set. Then the Dimension Approximation Conjecture holds for X.*

Proof. We recall that a map is called *light* if all its point preimages are 0-dimensional. We show that the set of light maps $f' : X \to \mathbb{R}^n$ with

$$DIM(f'(X)) = DIM(X) \tag{1}$$

is dense G_δ in $C(X, \mathbb{R}^n)$. First we show that the set of light maps $f' : X \to \mathbb{R}^n$ satisfying (1) is dense in $C(X, \mathbb{R}^n)$. Let $f : X \to \mathbb{R}^n$ be given. We approximate f by a light map $\tilde{f}$ and take a compact, n-dimensional polyhedron $P \subset \mathbb{R}^n$ with $X \subset P$ and extend $\tilde{f}$ over P, i.e., get a map $\bar{f} : P \to \mathbb{R}^n$ such that $\bar{f}|_X = \tilde{f}$. Then we approximate $\bar{f}$ by a simplicial, general position map $g : P \to \mathbb{R}^n$. Then the restriction $g|_\Delta : \Delta \to \mathbb{R}^n$ is an embedding for every simplex Δ in P. Consider $f' = g|_X$. Clearly f' is a light map. Since

$$X = \cup\{X \cap \Delta \mid \Delta \subset P\},$$

it follows that

$$f'(X) = \cup\{f'(X \cap \Delta) \mid \Delta \subset P\}.$$

Then

$$\dim_G f'(X) = \max \dim_G f'(X \cap \Delta) = \max \dim_G X \cap \Delta = \dim_G X$$

for all abelian groups G.

Then we fix $H \in \sigma$ and show that the set S of maps $f' : X \to \mathbb{R}^n$ with

$$\dim_H f'(X) > \dim_H X = m_H$$

is F_σ. Since a CW complex $K(H, m_H)$ is an ANE (absolute neighborhood extensor), there is a countable family $\{B_i\}$ of finite subcomplexes in $\mathbb{R}^n$ such that every map $f : A \to K_j$ of a compact set $A \subset \mathbb{R}^n$ is extendable over a set $B_i^j \supset A$ for some i. We note that there are at most countably many homotopy classes of mappings between two finite complexes. Since H is countable there is a complex $K(H, m_H)$ which can be presented as the countable union of finite complexes:

$$K(H, m_H) = \cup K_j.$$

Therefore for every i there is a countable family of maps $\phi_i^j : B_i \to K(H, m_H)$ representing all homotopy classes. In view of the above and the Homotopy Extension Theorem we conclude that for every map $\phi : A \to K(H, m_H)$ of a comact subset $A \subset \mathbb{R}^n$ there are i and j with $A \subset B_i^j$ and with ϕ homotopic to $\phi_i^j |_A$. We define

$$S_i^j = \{f : X \to \mathbb{R}^n \mid \phi_i^j |_{B_i \cap f(X)} : B_i \cap f(X) \to K(H, m_H) \text{ is not extendable to } f(X)\}.$$

We claim that

$$S = \cup S_i^j.$$

Clearly $S_i^j \subset S$ for all i and j. If

$$\dim_H f'(X) > m_H,$$

there exists a closed subset $A \subset f'(X)$ and a map $\phi : A \to K(H, m_H)$ which is not extendable to $f'(X)$. By the construction of the family $\{\phi_i^j\}$ there are i and j such that $A \subset B_i$ and ϕ is homotopic to $\phi_i^j |_A$. By the Homotopy Extension Theorem the map

$$\phi_i^j |_{B_i \cap f'(X)} : B_i \cap f'(X) \to K(H, m_H)$$

is not extendable to $f'(X)$. Hence $f' \in S_i^j$. Thus the set of maps $f' : X \to \mathbb{R}^n$ with

$$DIM(f'(X)) \leq DIM(X)$$

is dense G_δ. Since the light maps (maps with 0-dimensional point preimages) do not lower cohomological dimension and they form a dense G_δ set, the set of light maps $f' : X \to \mathbb{R}^n$ with

$$DIM(f'(X)) = DIM(X)$$

is dense G_δ. $\qquad\qquad\qquad\qquad\qquad\qquad\qquad\qquad\qquad\qquad\qquad\qquad\qquad\square$

Theorem 45. *The Simple Approximation Conjecture implies the Dimension Approximation Conjecture.*

Proof. According to the Section 1, the inequalities $\beta_X \geq 1$ and $\beta_X \leq n - 2$ imply the inequalities

$$\beta_X^* \leq -1 \quad \text{and} \quad \beta_X^* \geq 2 - n.$$

These inequalities imply that $\alpha \geq 1$ and $\alpha \leq n - 2$ for the Bockstein function

$$\alpha = \beta_X^* + n - 1.$$

By the Realization Theorem —Theorem 31—, α can be represented by a compactum $Y_\alpha \subset \mathbb{R}^n$. By Proposition 44 and the Simple Approximation Conjecture —Conjecture 42— there is a countable dense subset H in the space $C(Y_\alpha, \mathbb{R}^n)$ with $h(Y_\alpha)$ $(n - 3)$-simple and

$$DIM(h(Y_\alpha)) = \alpha \quad \text{for all} \quad h \in H.$$

We note that every compactum Z lying in the complement to $\cup_{h \in H} h(Y_\alpha)$ is Y_α-negligible. Then by the Remark 39 we have

$$\dim(Y_\alpha \times Z) \leq n - 1.$$

This can be written as

$$(\alpha - n + 1) \wedge \beta_Z \leq 0 \iff \beta_X^* \wedge \beta_Z \leq 0.$$

Therefore,

$$\beta_Z \leq (\beta_X^*)^* = \beta_X.$$

Hence,

$$\dim_G Z \leq \dim_G X \quad \text{for all} \quad G.$$

Since $\dim(h(Y_\alpha) \times X) < n$ and $h(Y_\alpha)$ is $(n-3)$-simple, by the Negligibility Criterion a map $f : X \to \mathbb{R}^n$ can be approximated by maps f' with

$$f'(X) \subset \mathbb{R}^n \setminus \bigcup_{h \in H} h(Y_\alpha).$$

Thus, $\dim_G f'(X) \leq \dim_G X$ for all G. Since the set of light maps f' is dense G_δ and light maps cannot lower cohomological dimension, we may assume that

$$\dim_G f'(X) = \dim_G X$$

for all G. $\qquad\square$

Corollary 46. *The simple Approximation Conjecture —Conjecture 42— gives a positive solution to Problem 2.*

2.5. The codimension three case

If a compactum X has dimension

$$\dim X \leq n - 3$$

then the Simple Approximation Conjecture holds true for mappings $f : X \to \mathbb{R}^n$. This follows, for example, from Stanko's tame reimbedding theorem [39]. However the Dimension Approximation Conjecture does not follow for such X. Indeed, the argument for Theorem 45 shows that we need the Simple Approximation Conjecture for Y_α with

$$\alpha = \beta_X^* + n - 1.$$

Proposition 47. *Suppose that a compactum X satisfies the inequalities*

$$\dim X \leq n - 3 \quad and \quad \dim_G X \geq 2 \quad for\ all \quad G \in \sigma.$$

Then the Approximation Conjecture holds for X.

Proof. The condition $\beta_X \geq 2$ implies that $\beta_X^* \leq -2$. The latter implies that

$$\beta_X^* + n - 1 \leq n - 3.$$

Hence by the above the Simple Approximation Theorem holds for Y_α and hence the Approximation Conjecture holds for X in view of the argument to Theorem 45. $\qquad\square$

The Problem 2 is solved in the codimension three case.

Theorem 48. *([10]) Suppose that compacta X and Y satisfy the inequalities*

$$\dim X, \ \dim Y \le n - 3 \quad and \quad \dim(X \times Y) < n.$$

Then every two maps $f : X \to \mathbb{R}^n$, $g : Y \to \mathbb{R}^n$ can be approximated by maps with disjoint images.

Proof. We replace Y with $Y \vee I^2$ to insure the inequality

$$\dim_G Y \ge 2.$$

Since $\dim X \le n - 3$, we still have the inequality

$$\dim(X \times Y) < n.$$

By Proposition 47 the Approximation Conjecture holds for Y. Hence we can approximate $g : Y \to \mathbb{R}^n$ arbitrary closely by $g' : Y \to \mathbb{R}^n$ with $(n - 3)$-simple $g'(Y)$ and with

$$DIM(g'(Y)) = DIM(Y).$$

By Theorem 30 and the Negligibility Criterion $g'(Y)$ is X-negligible. Therefore we can take an arbitrary close approximation f' of f with

$$f'(X) \cap g'(Y) = \varnothing.$$

This ends the proof.   □

Remark 49. The Casson finger move allows to prove the Simple Approximation Conjecture for $n = 4$.

References

1. Alexandroff, P.S., Dimensiontheorie, ein Betrag zur Geometrie der abgeschlossen Mengen, *Math. Ann.* **106** (1932), 161–238.
2. Bockstein, M.F., Homological invariants of topological spaces (Russian), *Trudy Moskov. Mat. Obshch.* **5** (1956), 3–80.
3. Bockstein, M.F., Homological invariants of topological spaces (Russian), *Trudy Moskov. Mat. Obshch.* **6** (1957), 3–133.
4. Boltyanskij, V.G., On dimensional full-valuedness of compacta, *Dokl. Akad. Nauk SSSR* **67** (1949), 773–777.
5. Cencelj, M., and Dranishnikov, A.N., Extension of maps to nilpotent spaces II and III, *Top. Appl.* **124** (2002), 77–83; and *Top. Appl.*, to appear.
6. Connes, A., Gromov, M., and Moscovici, H., Group cohomology with Lipschitz control and higher signatures, *GAFA* **3** (1993), 1–78.
7. Dranishnikov, A.N., Asymptotic topology, *Russian Math. Surveys* **55** (2000), 71–116.
8. Dranishnikov, A.N., Hypereuclidean manifolds and the Novikov conjecture, Preprint of MPIM 2000-65.

9. Dranishnikov, A.N., Homological dimension theory, *Russian Math. Surveys* **43** (1988), 11–63.

10. Dranishnikov, A.N., On the mapping intersection problem, *Pac. J. Maths.* **173** (1996), 403–412.

11. Dranishnikov, A.N., Extension of mappings into CW-complexes, *Matem. Sbornik* **182** (1991), 1300–1310.

12. Dranishnikov, A.N., On Intersections of Compacta in Euclidean Space, *Proc. AMS* **112** (1991), 267–275.

13. Dranishnikov, A.N., On the dimension of the product of two compacta and the dimension of their intersection in general position in Euclidean space, *Trans. Amer. Math. Soc.* **352** (2000), 559–5618.

14. Dranishnikov, A.N., Lipschitz cohomology, Novikov conjecture, and expanders, Preprint ArXiv, math. GT/0205172.

15. Dranishnikov, A.N., Cohomological dimension theory of compact metric spaces, in *Topology Atlas*, http://at.yorku.ca/topology/taic.html.

16. Dranishnikov, A.N, Ferry, S., and Weinberger, S., Large Riemannian manifolds which are flexible, *Ann. of Math.* **157** (2003), 919–938.

17. Dranishnikov, A.N., Gong, G., Lafforgue, V., and Yu, G., Uniform embeddings into Hilbert space and a question of Gromov, *Canad. Math. Bull.* **45** (2002), 60–70.

18. Dranishnikov, A.N., Repovš, D., and Schepin, E.V., On intersection of compacta of complementary dimensions in Euclidean space, *Topol. Appl.* **38** (1991), 237–253.

19. Dranishnikov, A.N., Repovš, D., and Schepin, E.V., An approximation and embedding problem for cohomological dimension, *Topol. Appl.* **55** (1994), 67–86.

20. Dranishnikov, A.N., Repovš, D., and Schepin, E.V., Transversal intersection formula for compacta, *Topol. Appl.* **85** (1998), 93–117.

21. Ferry, S., Ranicki, A., and Rosenberg, J., Editors, *Conjectures, Index Theorems and Rigidity, Vol. 1, 2*, London Math. Soc. Lecture Note Ser. 226, Cambridge Univ. Press, Cambridge, 1995.

22. Gromov, M., *Asymptotic Invariants of Infinite Groups*, Geometric Group Theory 2, Cambridge Univ. Press, Cambridge, 1993.

23. Gromov, M., Large Riemannian manifolds, in *Lecture Notes in Math. 1201*, pages 108–122, 1985.

24. Gromov, M., Positive curvature, macroscopic dimension, spectral gaps and higher signatures, in *Functional Analysis on the eve of the 21st century, Vol. 2*, pages 1–213, Progr. Math. 132, 1996.

25. Gromov, M., Spaces and questions, *GAFA* **Special Vol. I** (2000), 118–161.

26. Gromov, M., Random walk in random groups, *GAFA* **13** (2003), 73–146.

27. Higson, N., Bivariant K-theory and the Novikov conjecture, *GAFA* **10** (2000), 563–581.

28. Higson, N., and Roe, J., On the coarse Baum-Connes conjecture, in *Novikov Conjectures, Index Theorems and Rigidity, Vol. 1, 2*, pages 227–254, London Math. Soc. Lecture Note Ser. 226, Cambridge Univ. Press, Cambridge, 1995.

29. Higson, N., and Roe, J., Amenable action and the Novikov conjecture, *J. Reine Angew. Math.* **519** (2000), 143–153.

30. Lubotzky, A., *Discrete Groups, Expanding Graphs and Invariant Measures*, Birkhauser, Basel-Boston-Berlin, 1994.

31. Matoušek, J., On embedding expanders into l_p-spaces, *Israel J. Math.* **102** (1997), 189–197.
32. Mitchener, P., Coarse homology theories, *AGT* **1** (2001), 271–297.
33. Roe, J., *Coarse Cohomology and Index Theory for Complete Riemannian Manifolds*, Memoirs Amer. Math. Soc. 497, 1993.
34. Roe, J., *Index Theory, Coarse Geometry, and Topology of Manifolds*, CBMS Regional Conference Series in Mathematics 90, 1996.
35. Shchepin, E.V., Arithmetic of dimension theory, *Russian Math. Surveys* **53** (1998), 975–1069.
36. Skandalis, G., Tu, J.L., and Yu, G., Coarse Baum-Connes conjecture and groupoids, *Topology* **41** (2002), 807–834.
37. Spiez, S., On pairs of compacta with $\dim(X \times Y) < \dim X + \dim Y$, *Fund. Math.* **135** (1990), 213–222.
38. Spiez, S., and Toruńczyk, H., Moving compacta in $\mathbb{R}^m$ apart, *Top. Appl.* **41** (1991), 193–204.
39. Štanko, M.A., Approximation of compacta in E^n in codimension greater than two (Russian), *Mat. Sb.* **90** (1973), 625–636.
40. Tu, J.L., Remarks on Yu's property A for discrete metric spaces and groups, *Bull. Soc. Math. de France* **129** (2001), 115–139.
41. Walsh, J.J., Dimension, cohomological dimension, and cell-like mappings, in *Lecture Notes in Math. 870*, pages 105–118, Springer, 1981.
42. Yu, G., The coarse Baum-Connes conjecture for groups which admit a uniform embedding into Hilbert space, *Inv. Math.* **139** (2000), 201–240.
43. Yu, G., The Novikov conjecture for groups with finite asymptotic dimension, *Annals of Math.* **147** (1998), 325–355.

Ten Mathematical Essays on Approximation in Analysis and Topology
J. Ferrera, J. López-Gómez, F. R. Ruiz del Portal, Editors

Eigenvalues and Perturbed Domains

J. K. Hale

School of Mathematics, Georgia Institute of Technology,
Atlanta, Georgia 30332, USA

Abstract

For elliptic partial differential equations on a bounded domain, we present a survey of some results on the dependence of eigenvalues and eigenfunctions on smooth and nonsmooth perturbations of the domain.

Key words: elliptic operators, eigenvalues, eigenfunctions, domain variation

1. Introduction

In the study of the global dynamics of certain types of partial differential equations, the stability properties of equilibria play a very important role. These properties often are closely related to the eigenvalues and eigenfuntions of a linear partial differential equation given by the linear variation from an equilibrium. If the partial differential equation is defined on a bounded domain, then one must investigate the dependence of the eigenvalues and eigenfunctions on the boundary conditions and perturbations of the domain. The purpose of these notes is to survey some of the results dealing with this latter problem for second order elliptic operators.

The first situation deals with regular perturbations; that is, the boundary of the original domain and the perturbed domain are C^k-close for some $k \geq 1$. By a change of coordinates onto the original domain, the regularity properties of eigenvalues and eigenfunctions are reduced to the study of the dependence of these quantities on variations in coefficients in the equation and in the boundary conditions. A differential calculus with respect to the domain is needed to discuss nonlinear problems; for example, bifurcation theory, generic hyperbolicity and transversality of stable and unstable manifolds with respect to the domain, maximization of functions over a domain with fixed volume, etc. Problems of this type are discussed in Section 1.

If the domain is irregular (that is, not regular), then the definition of eigenvalues and eigenfunctions for some boundary conditions is a nontrivial task. For example, if the domain is irregular and the boundary conditions are Dirichlet, then one must give first a precise definition of Dirichlet boundary conditions as well as what is meant by eigenvalues. This is discussed in Section 2 for the first eigenvalue and eigenfunction and it is stated that a maximum principle holds if the first eigenvalue is positive. This definition depends upon the domain and then it becomes important to give a topology on the domains in order to know the maximum principle remains true under small perturbations in this topology. It also is necessary to do the same thing for the complete spectrum of the operator.

In Section 3, for the Laplacian with Neumann boundary conditions, we present some results on the dependence of eigenvalues and eigenfunctions on exterior irregular perturbations of the domain, including perturbations near points on the boundary of the domain, dumbbell shaped domains, thin domains and more general perturbations.

2. Regular domain perturbations

Let $\Omega_0 \subset \mathbb{R}^n$ be a bounded domain and let $\mathcal{K}_k(\Omega_0)$ be the collection of all regions Ω which are C^k-diffeomorphic to Ω_0. We introduce a topology by defining a sub-basis of the neighborhoods of a given Ω as

$$\{h \text{ is a small } C^k(\Omega, \mathbb{R}^n) - \text{neighborhood of the inclusion } i_\Omega : \Omega \subset \mathbb{R}^n \}.$$

When $\|h - i_\Omega\|_{C^k}$ is small, h is a C^k-imbedding of Ω into $\mathbb{R}^n$; that is, a C^k-diffeomorphism to its range $h(\Omega)$. Micheletti [42] has shown that this topology is metrizable and that $\mathcal{K}_k(\Omega_0)$ may be considered a separable complete metric space. We say that $\Omega \in \mathcal{K}_k(\Omega_0)$ is a C^k-*regular perturbations* (or, sometimes, simply a *regular perturbation*) of a given domain Ω_0 if h is a small perturbation in $C^k(\Omega_0, \mathbb{R}^n)$ of the inclusion i_{Ω_0}.

Courant and Hilbert [15] studied the effect of regular perturbations of the domain on the eigenvalues and eigenfunctions of boundary value problems for PDE. For example, if $\Omega_\epsilon \in \mathcal{C}_r(\Omega_0)$ is a continuous family of domains converging to Ω_0 as $\epsilon \to 0$ and $\lambda_k(\Omega_\epsilon)$, $k \geq 1$, denotes the ordered set of eigenvalues of the Laplacian $-\Delta_{BC}^{\Omega_\epsilon}$ with some boundary conditons BC and $\{\varphi_{k,\epsilon}\}$ is a set of normalized eigenvectors, they proved that the eigenvalues $\lambda_k(\Omega_\epsilon)$ and eigenfunctions $\{\varphi_{k,\epsilon}\}$ converge to those of $-\Delta_{BC}^{\Omega_0}$. The proof consisted of constructing the family of diffeomorphisms h_ϵ which map Ω_ϵ onto Ω_0 and reduce the problem to the study of a family of operators L_ϵ on Ω_0.

This result is very interesting, but it is desirable to have more information about the eigenvalues and eigenfunctions. For example, if the eigenvalues and eigenfunctions are smooth functions of ϵ, what is the Taylor series in ϵ? Other important problems arise which are concerned with the determination of those domains which are critical values of some function of the domain such as maximization of torsional rigidity with fixed area of the domain, minimization of the principle eigenvalue of the Laplacian with Dirichlet boundary conditions over domains with fixed volume, etc. To discuss such questions we need to have a differential calculus of boundary perturbations. Many people have been concerned about questions of this type. We are going to present the approach of Henry [29], [30], [31] and refer the reader to [31] for extensive references.

If $F : \mathcal{K}_k(\Omega_0) \to Y$, Y a Banach space, then we can define the smoothness of F at Ω_0 in the following way. For any $\Omega \in \mathcal{K}_k(\Omega_0)$ which is close to Ω_0, there is an $h \in C^k(\Omega_0, \mathbb{R}^n)$ which is close to the inclusion i_{Ω_0} such that

$$\Omega = h(\Omega_0).$$

Therefore,

$$F(\Omega) = F(h(\Omega_0)) \equiv (F \circ h)(\Omega_0).$$

We say that F is C^r (resp. C^∞) (resp. analytic) if the map $h \mapsto F \circ h$ is C^k (resp. C^∞) (resp. analytic). In this sense, problems of perturbation of the boundary (or, of the domain of definition) of a boundary value problem is reduced to differential calculus in Banach spaces.

Consider a non-linear formal differential operator

$$F_\Omega(u)(x) = f(x, Lu(x)), \quad x \in \Omega,$$

where L is a constant coefficient linear differential operator depending upon

$$u, \quad u_{x_j}, \ 1 \le j \le n, \quad u_{x_j x_k}, \ 1 \le j, k \le n, \dots$$

and $f(x, \zeta)$ is a given smooth function. We may consider F_Ω as a map from $C^m(\Omega)$ to $C^0(\Omega)$ [or from $W_p^m(\Omega)$ to $L_p(\Omega)$] under appropriate hypotheses.

If $h : \Omega \to h(\Omega) \subset \mathbb{R}^n$ is a C^m-embedding, then a basic problem is concerned with the manner in which the function $F_{h(\Omega)}(u)$ depends upon h and in exhibiting explicit formulas for the derivatives with respect to h if the derivatives exist. Obtaining derivatives with respect to h by working directly on $F_{h(\Omega)}(u)$ leads to many difficulties. As Henry points out, the computation working directly on $F_{h(\Omega)}(u)$ is analogous to treating continuum mechanics with the Lagrange description where particles are labeled as to position at a given time (and in different coordinate systems). The Eulerian description in continuum mechanics labels the particles by a velocity function of position and time in a fixed coordinate system. This suggests discussing properties of $F_{h(\Omega)}$ by considering functions which depend only upon the original domain Ω. This can be accomplished in the following way.

Any C^m-embedding $h : \Omega \to h(\Omega) \subset \mathbb{R}^n$ induces an isomorphism (pull-back)

$$h^* : C^r(h(\Omega)) \to C^r(\Omega)$$

[or $h^* : W_p^r(h(\Omega)) \to W_p^r(\Omega)$] for $0 \le r \le m$ by

$$h^* \varphi = \varphi \circ h.$$

Henry observed that the function (analogous to the Eulerian description in continuum mechanics)

$$h^* F_{h(\Omega)} h^{*-1} : C^m(\Omega) \to C^0(\Omega) \ [\text{or} \ W_p^m(\Omega) \to L_p(\Omega)]$$

acting in spaces which are independent of h could be of great assistance in the differential calculus with respect to the domain. The symbol h^{*-1} denotes $(h^{-1})^*$.

Let Ω be in $C_r(\Omega_0)$, $\alpha = (\alpha_1, \ldots, \alpha_n)$, $\alpha_j \geq 0$, integers, a_α be functions on Ω,

$$|\alpha| = \alpha_1 + \ldots + \alpha_n, \qquad (\partial/\partial y)^\alpha = \prod_{j=1}^{n} (\partial/\partial y_j)^{\alpha_j},$$

and define the linear operator

$$A_\Omega = \sum_{|\alpha| \leq m} a_\alpha(y) \left(\frac{\partial}{\partial y}\right)^\alpha.$$

If we suppose that $\Omega(t)$ is a C^1-curve of domains in $C_r(\Omega_0)$ represented by the diffeomorphisms

$$h(t, \cdot) = i_C + Vt + o(t)$$

as $t \to 0$ and apply the above calculus to A acting on functions $u(t, \cdot)$, then several computations yield

$$\frac{\partial}{\partial t} (h^*(t, \cdot) A_{h(t,\Omega)} h^{*-1}(t, \cdot) u)|_{t=0} = A \frac{\partial u}{\partial t} + [V \cdot \nabla, A] u,$$

where

$$A = A_{\Omega_0}, \qquad [V \cdot \nabla, A] = V \cdot \nabla A - A V \cdot \nabla$$

is the commutator of $V \cdot \nabla$ and A.

If the chain rule for differentiation were applied directly to the corresponding function $F_{\Omega(t)} u$, the special commutator structure as well as the higher order derivatives would not be easily recognizable. This seems to be the advantage of computing derivatives on a fixed domain. This point was noted by Peetre [52] who related it to a Lie derivative. For operators in variational form, Courant and Hilbert [15]) obtained an equivalent formula.

Henry has given many applications of this calculus to prove differentiability with respect to the domain for various quantities associated with a boundary value problem and to obtain explicit formulas for the first and sometimes second derivatives. We mention only a few without any proofs since they are very technical.

2.1. Torsional rigidity

The resistance to torsion of a cylindrical rod depends not only on the elastic constants of the material, but also on the geometry of the cross section $\Omega \subset \mathbb{R}^2$ through the torsional rigidity

$$R(\Omega) \equiv \int_\Omega |\nabla u|^2 = \int_\Omega u \, dx$$

where $u : \Omega \to \mathbb{R}$ is the solution of

$$\Delta u = -1, \quad u = 0 \text{ in } \partial\Omega.$$

Suppose that Ω is $C^{m,\alpha}$ regular, $m \geq 2, 0 < \alpha < 1$, and

$$t \mapsto h(t, \cdot) \in C^{m,\alpha}(\Omega, \mathbb{R}^2)$$

is a C^1 curve near $t = 0$ with

$$h(0, \cdot) = i_\Omega, \qquad \dot{h}(0, \cdot) = V.$$

Let $v(t, y)$ be the solution of

$$\Delta_y v(t, y) = -1 \quad \text{in } \Omega(t) = h(t, \Omega),$$
$$v(t, y) = 0 \quad \text{in } \partial\Omega(t) = h(t, \partial\Omega).$$

After several computations, Henry shows that the following results are true:

$$\frac{d}{dt} R(\Omega(t)) = \int_{\partial\Omega(t)} V \cdot N_{\Omega(t)} \left(\frac{\partial v}{\partial N_{\Omega(t)}} \right)^2, \tag{1}$$

$$\frac{d^2}{dt^2} R(\Omega(t)) = \int_{\partial\Omega(t)} \left(\frac{\partial v}{\partial N_{\Omega(t)}} \right)^2 \left(\frac{\partial \sigma}{\partial t} + \sigma \frac{\partial \sigma}{\partial N_{\Omega(t)}} + H\sigma^2 \right)$$
$$+ \int_{\partial\Omega(t)} 2\sigma \frac{\partial v}{\partial N_{\Omega(t)}} \left(\sigma \frac{\partial^2 v}{\partial N_{\Omega(t)}^2} + \frac{\partial \dot{v}}{\partial N_{\Omega(t)}} \right), \tag{2}$$

where $N_{\Omega(t)}$ is the unit outward normal to $\Omega(t)$,

$$\sigma = V \cdot N_{\Omega(t)}, \qquad H = \operatorname{div} N_{N(t)}$$

and

$$\dot{v} = \partial v / \partial t = -\sigma (dv/dN) \quad \text{on } \Omega(t).$$

Fixing the area of $\Omega(t)$ as a constant independent of t and evaluating (1) at $t = 0$, one can deduce that the disk D_2 is a critical point of $R(\Omega)$.

If $\Omega = D_2$ and the area of $\Omega(t)$ is a constant, then it is possible to show that

$$\frac{d^2}{dt^2} R(\Omega(t))|_{t=0} = -2\pi \sum_{k=2}^{\infty} |\sigma_k|^2 (k - 1)$$

where

$$\sigma = \sum_{k=-\infty}^{\infty} \sigma_k e^{ik\theta}, \qquad 0 \le \theta \le 2\pi,$$

in polar coordinates. Therefore, the disk D_2 is a maximum of the rigidity under the restriction that the area of Ω is constant. Serrin [58] has shown that the disk is the only critical point in the class of connected, bounded C^2-regions.

2.2. Eigenvalues

The calculus also leads directly to formulas for the manner in which eigenvalues of boundary value problems depend upon the domain. For example, consider the eigenvalue problem

$$\Delta u + \lambda u = 0 \text{ in } \Omega, \qquad u = 0 \text{ in } \partial\Omega, \tag{3}$$

where Ω is a C^2-regular domain. Suppose that $\lambda_0 = \lambda_0(\Omega)$ is a simple eigenvalue of (3) and let u_0 be the corresponding eigenfunction with

$$\int_\Omega u_0^2 = 1.$$

Using the implicit function theorem and the above calculus, it is shown that, for any $h \in C^2(\Omega, \mathbb{R}^n)$ in some C^2-neighborhood of the inclusion map $i_\Omega : \Omega \to \mathbb{R}^n$, there is a simple eigenvalue $\lambda(h(\Omega))$ near λ_0, the map $h \mapsto \lambda(h(\Omega))$ is analytic, and, if $h(t, \cdot)$ is a C^1-family of maps in this neighborhood with $h(0, \cdot) = i_\Omega$, $\dot{h}(0, \cdot) = V$, then

$$\frac{d}{dt}\lambda(h(t, \Omega))_{t=0} = - \int_{\partial\Omega} V \cdot N_\Omega (\frac{\partial u_0}{\partial N_\Omega})^2. \tag{4}$$

This derivative was computed formally, in some special cases, by Rayleigh [56] (in the edition of Dover (1945, p.338, eq. 11)) and, for general two dimensional regions by Hadamard [24]. Garabedian and Schiffer [23] did the general case.

Using (4) for $t = 0$, one deduces that, in the class of connected domains of fixed volume, if $\lambda(\Omega)$ is the principal eigenvalue, then the ball is the only critical point of $\lambda(\Omega)$ and is a minimizer.

Higher order derivatives may be computed but explicit expressions can only be given for special types of perturbations and special Ω. For an ellipse $\Omega(t)$ in $\mathbb{R}^2$ with semi-axes e^t, e^{-t}, the map

$$t \mapsto \lambda(\Omega(t))$$

is an even function and

$$\lambda(\Omega(t)) = \lambda_0 + \frac{1}{2}\lambda_0(\lambda_0 - 2)t^2 + O(t^4).$$

The eccentricity of the ellipse is the solution ϵ of the equation

$$\sqrt{1 - \epsilon^2} = e^{-2|t|}.$$

In terms of ϵ, we have

$$\lambda(\Omega(t(\epsilon))) = \lambda_0 + \frac{1}{32}\lambda_0(\lambda_0 - 2)(\epsilon^4 + \epsilon^6) + O(\epsilon^8)$$

as the eigenvalue near λ_0 in the ellipse of area π and eccentricity ϵ. Joseph [38] has computed the series for this latter case, but obtained $(3/2)\lambda_0 - 5$ in place of $\lambda_0 - 2$. This does not change the qualitative properties of his results.

Other boundary conditions can be considered. For example, if we change the boundary conditions to Robin conditions

$$\frac{\partial u}{\partial N} + \beta(x)u = 0 \quad \text{in } \partial\Omega, \tag{5}$$

and assume that the Laplacian with conditions (5) has a simple eigenvalue λ_0 with normalized eigenfunction u_0 and assume perturbations of the domain Ω as above, then there is a unique eigenvalue $\lambda(\Omega(t))$ near λ_0 and

$$\frac{d}{dt}\lambda(\Omega(t))|_{t=0} = \int_{\partial\Omega} V \cdot N[|\nabla_{\partial\Omega} u_0|^2 - (\lambda_0 + \beta^2 - H\beta - \frac{\partial\beta}{\partial N_\Omega})u_0^2],$$

where

$$H = \operatorname{div} N_\Omega$$

is the mean curvature of $\partial\Omega$ and $\nabla_{\partial\Omega} u_0$ is the tangential component of the gradient of u_0.

Henry also has shown that the spectral projections on eigenspaces corresponding to multiple eigenvalues are smooth functions of the perturbation. This implies that the complete set of eigenvalues and eigenfunctions for a domain $h(\Omega)$ will converge to those of Ω as $h \to i_\Omega$ in the C^r-topology.

Many other applications of the calculus with respect to the boundary of the domain are given by Henry. For example, explicit formulas are given for the first few terms in the expansion of capacity in terms of the variations in the boundary. Similar results are given for Green's function for second order differential operators.

2.3. Bifurcation and generecity

For second order differential operators, Henry considers bifurcation problems near equilibrium points for nonlinear problems. In the case where the bifurcation corresponds to a simple zero eigenvalue with the first nontrivial term being quadratic (a codimension one singularity), he gives a complete description of the manner in which the number of solutions changes from zero to two leading to an unfolding of the singularity. In the case of a zero eigenvalue and the first nontrivial nonlinear term being a cubic (a codimension two singularity), he describes the number of solutions as a function of the domain together with another parameter in the differential equation leading to an unfolding of the singularity.

Recall that, in a complete metric space X, a property is said to be generic if it holds on a residual set; that is, on a set which is the countable intersection of open dense sets. In differential equations, it is very important to be able to assert that a given property is generic with respect to some parameters (which may be the vector field, the domain, etc.). Transversality theorems are the normal way to obtain such results. The classical transversality theorem deals with Fredholm maps of finite index (see, for example, Abraham and Robbin [1]). In many of the problems dealing with perturbation of the boundary, the operator is Fredholm but has index $-\infty$. Therefore, to use transversality theory to discuss generic properties with respect to the boundary, a generalized transversality theorem is needed. An elegant and appropriate generalization has been given by Henry. We do not state the theorem and only mention some of the results that have been obtained by using this theorem.

Using his transversality theorem, Henry proved that, for a residual set of h near i_Ω, the eigenvalues of the Laplacian with Dirichlet, Robin or Neumann boundary conditions are simple. In the Neumann case, if the domain is not connected, then zero is a multiple eigenvalue. In this case, the simplicity of eigenvalues refers to all eigenvalues except zero. The case of Dirichlet boundary conditions also was proved by Uhlenbeck [60]. Henry has similar results for more general differential operators which generalize those of Micheletti [44].

When the domain Ω enjoys some symmetry properties and the perturbations are in some symmetry class, then it may or may not be possible to make perturbations which make the eigenvalues simple. On the other hand, it is reasonable to conjecture that there is an integer p, determined by the symmetries, for which perturbation in a residual set will yield eigenvalues of dimension at most p. Henry has an example with reflection symmetry for which $p = 1$; that is, the eigenvalues are still simple. Pereira [49], [51], [50] has discussed more general situations for which the symmetry class generically has eigenvalues of multiplicity ≥ 2.

Let us now consider the equation

$$\Delta u + f(x, u, \nabla u) = 0 \text{ in } \Omega, \quad u = 0 \text{ in } \partial\Omega, \tag{6}$$

where f is a C^2-function.

Using the generalized transversality theorem, Henry proved that there is a residual set of domains Ω for which all solutions of (6) are simple; that is, the linear variational equation operator about any solution considered as a map from $H^2(\Omega) \cap H_0^1(\Omega)$ to $L^2(\Omega)$ is an isomorphism. For the case in which $f(x, 0, 0) \equiv 0$, one can use the standard transversality theorem to prove this result. This latter case was considered by Saut and Temam [57], but they inadvertently forgot to say that $f(x, 0, 0) \equiv 0$. Without this latter condition, the standard transversality theorem does not apply.

Henry also has some similar results for domains $\Omega \subset \mathbb{R}^n$, $n \geq 2$, for the case where the function f in (6) does not depend upon ∇u; that is, $f = f(x, u)$. Under some conditions on f to be listed below and for homogeneous Neumann boundary conditions (he also allows nonlinear Neumann boundary conditions), he proves, generically in Ω, that the equilibrium solutions are simple. Since the system in this case is gradient, this is the same as saying that the equilibrium points are hyperbolic. The conditions on f are

(i) There is a discrete set $\{c_j\} \subset \mathbb{R}$, possible empty, such that $f(x, c_j) = 0$ for all x and, at each such c_j, $f_u(x, c_j) \neq 0$ on a dense set of $\mathbb{R}^n$.

(ii) For any $c \in \mathbb{R} \setminus \{c_j\}$, the set

$$\{x \in \mathbb{R}^n : f = 0, f_x = 0, f_{xx} = 0, \text{ at } (x, c)\}$$

has dimension $< n - 1$.

(iii) The set

$$\{(x, u) : u \notin \{c_j\}, f = 0, f_x = 0, f_{xx} = 0, f_u = 0, f_{xu} = 0, \text{ at } (x, u)\}$$

has dimension $< n - 1$.

An example is

$$f(x, u) = r(u)s(x),$$

where $s(x) \neq 0$ on a dense set and each zero of r is simple.

There also are some results in Henry which deal with the simplicity of solutions of nonlinear equations for which the boundary conditions are nonlinear.

3. Irregular domains and Dirichlet boundary conditions

In the previous section, given an elliptic PDE on a regular domain with specified boundary conditions, we have discussed the effect on eigenvalues and eigenfunctions of regular perturbations of the domain. On the other hand, there are problems in PDE for which the original domain is irregular. If the original domain is irregular, then there first is the problem of the existence of solutions with specified boundary conditions. In this section, we discuss the work of Berestycki, Nirenberg and Varadhan [12] in which they give a definition of the solution of an elliptic equation with Dirichlet boundary conditions on an irregular domain, as well as a definition of the first eigenvalue and show that the maximum principle holds if this eigenvalue is positive. Since this eigenvalue depends upon the domain, it is important to have a topology on the irregular domains which will imply the first eigenvalue remains positive under perturbations of the original domain and, therefore, conclude that the maximum principle holds on the perturbed domain. Such a topology has been given by Arrieta [5], [6].

3.1. *General elliptic operators*

Suppose that

$$a^{ij} \in C(\mathbb{R}^n), \qquad b^i, c \in L^\infty(\mathbb{R}^n), \quad 1 \le i, j \le n,$$

with

$$(\sum_i |b^i|^2)^{1/2} < \infty, \quad |c| \le b_0,$$

for some constant $b_0 \ge 0$, and

$$c_0|\xi|^2 \le \sum a^{ij}\xi_i\xi_j \le C_0|\xi|^2$$

for some positive constants c_0, C_0, and define the elliptic differential operator

$$L = \sum_{i,j} a^{ij}\partial_{x_i x_j} + \sum_i b^i\partial_{x_i} + c. \tag{7}$$

If Ω is a bounded domain, we want to consider the Dirichlet problem

$$Lv = f \qquad \text{in } \Omega,$$
$$v = 0 \qquad \text{in } \partial\Omega, \tag{8}$$

where $f \in L^n(\Omega)$ and $v = 0$ in $\partial\Omega$ means that

$$\lim_{x \to \partial\Omega} v(x) = 0.$$

It is known that there exist domains Ω and operators L for which (8) has no solution. To overcome this difficulty, one must relax the manner in which the boundary conditions are to be satisfied. Berestycki, Nirenberg and Varadhan [12] proceeded in the following very interesting way. Define the differential operator

$$M = L - c \tag{9}$$

and let H_j, $j \geq 1$, be a family of smooth domains,

$$\bar{H}_j \subset H_{j+1} \subset \Omega, \qquad j \geq 1,$$

such that

$$\bigcup_{j \geq 1} H_j = \Omega.$$

Since H_j is smooth, there is a unique solution u_j of the problem

$$Mu_j = -1 \text{ in } H_j, \quad u_j = 0 \text{ in } \partial H_j. \tag{10}$$

It can be shown that u_j is a nondecreasing sequence which converges to a function $u^{\Omega M}$ weakly in $W^{2,p}(J)$ and strongly in $C^1(J)$ for any compact set $J \subset \Omega$. Moreover, $u^{\Omega M}$ is a strong solution of $Mu = -1$ in Ω, $u^{\Omega M} > 0$ in Ω and only depends upon the domain Ω and the operator M and not on the sets H_j.

The Dirichlet condition $v = 0$ in (8) is replaced by the condition $v = 0(u^{\Omega M})$ where this means that $v(x^{(j)}) \to 0$ for any sequence $\{x^{(j)}\} \subset \Omega$ for which $x^{(j)} \to \partial \Omega$ and $u^{\Omega M}(x^{(j)}) \to 0$ as $j \to \infty$. Let us also use the notation $x^{(j)} \to \partial \Omega(u^{\Omega M})$ to denote that $x^{(j)} \to \partial \Omega$ and $u^{\Omega M}(x^{(j)}) \to 0$ as $j \to \infty$.

We say that the operator L satisfies the *Refined Maximum Principle (RMP) in Ω* if the condition, $Lw \geq 0$ in Ω for w bounded above and

$$\limsup_{j \to \infty} w(x^{(j)}) \leq 0 \text{ if } x^{(j)} \to \partial \Omega(u^{\Omega M}),$$

implies that $w \leq 0$ in Ω.

The principal eigenvalue $\lambda(L, \Omega)$ of the operator L in a domain Ω is defined as

$$\lambda(L, \Omega) = \sup\{\mu : \exists \varphi > 0 \text{ in } \Omega, (L + \mu)\varphi \leq 0 \text{ in } \Omega\}. \tag{11}$$

The following very interesting results have been proved by Berestycki, Nirenberg and Varadhan [12].

Theorem 1. *RMP holds for L if and only if $\lambda(L, \Omega) > 0$.*

Theorem 2. *If $\lambda(L, \Omega) > 0$, then there is a positive constant*

$$A = A(\Omega, c_0, C_0, b, \lambda(L, \Omega))$$

such that, for any $f \in L^n(\Omega)$, there is a unique solution v of

$$\begin{aligned} Lv &= f & \text{in } \Omega, \\ v &= 0(u^{\Omega M}) & \text{in } \partial \Omega, \end{aligned} \tag{12}$$

and

$$\|v\|_{L^\infty(\Omega)} \leq A\|f\|_{L^n(\Omega)}. \tag{13}$$

Both of these results depend upon knowing that $\lambda(L, \Omega) > 0$. If we know that this condition is satisfied for Ω, how do we characterize the class of perturbations of Ω for which it will still be true? Arrieta [5] has introduced a complete metric space of equivalence classes of bounded open sets in which $\lambda(L, \Omega)$, as well as the solution of (12), is continuous in Ω. We now describe this result.

Let

$$\Theta = \{\Omega \subset B_1 \subset \mathbb{R}^n : \Omega \text{ is open}\},$$

where B_1 is the unit ball with center zero. For $\Omega \in \Theta$, if

$$\Gamma_{\Omega M} = \{x \in \partial\Omega : \exists \{x^{(j)}\} \subset \Omega, x^{(j)} \to x, u^{\Omega M}(x^{(j)}) \to 0 \text{ as } j \to \infty\},$$

then the set

$$\Omega^{*M} = \bar{\Omega} \setminus \Gamma_{\Omega M}$$

is open. We say that $\Omega_1, \Omega_2 \in \Theta$ *are equivalent relative to the operator M and Dirich-let boundary conditions*, $\Omega_1 \sim_M \Omega_2$, if $\Omega_1^{*M} = \Omega_2^{*M}$. With this equivalence relation, following Arrieta [5], we define

$$\tilde{\Theta}^M = \Theta / \sim_M$$

and the metric

$$d_{L^\infty}^M : \tilde{\Theta}^M \times \tilde{\Theta}^M \longrightarrow \mathbb{R}$$
$$(\Omega_1, \Omega_2) \mapsto d_{L^\infty}^M(\Omega_1, \Omega_2) = \|u^{\Omega_1 M} - u^{\Omega_2 M}\|_{L^\infty(B_1)}$$

Arrieta [5] shows that $(\tilde{\Theta}^M, d_{L^\infty}^M)$ is a complete metric space and also proves the following result.

Theorem 3. *If $\lambda(L, \Omega)$ is defined as in (11), then $\lambda(L, \Omega)$ and the corresponding eigen-function are continuous in the metric $d_{L^\infty}^M$. If $\lambda(L, \Omega) > 0$, then so is the unique solution of* (12).

This result shows that, if the conditions of Theorems 1 and 2 hold for a given domain Ω_0, then they hold for an open neighborhood of Ω_0 in the space $(\tilde{\Theta}^M, d_{L^\infty}^M)$.

3.2. *Operators in divergence form*

Several important questions arise with respect to the above metric imposed on the domains.

(1) Is it possible to show that the equivalence relation $\sim_M$ is independent of M for M in some class?

(2) In the definition of the metric, is it possible to replace $L^\infty(B_1)$ by $L^p(B_1)$ or $H^1(B_1)$?

(3) Is there a class of operators for which the metric for M and M^* are equivalent if they belong to this class?

(4) In these metrics for M in some class of operators, is it possible to obtain continuity of all of the spectrum?

Arrieta [6] shows that the answers to these questions are mostly affirmative in the class of operators which can be described in divergence form. To describe the results, we need some notation. For a fixed constant $\nu > 0$, let

$$
\mathcal{D} = \left\{ L = \sum_{i,j} \partial_{x_i}(a^{ij}\partial_{x_j}) + \sum_i b^i \partial_{x_i} + c, \right.
$$
$$
a^{ij} \in C^{0,1}(\mathbb{R}^n), \ b^i, c \in L^\infty(\mathbb{R}^n), \ 1 \le i,j \le n,
$$
$$
\left. \sum_{i,j=1}^n a^{ij}\xi_i\xi_j \ge \nu|\xi|^2 \right\}
$$
$$
\mathcal{D}_0 = \{L \in \mathcal{D} : c = 0\},
$$
$$
\mathcal{D}_{00} = \{L \in \mathcal{D}_0 : b_i = 0, 1 \le i \le n\}
$$

Proposition 4. *The equivalence relation $\sim_M$ is independent of the operator $M \in \mathcal{D}_0$; that is,*
$$
\Omega^{*M} = \Omega^{*M^*}
$$
for every $M, M^* \in \mathcal{D}_0$.

From Proposition 4, we can define
$$
\Omega^* = \Omega^{*M} \quad \text{and} \quad \tilde{\Theta} = \tilde{\Theta}^M
$$
for any $M \in \mathcal{D}_0$. From now on, when an open set Ω is considered, we can suppose that
$$
\Omega = \Omega^*
$$
since the properties of an operator $L \in \mathcal{D}_0$ are the same on Ω and Ω^*. As we did for the metric $d_{L^\infty}^M$, we can define the metrics $d_{L^p}^M$, $1 \le p < \infty$, and $d_{H^1}^M$ on $\tilde{\Theta}$.

The metric $d_{L^p}^M$ is strictly weaker that the metric $d_{L^\infty}^M$. On the other hand, as noted by Arrieta (1997), even though the space $(\tilde{\Theta}, d_{L^\infty}^M)$ is complete, the space $(\tilde{\Theta}, d_{L^p}^M)$ is not complete for any $1 \le p < \infty$. As compensation, we have the following

Proposition 5. *For any $M \in \mathcal{D}_{00}$, the metrics $d_{L^p}^M$, $1 \le p < \infty$, and $d_{H^1}^M$ on $\tilde{\Theta}$ are equivalent.*

For any linear operator L, we let $\sigma(L)$ denote the spectrum of L and $\rho(L)$ the resolvent set of L. Regarding the convergence of the spectrum of an operator L in the metric $d_{L^p}^{L-c}$, Arrieta [6] proves the following

Theorem 6. *Let*
$$
L \in \mathcal{D}, \qquad M = L - c \in \mathcal{D}_{00},
$$
and suppose that Ω_k, $k \ge 0$, is a sequence in $\tilde{\Theta}$ and define L_k to be the operator L with Dirichlet boundary conditions on Ω_k. If

$$\lim_{k\to\infty} d_{L^2}^M(\Omega_k, \Omega_0) = 0,$$

then the following statements are true:

(i) *For any C^1-Jordan curve Γ in the complex plane such that $\Gamma \cap \sigma(L_0) = \varnothing$, there exists a $k_0 = k_0(\Gamma)$ such that $\Gamma \cap \sigma(L_k) = \varnothing$ for $k \geq k_0$. Moreover, if P_{Γ, L_k} is the spectral projection over the part of the spectrum inside Γ, then*

$$\lim_{k\to\infty} \|P_{\Gamma, L_k} - P_{\Gamma, L_0}\|_{\mathcal{L}(L^2(B_1), H_0^1(B_1))} = 0.$$

(ii) *If $R(\lambda, L_k)$ is the resolvent of L_k, then*

$$\lim_{k\to\infty} \|R(\lambda, L_k) - R(\lambda, L_0)\|_{\mathcal{L}(L^2(B_1), H_0^1(B_1))} = 0$$

and the convergence is uniform in any compact $\Gamma \subset \rho(L_0)$.

Since the Laplace operator Δ is the simplest second order elliptic differential operator, it is natural to define a canonical metric d_2 by

$$d_2(\Omega_1, \Omega_2) = \|u^{\Omega_1 \Delta} - u^{\Omega_2 \Delta}\|_{L^2(B_1)}.$$

With this notation, Arrieta [6] obtains the following interesting result.

Theorem 7. *Suppose that Ω_k, $k \geq 0$, is a sequence of domains in $\tilde{\Theta}$ and let $u^k = u^{\Omega_k \Delta}$. For any $L \in \mathcal{D}$, let L_k be the operator L_k with Dirichlet boundary conditions acting on Ω_k. For the following statements:*

(i) *$d_2(\Omega_k, \Omega_0) \to 0$ as $k \to \infty$,*

(ii) *The spectrum of L_k aprroaches the spectrum of L_0 and the spectral projections of L_k approach the spectral projections of L_0 in $\mathcal{L}(L^2(B_1), H_0^1(B_1))$ as $k \to \infty$,*

we have (i) implies (ii). Moreover, if L is self-adjoint, then both statements are equivalent.

Micheletti [42]-[45] has given results about the convergence of the spectrum of operators in the case of regular perturbations of the domain in the Courant metric. The Courant metric is stronger than the d_2 metric and therefore we can have convergence of the spectrum for more general domains.

Most of the results in the literature related to the behavior of the spectrum of an operator when the domain is perturbed put the emphasis on geometric conditions on the perturbations of the domain to guarantee the continuity of the spectrum (see the previous and the next section). For Dirichlet boundary condtions, the conditions in Theorem 7 are different from the conditions being imposed on the convergence properties of solutions of the simplest nontrivial elliptic equation $\Delta u = 1$ in the perturbed domains.

It is clear that it would be interesting to characterize, in some more analytic way, large classes of domains for which the condition (i) in Theorem 7 is satisfied. It would also be interesting to see if some similar theory is valid for other types of boundary conditions.

4. Neumann conditions and irregular perturbations

If the perturbed domain depends upon a parameter ϵ in a metric space containing zero, then a family of domains Ω_ϵ is said to be an *irregular perturbation* of the domain Ω_0 if the measure of $\Omega_\epsilon \setminus \Omega_0$ approaches zero as $\epsilon \to 0$. The set of irregular perturbations contains but is more general that the set of regular perturbation of Ω_0 as defined in Section 1. For example, the domain Ω_ϵ could be a perturbation of Ω_0 which introduces an irregular bump at a point on the boundary of Ω_0. Another example could be a dumbbell shaped domain for which the connecting bar degenerates to a curve as $\epsilon \to 0$. A domain $\Omega_\epsilon \subset \mathbb{R}^n$ which degenerates to a domain $\Omega_0 \subset \mathbb{R}^m$ with $m < n$ (*thin domain*) also is an irregular perturbation.

In this section, we study the properties of eigenvalues and eigenfunctions of elliptic operators with Neumann boundary conditions as a function of external irregular perturbations of a bounded domain.

Problems of this type have independent interest and also play an important role in the dynamics of nonlinear equations. For example, if the nonlinear system is gradient, then the compact global attractor (that is, the maximal compact invariant set which attracts bounded sets uniformly) consists of the union of the unstable sets of the equilibrium. Knowing convergence properties of the eigenfunctions and eigenfunctions with respect to the domain leads, without too much difficulty, to results on the upper semicontinuity of attractors at the limit domain for parabolic equations. For hyperbolic equations, the upper semicontinuity is more difficult to prove. In some cases (for example, the variational case), it is easier to show upper semicontinuity directly. If each equilibrium is hyperbolic, then one can deduce continuity properties of the unstable manifolds and, as a consequence, deduce that the compact global attractors are Hausdorff continuous at the limit domain. We do not discuss this problem and refer the reader to Hale and Raugel [25], [26], Raugel [55], Arrieta [7].

In this section, we concentrate on Neumann boundary conditions for these types of perturbations. However, we begin with a few remarks about other types of boundary conditions.

If we assume Dirichlet boundary conditions, then it is possible to prove very general results. In fact, Babuška and Vyborny [9] proved that the eigenvalues and eigenfunctions converge for a general $2m$-order elliptic operator with Dirichlet boundary conditions when the domains Ω_ϵ satisfy the following conditions:

(i) For all compact sets $K \subset \Omega_0$, there exists $\epsilon(K) \in (0, \epsilon_0)$ such that $K \subset \Omega_\epsilon$ for $\epsilon \in (0, \epsilon(K))$.

(ii) For each open set U with $\bar{\Omega}_0 \subset U$, there exists $\epsilon(U) \in (0, \epsilon_0)$ such that $\Omega_\epsilon \subset U$ for $\epsilon \in (0, \epsilon(U))$.

Other references dealing with these problems for Dirichlet boundary conditions are Courant and Hilbert [15], Dancer [17], [18], [19], Daners [21], López-Gómez [41].

We will not discuss Robin boundary conditions and only mention that some references for this case are Dancer and Daners [20], Daners [21], Ozawa [47], Ozawa and Roppongi [48], Stummel [59], Ward, Henshaw and Keller [62], Ward and Keller [63], [64]. Results related to convergence of eigenvalues and eigenvectors are more closely related to the Dirichlet problem than to the Neumann problem.

It was shown by an example in Courant and Hilbert [15] that the eigenvalues of the Laplacian with Neumann boundary conditions may not be continuous if the perturbation of the domain is irregular. In the last few years, Neumann problems have received considerable attention by Arrieta [2], [3], [4], [5], [6], Arrieta, Hale and Han [8], Beale [10], Brown, Hislop and Martinez [13], Chavel and Feldman [14], Ciuperca [16], Hale and Raugel [25], [26], Hale and Vegas [27], Hempel, Seco and Simon [28], Hislop and Martinez [32], Jimbo [33], [34], [35], [36], Jimbo and Morita [37], Lobo-Hidalgo and Sanchez-Palencia [40], Rauch and Taylor [54], Raugel [55], Vegas [61], as well as others contained in the references of the above papers.

4.1. *Perturbations near boundary points*

It is instructive to begin with an example of Courant and Hilbert [15]. Let

$$\Omega_\sigma = \{(x,y) : |x| < \sigma/2, \ |y| < \sigma/2\}$$

of area $\sigma^2/4$ with center $(0,0)$. For any $\epsilon > 0, \tau > 0$, let

$$R_{\epsilon,\tau} = \{(x,y) : 0 < x < \epsilon, \ |y| < \tau/2\}$$

and define

$$\Omega_{\epsilon,\tau} = \Omega_1 \cup (R_{\epsilon,\tau} + (1/2,0)) \cup (\Omega_\epsilon + (1/2 + \epsilon, 0)).$$

For $\tau = \epsilon^4$, the domain $\Omega_{\epsilon,\epsilon^4}$ can be viewed as a C^0-perturbation of Ω_0, but not a C^1-perturbation. Let

$$A_\epsilon : \mathcal{D}(A_\epsilon) \subset L^2(\Omega_{\epsilon,\epsilon^4}) \to L^2(\Omega_{\epsilon,\epsilon^4}),$$
$$\mathcal{D}(A_\epsilon) = \{\varphi \in H^2(\Omega_{\epsilon,\epsilon^4}), \ \partial\varphi/\partial n = 0 \text{ on } \partial\Omega_{\epsilon,\epsilon^4}\},$$
$$A_\epsilon\varphi = -\Delta\varphi, \qquad \varphi \in \mathcal{D}(A_\epsilon).$$

For all $\epsilon \geq 0, 0$ is an eigenvalue of A_ϵ. If λ_k^ϵ are the ordered eigenvalues of A_ϵ, then $\lambda_2^\epsilon > 0$ for $\epsilon > 0$. It is shown in Courant and Hilbert [15] that $\lambda_2^\epsilon \to 0$ as $\epsilon \to 0$. The eigenvalues exhibit a singular behavior at $\epsilon = 0$ in the sense that the second eigenvalue for $\epsilon > 0$ is not close to the second eigenvalue for $\epsilon = 0$.

For some cases, a C^0-perturbation does not yield singular behavior of the eigenvalues of A_ϵ. In the example of Courant and Hilbert [15], if $\tau = \epsilon^\beta$ and β is too small, this will be the case. A more trivial example can be obtained by eliminating the retangular square of size ϵ from the perturbation.

Arrieta, Hale and Han [8] have given a complete description of the behavior of the eigenvalues and eigenfunctions of the Laplacian with Neumann boundary conditions for a general class of perturbations including the example above of Courant and Hilbert [15]. As we will see, the singular behavior of the eigenvalues relies on the way in which the original domain is perturbed as well as the relative sizes of the domains used as perturbation, but the shape of the perturbation is of no importance.

We now give a precise definition of the domain considered in Arrieta, Hale and Han [8]. Let Ω_0, D_1 be bounded, connected smooth domains such that

(H.1) There exist positive constants α, β such that

$$\{(x,y) \in \mathbb{R} \times \mathbb{R}^{n-1} : |x| < \alpha, |y| < \beta\} \cap \Omega_0$$
$$= \{(x,y) : -\alpha < x < 0, |y| < \beta\},$$

$$\{(x,y) \in \mathbb{R} \times \mathbb{R}^{n-1} : 0 < x < 2\alpha, |y| < \beta\} \cap D_1$$
$$= \{(x,y) : \alpha < x < 2\alpha, |y| < \beta\},$$

$$((0,\alpha) \times (-\beta,\beta)) \cap (\Omega_0 \cup D_1) = \varnothing,$$

$$\{0\} \times (-\beta,\beta) \subset \partial\Omega_0, \quad \{\alpha\} \times (-\beta,\beta) \subset \partial D_1,$$

(H.2) $\bar{\Omega}_0 \cap \bar{D}_1 = \varnothing.$

(H.3) For any connected set

$$R_1 \subset \{(x,y) \in \mathbb{R} \times \mathbb{R}^{n-1} : 0 \le x \le \alpha, |y| < \beta\},$$

the set $\Omega_0 \cup D_1 \cup R_1$ is a bounded connected smooth domain in $\mathbb{R}^n$. Also, if $\Gamma_1^1 = \partial\Omega_0 \cap \partial R_1$, then $R_1 \cap \Gamma_1^1 \ne \varnothing$.

The set $(R_1 \setminus \Gamma_1^1) \cap D_1$ is a bounded connected domain with smooth boundary except probably at some points of Γ_1^1. Let $\eta > 0$ be a constant which will be fixed later. For $\epsilon > 0$ small, let

$$R_{\epsilon,\eta} = \{(\epsilon x, \epsilon^\eta y) : (x,y) \in R_1\},$$
$$D_\epsilon = \{(\epsilon x, \epsilon y) : (x,y) \in D_1\}. \tag{14}$$

There is an $\epsilon_0 > 0$ such that, for each $\epsilon \in (0, \epsilon_0)$, we have

$$\bar{\Omega}_0 \cap \bar{D}_\epsilon = \varnothing$$

and

$$\bar{R}_{\epsilon,\eta} \cup \bar{D}_\epsilon \subset \{(x,y) : 0 \le x < \alpha, |y| \le \beta\}.$$

The set

$$\Omega_\epsilon = \Omega_0 \cup R_{\epsilon,\eta} \cup D_\epsilon$$

is a bounded open connected smooth domain.

Remark 8. The fact that $\partial\Omega_0$ is a piece of a hyperplane near $(0,0)$ is merely technical. It is shown in Arrieta, Hale and Han [8] how to attach the perturbation near a point for arbitrary smooth domains Ω_0 and so all of the results below will be valid.

For each fixed $\epsilon_1 \in (0, \epsilon_0)$, the domain Ω_ϵ, for ϵ close to ϵ_1, is a C^1-perturbation of Ω_{ϵ_1}. Although this is not true at $\epsilon = 0$, we do have

$$\mu(\Omega_\epsilon \setminus \Omega_0) \to 0 \quad \text{as} \;\; \epsilon \to 0,$$

where μ is Lebesgue measure. Let us also introduce the set S_γ by the relation

$$S_\gamma = \{(x, y) \in \mathbb{R} \times \mathbb{R}^{n-1} : x^2 + |y|^2 \leq \gamma^2\} \cap \bar{\Omega}_0.$$

There is a γ_0 such that, for $0 < \gamma < \gamma_0$, we have

$$S_\gamma \subset \{(x, y) \in \mathbb{R} \times \mathbb{R}^{n-1} : -\alpha < x \leq 0, \; |y| \leq \beta\}.$$

For $0 \leq \epsilon < \epsilon_0$, we denote by

$$\{\omega_m^\epsilon, 1 \leq m < \infty\},$$

a set of orthonormal eigenvectors corresponding to the ordered set of eigenvalues

$$\{\lambda_m^\epsilon, 1 \leq m < \infty\}$$

of the Laplacian on Ω_ϵ with Neumann boundary conditions.

The following result regarding the second eigenvalue and eigenfunction is due to Arrieta, Hale and Han [8].

Theorem 9. *Let*

$$\Omega_\epsilon = \Omega_0 \cup R_{\epsilon,\eta} \cup D_\epsilon$$

with $R_{\epsilon,\eta}, D_\epsilon$ defined by (14). For $\eta > (n+1)/(n-1)$, the following conditions hold:

$$\lim_{\epsilon \to 0} \lambda_2^\epsilon = 0,$$

$$\lim_{\epsilon \to 0} \|\omega_2^\epsilon\|_{H^1(\Omega_0)} = 0,$$

$$\lim_{\epsilon \to 0} \|\omega_2^\epsilon\|_{H^2(R_{\epsilon,\eta})} = 0,$$

$$\lim_{\epsilon \to 0} \|\omega_2^\epsilon\|_{L^2(D_\epsilon)} = 1,$$

$$\lim_{\epsilon \to 0} \frac{1}{\mu(D_\epsilon)} \left(\int_{D_\epsilon} \omega_2^\epsilon \right)^2 = 1.$$

Furthermore, if Ω_0 is a C^∞-domain, then, for any integer $\ell \geq 1$ and any $\gamma \in (0, \gamma_0)$,

$$\lim_{\epsilon \to 0} \|\omega_2^\epsilon\|_{H^\ell(\Omega_0 \setminus S_\gamma)} = 0.$$

Therefore, for any $\gamma \in (0, \gamma_0)$, the function ω_2^ϵ together with all derivatives up to order ℓ converge to zero pointwise in Ω_0 and uniformly in $\bar{\Omega}_0 \setminus S_\gamma$ as $\epsilon \to 0$.

The limit properties of the remainder of the eigenvalues and eigenfunctons is given in the following result.

Theorem 10. *Let*

$$\Omega_\epsilon = \Omega_0 \cup R_{\epsilon,\eta} \cup D_\epsilon$$

with $R_{\epsilon,\eta}, D_\epsilon$ defined by (14). For $\eta > (n+1)/(n-1)$, the following conditions hold:

$$\lim_{\epsilon \to 0} \lambda_m^\epsilon = \lambda_{m-1}^0 \text{ for } m \geq 3.$$

The corresponding eigenvectors can be chosen so that, for any sequence of positive numbers $\{\epsilon_k, 1 \leq k < \infty\}$ with $\epsilon_k \to 0$ as $k \to \infty$, there is a subsequence $\{\delta_k, 1 \leq k < \infty\}$ such that, for each $m \geq 3$, we have

$$\lim_{k \to \infty} \|\omega_m^{\delta_k} - \omega_{m-1}^0\|_{H^1(\Omega_0)} = 0,$$

$$\lim_{k \to \infty} \|\omega_m^{\delta_k}\|_{H^1(R_{\delta_k} \cup D_{\delta_k})} = 0.$$

Furthermore, if Ω_0 is a C^∞-domain, then, for any integer $\ell \geq 1$ and any $\gamma \in (0, \gamma_0)$,

$$\lim_{k \to \infty} \|\omega_m^{\delta_k} - \omega_{m-1}^0\|_{H^\ell(\Omega_0 \setminus S_\gamma)} = 0$$

Therefore, for any $\gamma \in (0, \gamma_0)$, the function $\omega_m^\epsilon, m \geq 3$, together with all derivatives up to order ℓ converge to ω_{m-1}^0 pointwise in Ω_0 and uniformly in $\overline{\Omega}_0 \setminus S_\gamma$ as $\epsilon \to 0$.

It is worth making a few remarks about these results. If we ignore the set $R_{\epsilon,\eta}$ and consider the eigenvalue problem on $\Omega_0 \cup D_\epsilon$, then there is no singular behavior in the eigenvalues. This is due to the fact that the only eigenvalue on the domain D_ϵ that remains bounded as $\epsilon \to 0$ is the eigenvalue zero. Theorems 9 and 10 assert that the double eigenvalue zero on the disconnected domain $\Omega_0 \cup D_\epsilon$ becomes two simple eigenvalues, zero and λ_2^ϵ with $\lambda_2^\epsilon \to 0$ as $\epsilon \to 0$ and the other eigenvalues converge to the eigenvalues of Ω_0 as $\epsilon \to 0$ provided that they remain bounded. Of course, this is under the restriction that $\eta > (n+1)/(n-1)$. If η is too small, then the eigenvalue problem on Ω_ϵ may not correspond so well to the one on the disconnected domain $\Omega_0 \cup D_\epsilon$.

Remark 11. The above result could have been stated in terms of spectral projections and then it would not be necessary to make a choice for the eigenfunctions.

Remark 12. Mixed boundary conditions as well as perturbations at a finite number of points also are discussed in Arrieta, Hale and Han [8].

4.2. Dumbbell shaped domains

Let us now turn to the disucussion of dumbbell shaped domains. Jimbo [33], [34], [35] seems to have been the first to discuss this problem in some generality for some special smooth domains in $\mathbb{R}^2$. For example, suppose that

$$\Omega_\epsilon = \Omega_0^L \cup \Omega_0^R \cup R_\epsilon$$

is a smooth, connected domain in $\mathbb{R}^2$ for which $\Omega_0^L, \Omega_0^R, R_\epsilon$ are disjoint, Ω_0^L, Ω_0^R are smooth connected domains joined by a rectangular channel

$$R_\epsilon = L \times (0, \epsilon), \quad L = [0, 1].$$

Jimbo pointed out that the relevant limit problem should consist of the following three eigenvalue problems:

$$\Delta u = \mu u \text{ in } \Omega_0^R \cup \Omega_0^L, \quad \partial u / \partial n = 0 \text{ in } \partial \Omega_0^R \cup \Omega_0^L, \tag{15}$$

$$u_{xx} = \mu u \text{ in } L, \quad u = 0 \text{ in } \partial L. \tag{16}$$

We order the eigenvalues of the problems (15), (16) as

$$\mu_1^0 = \mu_2^0 = 0 > \mu_3^0 \geq \mu_4^0 \geq \ldots,$$

and let $\psi_1^0, \psi_2^0, \ldots$ be a corresponding set of normalized eigenfunctions. He proved the convergence of the eigenvalues and eigenfunctions on Ω_ϵ to those of (15), (16) as $\epsilon \to 0$.

Arrieta, Hale and Han [8] considered a more general type of dumbbell shaped domain for which the connecting channel R_ϵ could have a boundary which may not even be connected. Allowing this complicated type of channel is the main difference between this situation and the one considered by Jimbo [35] and Hale and Vegas [27].

We now give a precise definition of the perturbed domain. Let Ω_0^L, Ω_0^R be bounded connected smooth domains such that

(H.4) There exist positive constants α, β, γ such that

$$\{(x,y) \in \mathbb{R} \times \mathbb{R}^{n-1} : -\alpha < x < \gamma, |y| < \beta\} \cap \Omega_0^L$$
$$= \{(x,y) : -\alpha < x < 0, |y| < \beta\}$$

$$\{(x,y) \in \mathbb{R} \times \mathbb{R}^{n-1} : 0 < x < \gamma + \alpha, |y| < \beta\} \cap \Omega_0^R$$
$$= \{(x,y) : \gamma < x < \gamma + \alpha, |y| < \beta\}$$

(H.5) $\bar{\Omega}_0^L \cap \bar{\Omega}_0^R = \varnothing.$

(H.6) For any connected set

$$R_1 \subset \{(x,y) \in \mathbb{R} \times \mathbb{R}^{n-1} : 0 \leq x \leq \gamma, |y| < \beta\},$$

the set $\Omega_0^L \cup \Omega_0^R \cup R_1$ is a bounded connected smooth domain in $\mathbb{R}^n$.

For $\epsilon > 0$ small, if we let

$$R_\epsilon = \{(x, \epsilon y) : (x,y) \in R_1\} \tag{17}$$

and define

$$\Omega_\epsilon = \Omega_0^L \cup R_\epsilon \cup \Omega_0^R,$$

then Ω_ϵ is a bounded open connected smooth domain.

Remark 13. As noted in Remark 8, the fact that $\partial \Omega_0^L$ and $\partial \Omega_0^R$ are pieces of a hyperplane near $(0,0)$ is merely technical.

Let

$$\{\mu_1^\epsilon = 0 > \mu_2^\epsilon \geq \mu_3^\epsilon \geq \ldots\}$$

be the ordered set of eigenvalues of the Laplacian with Neumann boundary conditions on Ω_ϵ and let

$$\psi_1^\epsilon, \; \psi_2^\epsilon, \; \ldots$$

be a corresponding set of normalized eigenfunctions. Arrieta, Hale and Han [8] showed that

$$\mu_2^\epsilon \to 0 \quad \text{as} \quad \epsilon \to 0$$

and that μ_3^ϵ is negative and bounded away from zero, which generalized a result of Hale and Vegas [27]. The methods used there as well as refinements of Arrieta [2], [3] yield the following theorem.

Theorem 14. *If*

$$\Omega_\epsilon = \Omega_0^L \cup R_\epsilon \cup \Omega_0^R,$$

where R_ϵ is defined by (17), *then the following conclusions hold for any m:*

$$\lim_{\epsilon \to 0} \mu_m^\epsilon = \mu_m^0$$

and the corresponding eigenfunctions can be chosen so that

$$\lim_{\epsilon \to 0} \|\psi_m^\epsilon - \psi_m^0\|_{H^1(\Omega_0^L \cup \Omega_0^R)} = 0.$$

Remark 15. We remark there can be many different channels and many different open sets connected by these channels. The results will be the same except there are more eigenvalue problems for the limit as $\epsilon \to 0$.

4.3. *Thin domains*

Hale and Raugel [25] have considered some properties of the dynamics of reaction diffusion equations on thin domains and, as a byproduct of the investigation, also have given results on the convergence of eigenvalues and eigenfunctions of the Laplacian with mixed boundary conditions. We describe a special case of their results for a particular case of a thin domain over a line segment for Neumann boundary conditions. For a more complete and more general discussion, see Raugel [55].

Let

$$R_\epsilon = \{(x, y) \in \mathbb{R}^2 : 0 < x < 1, \; 0 < y < G(x, \epsilon)\}, \tag{18}$$

where the function $G \in C^1([0, 1] \times [0, \epsilon_0])$ and satisfies

$$G(x, 0) = 0, \quad G_0(x) = \frac{\partial G}{\partial \epsilon}(x, 0) > 0, \quad x \in [0, 1]. \tag{19}$$

Let $\{\lambda_m^\epsilon, m \geq 1\}$ be the ordered set of eigenvalues of $-\Delta$ with homogeneous Neumann boundary conditions and let $\{\varphi_m^\epsilon, m \geq 1\}$ be a corresponding set of normalized eigenfunctions. Hale and Raugel [25] show that the appropriate limit problem as $\epsilon \to 0$ is the eigenvalue problem

$$-\frac{1}{G_0(x)}(G_0(x)u_x)_x = \lambda u \quad \text{in} \ (0,1), \tag{20}$$

$$u_x = 0 \quad \text{at} \ x = 0,1. \tag{21}$$

If $\{\lambda_m^0, m \geq 1\}$ is the ordered set of eigenvalues of (20) and $\{\varphi_m^0, m \geq 1\}$ is a corresponding set of normalized eigenfunctions, they prove that the following statement is true.

Theorem 16. *For any integer m, there are positive constants $\epsilon_0(m)$, $C(m)$ such that, for every integer $n \leq m$, $0 < \epsilon \leq \epsilon_0$,*

$$|\lambda_n^\epsilon - \lambda_n^0| \leq C(m)|\epsilon|,$$

and

$$\|\varphi_n^\epsilon - (\varphi_n^\epsilon, \epsilon^{-1/2}\varphi_n^0)\epsilon^{-1/2}\varphi_n^0\|_{H^1(R_\epsilon)}^2 \leq C(m)|\epsilon|,$$

where $(\cdot, \cdot)$ is the L^2-inner product.

Results of this type permit the reduction of the two dimensional boundary value problem to a problem in one dimension.

Remark 17. If

$$\Gamma_\epsilon = \partial R_\epsilon \cap ((\{G(0,\epsilon)\} \times (0,1)) \cup (\{G(1,\epsilon)\} \times (0,1))),$$

and one assumes Dirichlet boundary conditions on Γ_ϵ with Neumann on $\partial R_\epsilon \setminus \Gamma_\epsilon$, then the same conclusions as in Theorem 16 hold if we suppose that the limit problem satisfies Dirichlet boundary conditions.

It also is possible to consider nonlinear equations on thin domains and relate the flow to a nonlinear equation on the limit domain. For example, consider the equation

$$u_t = \Delta u + f(u) \quad \text{in} \ R_\epsilon$$

with homogeneous Neuman boundary conditions. Under natural conditions on f, this equation defines a flow which has a compact global attractor $\mathcal{A}_\epsilon$. If $\mathcal{A}_0$ is the compact global attractor for the equation

$$v_t = \frac{1}{G_0(x)}(G_0(x)v_x)_x + f(v)$$

with Neuman boundary conditions, the set of global attractors

$$\{\mathcal{A}_\epsilon, \epsilon > 0\} \cup \mathcal{A}_0$$

is upper semicontinuous at $\epsilon = 0$.

Prizzi and Rybakowski [53] have considered thin domains R_ϵ for which the function $G(x, \epsilon)$ can be multivalued. More specifically, suppose that Ω is an arbitrary smooth domain in $\mathbb{R}^2$. For any $\epsilon > 0$, let

$$T_\epsilon : (x,y) \in \mathbb{R}^2 \mapsto (x, \epsilon y) \in \mathbb{R}^2$$

and define

$$R_\epsilon = T(\epsilon)\Omega.$$

The domain R_ϵ has a smooth boundary, but it need not be a graph over the x-axis and it may not be connected. However the domain converges to a line segment on the x-axis. Prizzi and Rybakowski [53] prove that the corresponding limit differential equation is a differential equation in one space variable over a graph with the boundary conditions at each point on the graph being uniquely determined.

With the function $G(x, \epsilon)$ satisfying (19), the domain degenerates to the line in a nice uniform way. It does not allow, for example, the boundary to oscillate rapidly as $\epsilon \to 0$. Such problems are of interest in the theory of homogenization (see, for example, Bensoussan, Lions and Papanicolaou [11], de Giorgi and Spagnolo [22], Kesavan [39]). Arrieta [2], [4] has allowed rapid oscillations as $\epsilon \to 0$. We now describe the results of Arrieta.

Consider a fixed function $a \in C^1([0, 1], (0, \infty))$, let

$$k \in C^1([0, 1] \times (0, \epsilon_0), (0, \infty)), \qquad k_\epsilon(x) = k(x, \epsilon),$$

such that

$$\lim_{\epsilon \to 0} \epsilon \left| \frac{dk_\epsilon}{dx}(x) \right| = 0 \ \text{ uniformly in } (0, 1),$$

define

$$g_\epsilon = a + k_\epsilon$$

and suppose that there are positive constants c_3, c_4 and a function $b \in C^1([0, 1], (0, \infty))$ such that

$$c_3 \le g_\epsilon(x) \le c_4, \quad x \in [0, 1],$$

and

$$\lim_{\epsilon \to 0} g_\epsilon = a \ \text{ weakly } L^2,$$

$$\lim_{\epsilon \to 0} \frac{1}{g_\epsilon} = b \ \text{ weakly } L^2.$$

With this notation, define the thin domain as

$$R_\epsilon = \{(x_1, x_2) \in \mathbb{R}^2 : 0 < x_1 < 1, \ 0 < x_2 < \epsilon g_\epsilon(x)\},$$

and let

$$\Gamma_\epsilon = \{(0, x_2) : 0 \le x_2 \le \epsilon g_\epsilon(0)\} \cup \{(1, x_2) : 0 \le x_2 \le g_\epsilon(1)\}.$$

Denote by $\{\lambda_n^\epsilon, n \ge 1\}$ the ordered set of eigenvalues and $\{\varphi_n^\epsilon, n \ge 1\}$ the corresponding eigenfunctions of the eigenvalue problem

$$\begin{cases} -\Delta\varphi = \lambda\varphi & \text{in } R_\epsilon, \\ \varphi = 0 & \text{in } \Gamma_\epsilon, \\ \dfrac{\partial\varphi}{\partial n} = 0 & \text{in } \partial R_\epsilon \setminus \Gamma_\epsilon. \end{cases}$$

Denote by $\{\mu_n^\epsilon, n \geq 1\}$ the ordered set of eigenvalues and $\{\psi_n^\epsilon, n \geq 1\}$ the corresponding eigenfunctions of the eigenvalue problem

$$\begin{cases} -\dfrac{1}{g_\epsilon}(g_\epsilon \psi_x)_x = \mu\psi & \text{in } (0,1) \\ \psi = 0 & \text{at } x = 0, 1. \end{cases} \tag{22}$$

The following result is due to Arrieta [2], [3].

Theorem 18. *Let* $(\lambda_n^\epsilon, \varphi_n^\epsilon)$, $(\mu_n^\epsilon, \psi_n^\epsilon)$ *be the eigenpairs defined above. If*

$$\eta_\epsilon = \sup\{\epsilon|g'_\epsilon(x)| : x \in [0,1]\},$$

then, for any integer m, *there are positive constants* $\epsilon_0(m), C(m)$ *such that, for* $n \leq m$, $\epsilon \in (0, \epsilon_0)$,

$$0 \leq \mu_n^\epsilon - \lambda_n^\epsilon \leq C(m)\eta_\epsilon^2,$$

and

$$\|\varphi_n^\epsilon - (\varphi_n^\epsilon, \epsilon^{-1/2}\psi_n^\epsilon)_{L^2(R_\epsilon)}\epsilon^{-1/2}\psi_n^\epsilon\|_{H^1(R_\epsilon)}^2 \leq C(m)\eta_\epsilon^2.$$

Theorem 18 gives the reduction of the problem to a one dimensional problem which still depends upon ϵ. It remains to find the appropriate limit problem for (22) as $\epsilon \to 0$. Arrieta [2], [4] shows that it should be the following eigenvalue problem

$$\begin{cases} -\dfrac{1}{a}(\dfrac{1}{b}\xi_x)_x = \nu\psi & \text{in } (0,1), \\ \xi = 0 & \text{at } x = 0, 1. \end{cases}$$

If we denote by $\{\nu_n, n \geq 1\}$ the ordered set of eigenvalues and $\{\xi_n, n \geq 1\}$ the corresponding normalized eigenfunctions of this eigenvalue problem, then we have the following result.

Theorem 19. *The following statements are true:*

$$\lim_{\epsilon \to 0} \mu_m^\epsilon = \nu_m, \quad m \geq 1,$$

$$\lim_{\epsilon \to 0}[\psi_m^\epsilon - D_\epsilon(\psi_m^\epsilon, \xi_m)\xi_m] = 0 \ \ \text{strongly in } L^2((0,1)), \ \text{weakly in } H^1((0,1)),$$

where

$$D_\epsilon(f, g) = \int_0^1 g_\epsilon f g.$$

An example for which the above result applies is the function

$$g_\epsilon(x) = 1 + \rho\sin(x\,e^{-\alpha}),$$

where $\rho \in (0,1)$, $\alpha > 0$. In this case, $a \equiv 1$ and

$$b \equiv (1 - \rho^2)^{-1/2} > 1$$

and the eigenvalues of the two dimensional problem are close to the eigenvalues of the operator $(-1/b)\partial_x^2$ with Dirichlet boundary conditions.

4.4. *General variations*

For Neumann boundary conditions, Lobo-Hidalgo and Sanchez-Palencia [40] proved that, if $\Omega_0 \subset \Omega_\epsilon$ and

$$m_n(\Omega_\epsilon \setminus \Omega_0) \to 0 \quad \text{as} \quad \epsilon \to 0,$$

where m_n is the Lebesgue measure in $\mathbb{R}^n$, then every point of the spectrum of $-\Delta_N^{\Omega_0}$ is approximated by points of the spectrum of $-\Delta_N^{\Omega_\epsilon}$, whereas the contrary statement is false in general; that is, there may be situations for which there are accumulation points of the spectrum of $-\Delta_N^{\Omega_\epsilon}$ which are not in the spectrum of $-\Delta_N^{\Omega_0}$.

Arrieta [4] has considered perturbed domains in this general setting and has given conditions for which one has convergence of eigenvalues and eigenfunctions. The conditions are stated in such a way as to lead to proofs of the results in this section as well as many more. To be specific, let $\Omega_0 \subset \Omega_\epsilon$ and

$$m_n(\Omega_\epsilon \setminus \Omega_0) \to 0 \quad \text{as} \quad \epsilon \to 0,$$

let

$$R_\epsilon = \Omega_\epsilon \setminus \bar{\Omega}_0, \qquad \Gamma_\epsilon = \partial\Omega_0 \cap \partial R_\epsilon.$$

For functions $V_\epsilon \in L^\infty(\Omega_\epsilon)$ with

$$\|V_\epsilon\|_{L^\infty(\Omega_\epsilon)} \leq C$$

for some $C > 0$ independent of ϵ, consider the Schrödinger operators

$$A_N^{\Omega_\epsilon} = -\Delta_N^{\Omega_\epsilon} + V_\epsilon$$
$$A_N^{\Omega_0} = -\Delta_N^{\Omega_0} + V_\epsilon$$
$$A_{D(\Gamma_\epsilon)N}^{R_\epsilon} = -\Delta_{D(\Gamma_\epsilon)N}^{R_\epsilon} + V_\epsilon$$

where the superscript denotes the domain on which the operator is applied, the subscript N denotes homogeneous Neumann boundary conditions on the domain and the subscript $D(\Gamma_\epsilon)N$ denotes homogeneous Dirichlet conditions on Γ_ϵ and Neumann conditions on the remainder of the domain. We let $H^1_{\Gamma_\epsilon}(R_\epsilon)$ denote the space of H^1 functions which respect the Dirichlet boundary conditions on Γ_ϵ.

Without loss of generality, we may suppose that

$$V_\epsilon \geq 0.$$

The objective is to show that eigenvalues and eigenfunctions of $A_N^{\Omega_\epsilon}$ behave as the eigenvalues and eigenfunctions of $A_N^{\Omega_0}$ and $A_{D(\Gamma_\epsilon)N}^{R_\epsilon}$. To achieve this, the following hypothesis is assumed:

(H) If $u_\epsilon \in H^1(\Omega_\epsilon)$ with

$$\|u_\epsilon\|_{H^1(\Omega_\epsilon)} \leq C_1$$

for some positive constant C_1 independent of ϵ, then there exists $\bar{u}_\epsilon \in H^1_{\Gamma_\epsilon}(R_\epsilon)$ such that

$$\lim_{\epsilon \to 0} \|u_\epsilon - \bar{u}_\epsilon\|_{L^2(R_\epsilon)} = 0,$$
$$\|\nabla\bar{u}_\epsilon\|_{L^2(R_\epsilon)} \leq \|\nabla u_\epsilon\|_{L^2(\Omega_\epsilon)} + o(1).$$

Let
$$\{\lambda_{m,\Omega_\epsilon}, m \geq 1\}, \quad \{\lambda_m(\Omega_0,\epsilon), m \geq 1\}, \quad \{\tau_m(R_\epsilon), m \geq 1\},$$
be the ordered set of eigenvalues counting multiplicity of the operators
$$A_N^{\Omega_\epsilon}, \quad A_N^{\Omega_0}, \quad A_{D(\Gamma_\epsilon)N}^{R_\epsilon},$$
and let
$$\{\varphi_{m,\Omega_\epsilon}, m \geq 1\}, \quad \{\psi_m(\Omega_0,\epsilon), m \geq 1\}, \quad \{\psi_m(R_\epsilon), m \geq 1\},$$
be a corresponding set of orthonormal eigenvectors. Let
$$\{\lambda_m^\epsilon, m \geq 1\} = \{\lambda_m(\Omega_0,\epsilon), m \geq 1\} \cup \{\tau_m(R_\epsilon), m \geq 1\}$$
be ordered (counting multiplicity), and define
$$\varphi_m^\epsilon = \psi_i(\Omega_0,\epsilon) \text{ in } \Omega_0, \ = 0 \text{ in } R_\epsilon \text{ if } \lambda_m^\epsilon = \lambda_i(\Omega_0,\epsilon),$$
$$\varphi_m^\epsilon = 0 \text{ in } \Omega_0, \ \psi_j(R_\epsilon) \text{ in } R_\epsilon, \text{ if } \lambda_m^\epsilon = \tau_j(R_\epsilon).$$
Obviously, we have
$$\varphi_m^\epsilon \in H^1(\Omega_0) \cup H^1_{\Gamma_\epsilon}(R_\epsilon).$$
We say that $\sigma_\epsilon > 0$ divides the spectrum if there are positive constants δ, M, N such that, for $\epsilon \in (0,\epsilon_0)$, we have
$$[\sigma_\epsilon - \delta, \sigma_\epsilon + \delta] \cap \{\lambda_m^\epsilon, m \geq 1\} = \varnothing, \qquad \sigma_\epsilon \leq M,$$
and
$$N(\sigma_\epsilon) \equiv \text{Card}\{\lambda_i^\epsilon : \lambda_i^\epsilon \leq \sigma_\epsilon\} \leq N.$$
If σ_ϵ divides the spectrum, then we can define the projection operator
$$P_{\sigma_\epsilon} : L^2(\Omega_\epsilon) \longrightarrow [\varphi_1^\epsilon, \ldots, \varphi_{N(\sigma_\epsilon)}^\epsilon]$$
$$g \mapsto \sum_{i=1}^{N(\sigma_\epsilon)} (g, \varphi_i^\epsilon)_{L^2(\Omega_\epsilon)} \varphi_i^\epsilon.$$
With this notation, Arrieta [4] proved the following result.

Theorem 20. *If condition (H) is satisfied, then*

(i) $\lim_{\epsilon \to 0}(\lambda_{m,\Omega_\epsilon} - \lambda_m^\epsilon) = 0, \quad m \geq 1,$

(ii) $\lim_{\epsilon \to 0} \|\varphi_{r_\epsilon,\Omega_\epsilon} - P_{\sigma_\epsilon}\varphi_{r_\epsilon,\Omega_\epsilon}\|_{H^1(\Omega_\epsilon \cup R_\epsilon)} = 0, \quad r_\epsilon = 1, 2, \ldots, N(\sigma_\epsilon), \text{ for any } \sigma_\epsilon$
which divides the spectrum.

If we impose conditions on Ω_0 and Ω_ϵ which will ensure that the eigenvalues λ_m^ϵ do not accumulate at any finite point in $(0, \infty)$, then more can be proved. In fact, Arrieta [4] obtained the following result.

Theorem 21. *If the condition*

(C) For any $k \geq 1$, there exists an integer L_k such that
$$\text{Card}\{\lambda_m^\epsilon : \lambda_m^\epsilon \leq k\} \leq L_k,$$

*is satisfied, then condition (**H**) is equivalent to the statements (i) and (ii) of Theorem 20.*

Arrieta [4] shows that these results include all of the examples of the previous subsections.

References

1. Abraham, R., and Robbin, J., *Transversal Mappings and Flows*, Benjamin, 1967.
2. Arrieta, J.M., *Spectral Properties of Schrödinger Operators under Perturbation of the Domain*, Ph.D. Dissertation, Georgia Tech., 1991.
3. Arrieta, J.M., Neumann eigenvalue problems on exterior prturbations of the domain, *J. Diff. Eqns.* **118** (1995), 54–103.
4. Arrieta, J.M., Rates of eigenvalues on a dumbbell domain. Simple eigenvalue case, *Trans. Amer. Math. Soc.* **347** (1995), 3503–3531.
5. Arrieta, J.M., Elliptic equations, principal eigenvalues and dependence on the domain, *Comm. PDE* **21** (1996), 971–991.
6. Arrieta, J.M., Domain dependence of elliptic operators in divergence form, *Resenhas IME-USP* **3** (1997), 107–123.
7. Arrieta, J.M., Spectral behavior and upper semicontinuity of attractors, in *Differential Equations Vol.1* (Fiedler, B., Gröger, K., Sprekels, J., Eds.), pages 11–196, World Scientific, 2000.
8. Arrieta, J.M., Hale, J.K., and Han, W., Eigenvalue problems for nonsmoothly perturbed domains, *J. Diff. Eqns.* **91** (1991), 24–52.
9. Babuška, I., and Vyborny, R., Continuous dependence of eigenvalues on the domains, *Czech. Math. J.* **15** (1965), 169–178.
10. Beale, J.T., Scattering frequencies of resonators, *Pure Appl. Math.* **26** (1973), 549–563.
11. Bensoussan, A., Lions, J.L., and Papanicolaou, G., *Asymptotic Methods in Periodic Structures*, North-Holland, 1978.
12. Berestycki, H., Nirenberg, L., and Varadhan, S.R.S., The principal eigenvalue and maximum principle for second-order elliptic operators in general domains, *Comm. Pure. Appl. Math.* **47** (1994), 47–92.
13. Brown, R., Hislop, P.D., and Martinez, A., Eigenvalues and resonances for domains with tubes: Neumann boundary conditions, *J. Diff. Eqns.* **115** (1995), 458–476.
14. Chavel, I., and Feldman, E.A., Spectra of domains in compact manifolds, *J. Funct. Anal.* **30** (1978), 196–222.
15. Courant, R., and Hilbert, D., *Methods of Mathematical Physics, Vol. 1*, Wiley-Interscience, New York, 1953. Translation from the German edition, 1937.
16. Ciuperca, I.S., Spectral properties of Schrödinger operators on domains with varying order of thinness, *J. Dyn. Diff. Eqns.* **10** (1998), 73–108.
17. Dancer, E.N., The effect of domain shape on the number of positive solutions of certain nonlinear equations, *J. Diff. Eqns.* **74** (1988), 120–156.
18. Dancer, E.N., The effect of domain shape on the number of positive solutions of certain nonlinear equations, II, *J. Diff. Eqns.* **87** (1990), 316–339.

19. Dancer, E.N., Domain variation for certain sets of solutions with applications, *Top. Meth. Nonlinear Anal.* Preprint.

20. Dancer, E.N., and Daners, D., Domain perturbation of elliptic equations subject to Robin boundary conditions, *J. Diff. Eqns.* **74** (1997), 86–132.

21. Daners, D., Domain perturbation for linear and nonlinear parabolic equations, *J. Diff. Eqns.* **129** (1996), 358–402.

22. de Giorgi, E., and Spagnolo, S., Sulla convergenzia degli integrali dell'energia per operatori ellitici del 2 deg ordine, *Boll. U. Mat. Ital.* **8** (1973), 391–411.

23. Garabedian, P., and Schiffer, M., Convexity of domain functionals, *J. Analyse Math.* **2** (1952), 281–368.

24. Hadamard, J., Mémoire sur le problème d'analyse relatif à l'équilibre des plaques élastiques encastrées, in *Œuvres de J. Hadamard Vol. 2*, 1968.

25. Hale, J.K., and Raugel, G., Convergence in gradient-like systems and applications, *ZAMP* **43** (1992), 63–124.

26. Hale, J.K., and Raugel, G., A reaction-diffusion equation on a thin *L*-shaped domain, *Proc. Roy. Soc. Edinburgh* **125A** (1995), 283–327.

27. Hale, J.K., and Vegas, J.M., A nonlinear parabolic equation with varying domain, *Arch. Rat. Mech. Anal.* **2** (1984), 99–123.

28. Hempel, R., Seco, L.A., and Simon, B., The essential spectrum of Neumann Laplacians on some bounded singular domains, *J. Funct. Anal.* **102** (1991), 448–483.

29. Henry, D., *Perturbation of the Boundary for Boundary Value Problems for Partial Differential Equations*, Sem. Brasileiro Anal. ATS 22, 1985.

30. Henry, D., Generic properties of equilibrium solutions by perturbation of the boundary, in *Dynamics of Infinite Dimensional Systems* (Chow, S.N., Hale, J.K., Eds.), pages 129–139, NATO ASI Series F 37, Springer-Verlag, 1987.

31. Henry, D., *Perturbation of the Boundary in Partial Differential Equations*, Lecture Notes (1996). To appear in Cambridge University Press.

32. Hislop, P.D., and Martinez, A., Scattering resonances of a Helmholtz resonator, *Indiana U. Math. J.* **40** (1991), 767–788.

33. Jimbo, S., Singular perturbation of domains and the semilinear elliptic equation, *J. Fac. Sci. Univ. Tokyo* **35** (1988), 27–76.

34. Jimbo, S., Singular perturbation of domains and the semilinear equation, II, *J. Diff. Eqns.* **75** (1998), 264–289.

35. Jimbo, S., The singularly perturbed domain and the characterization for the eigenfunctions with Neumann boundary conditions, *J. Diff. Eqns.* **77** (1989), 322–350.

36. Jimbo, S., Perturbation formula of eigenvalues in a singularly perturbed domain, *J. Math. Soc. Japan* **45** (1993), 339–356.

37. Jimbo, S., and Morita, Y., Remarks on the behavior of certain eigenvalues on a singularly perturbed domain with several thin channels, *Comm. PDE* **17** (1992), 523–552.

38. Joseph, D., Parameter and domain dependence, *Arch. Rat. Mech. Anal.* **24** (1967), 325–351.

39. Kesavan, S., Homogenization of elliptic eigenvalue problems, I, *Appl. Math. Optim.* **5** (1979), 153–167.

40. Lobo-Hidalgo, M., and Sanchez-Palencia, E., Sur certaines propriétés spectrales des perturbations du domaine dans les problèmes aux limites, *Comm. PDE* **4** (1979), 1085–1098.

41. López-Gómez, J., The maximum principle and the existence of principal eigenvalues for some linear weighted boundary value problems, *J. Diff. Eqns.* **127** (1996), 263–294.

42. Micheletti, A.M., Perturbazione dello spettro dell'operatore di Laplace in relazione ad una variazione del campo, *Ann. Scuola Norm. Sup. Pisa* **XXVI Fasc. I** (1972) 151–169.

43. Micheletti, A.M., Metrica per famiglie de domini limitati e proprietá generiche degli autovalori, *Ann. Scuola Norm. Sup. Pisa* **XXVI Fasc. III** (1973), 633–694.

44. Micheletti, A.M., Perturbazione dello spettro di un operatoire ellittico tipo variazionale, in relazione ad una variazione del campo, *Annali Mat. Pura Appl.* **XCVII, Fasc. IV** (1973), 261–281.

45. Micheletti, A.M., Perturbazione dello spettro di un operatore ellittico di tipo variazionale, in relazione ad una variazione del campo, II, *Recerche Mat.* **25** (1976), 187–200.

46. Ozawa, S., Singular variations of domains and eigenvalues of the Laplacian, *Duke Math. J.* **48** (1981), 769–778.

47. Ozawa, S., Singular variation of domain and spectra of the Laplacian with small Robin conditional boundary, I, *Osaka J. Math.* **29** (1992), 837–850.

48. Ozawa, S., and Roppongi, S., Singular variation of domains and continuity property of eigenfunction for some semi-linear elliptic equations, II, *Kodai Math. J.* **18** (1995), 315–327.

49. Pereira, A.L., *Auto valores do Laplaciano im regiõ es simé tricas*, Ph.D. Thesis, Univ. São Paulo, Brasil, 1989.

50. Pereira, A.L., Appendix to Henry, D., [31], 1996.

51. Pereira, A.L., Eigenvalues of the Laplacian on symmetric regions, *NoDEA* **2** (1995), 63–109.

52. Peetre, J., On Hadamard's variational formula, *J. Diff. Eqns.* **36** (1980), 335–346.

53. Prizzi, M., and Rybakowski, K.P., The effect of domain squezzing upon the dynamics of reaction-diffusion equations, *J. Diff. Eqns.* **173** (2001), 271–320.

54. Rauch, J., and Taylor, M., Potential and scattering theory on wildly perturbed domains, *J. Funct. Anal.* **18** (1975), 27–59.

55. Raugel, G., Dynamics of partial differential equations on thin domains, in *Dynamical Systems* (Johnson, R., Ed.), pages 208–315, Lectures Notes in Mathematics 1609, Springer, 1995.

56. Rayleigh, J.W.S., *Theory of Sound*, Dover, 1945.

57. Saut, J.C., and Temam, R., Generic properties of nonlinear boundary value problems, *Comm. PDE* **4** (1979), 293–319.

58. Serrin, J., A symmetry problem of potential theory, *Arch. Rat. Mech. Anal.* **43** (1971), 304–318.

59. Stummel, F., Perturbation of domains in elliptic boundary value problems, in *Applications of Methods of Functional Analysis to Problems in Mechanics* (Germain, P., and Nayroles, B., Eds.), pages 110–135, Lecture Notes Mathematics 503, 1976.

60. Uhlenbeck, K., Eigenfunctions of Laplace operators, *Amer. J. Math.* **98**, (1972), 1073–1076.

61. Vegas, J.M, A functional-analytic framework for the study of elliptic operators on variable domains, *Proc. Roy. Soc. Edinburgh* **116A** (1990), 367–380.
62. Ward, M.J., Henshaw, W.D., and Keller, J.B., Summing logarithmic expansions for singularly perturbed eigenvalue problems, *SIAM J. Appl. Math.* **53** (1993), 799–828.
63. Ward, M.J., and Keller, J.B., Nonlinear eigenvalue problems under strong localized perturbations with applications to chemical reactors, *Stud. Appl. Math.* **85** (1991), 1–28.
64. Ward, M.J., and Keller, J.B., Strong localized perturbations of eigenvalue problems, *SIAM J. Appl. Math.* **53** (1993), 770–798.

Ten Mathematical Essays on Approximation in Analysis and Topology
J. Ferrera, J. López-Gómez, F. R. Ruiz del Portal, Editors

125

Monotone Approximations and Rapid Convergence

V. Lakshmikantham

Florida Institute of Technology, Department of Mathematical Sciences,
Melbourne, FL 32901 USA

Abstract

Starting from the classical method of successive approximations, in a general set up, this paper describes in detail, monotone iterative technique and the method of generalized quasilinearization. The paper demonstrates how monotone approximation techniques cover a broad range of nonlinear problems in a variety of situations. The paper also traces the research excursions of the author for the benefit of the readers.

Key words: monotone approximations, rapid convergence, nonlinear differential equations

1. Introduction

One of the objectives in approximation theory is the investigation of the following problem. If E is a Banach space, a linear subspace S is prescribed, whose elements $s \in S$ are utilized for approximating $u \in E$. The interesting problem, known as best approximation, is to find, for a fixed $u \in E$, an element $s \in S$, for which $\|u - s\|$ is as small as possible. To be precise, if the infimum is attained by one or more elements of S, in the relation

$$d(u, s) = \inf_{s \in S} \|u - s\|,$$

then these elements of S, are called best approximations of u in S. The classical approximation theory of functions of one variable has dealt with such a problem. Multivariate approximation theory concerns with the approximation of functions of several variables and therefore is more complicated, which is clear by experience in solving partial differential equations. In this contribution, we shall investigate monotone flows which approximate the solutions of various nonlinear dynamic equations starting from the classical successive approximations of ordinary differential equations.

2. Successive approximations

Consider the Cauchy problem

$$x' = f(t, x), \quad x(t_0) = x_0, \quad t_0 \in R_+, \tag{1}$$

where $f \in C[R_0, R^n]$, and

$$R_0 = [(t, x) : t_0 \leq t \leq t_0 + a \text{ and } \|x - x_0\| \leq b].$$

We shall begin with the following existence and uniqueness result, that is perhaps not well known, under more general assumption than Lipschitz. It shows, at the same time, the convergence of successive approximations to the solution of (1) as well as the generation of monotone flows of the comparison function when it is also monotone. Moreover, this result demonstrates the power of the comparison principle and also has an interesting history which we shall indicate later.

Theorem 1. *Assume that*

(i) $g \in C[[t_0, t_0 + a] \times [0, 2b], R_+],$

$$0 \leq g(t, u) \leq M_1 \quad on \quad [t_0, t_0 + a] \times [0, 2b],$$

$g(t, 0) \equiv 0,$ $g(t, u)$ *is nondecreasing in u for each t and $u \equiv 0$ is the unique solution of the scalar differential equation*

$$u' = g(t, u), \quad u(t_0) = 0 \tag{2}$$

on $[t_0, t_0 + a]$;

(ii) $\|f(t, x) - f(t, y)\| \leq g(t, \|x - y\|)$ *on R_0.*

Then the successive approximations defined by

$$x_{n+1}(t) = x_0 + \int_{t_0}^{t} f(s, x_n(s))ds, \quad n = 0, 1, 2, \ldots, \tag{3}$$

exist on $[t_0, t_0 + \alpha]$, where

$$\alpha = \min(a, \frac{b}{M}), \quad M = \max(M_0, M_1),$$

as continuous functions and converge uniformly to the unique solution $x(t)$ of the Cauchy problem (1) on $[t_0, t_0 + \alpha]$. Here M_0 is the bound of f on R_0.

Proof. It is easy to see, by induction, the successive approximations (3) are defined and continuous on $[t_0, t_0 + \alpha]$ and

$$\|x_{n+1}(t) - x_0\| \leq b, \quad n = 0, 1, 2, \ldots.$$

We shall now define the successive approximations for problem (2) as follows:

$$\begin{cases} u_0(t) = M(t - t_0) \\ u_{n+1}(t) = \displaystyle\int_{t_0}^{t} g(s, u_n(s))ds, \ t_0 \le t \le t_0 + \alpha. \end{cases} \tag{4}$$

An easy induction proves that the successive approximations (4) are well defined and satisfy

$$0 \le u_{n+1}(t) \le u_n(t), \quad t_0 \le t \le t_0 + \alpha.$$

Since $|u'_n(t)| \le M_1$, we conclude by Ascoli-Arzelá theorem and the monotonicity of the sequence $\{u_n(t)\}$ that

$$\lim_{n \to \infty} u_n(t) = u(t)$$

uniformly on $[t_0, t_0 + \alpha]$. It is also clear that $u(t)$ satisfies (2). Hence, by (i), $u(t) \equiv 0$ is $[t_0, t_0 + \alpha]$.

Now,

$$\|x_1(t) - x_0\| \le \int_{t_0}^{t} \|f(s, x_0)\|ds \le M(t - t_0) \equiv u_0(t).$$

Assume that

$$\|x_k(t) - x_{k-1}(t)\| \le u_{k-1}(t)$$

for a given k. Since

$$\|x_{k+1}(t) - x_k(t)\| \le \int_{t_0}^{t} \|f(s, x_k(s)) - f(s, x_{k-1}(s))\|ds,$$

using the nondecreasing character of $g(t, u)$ in u and the assumption (ii), we get

$$\|x_{k+1}(t) - x_k(t)\| \le \int_{t_0}^{t} g(s, u_{k-1}(s))ds \equiv u_k(t).$$

Thus, by induction, the inequality

$$\|x_{n+1}(t) - x_n(t)\| \le u_n(t), \quad t_0 \le t \le t_0 + \alpha,$$

is true for all n. Also

$$\begin{aligned} \|x'_{n+1}(t) - x'_n(t)\| &\le \|f(t, x_n(t)) - f(t, x_{n-1}(t))\| \\ &\le g(t, \|x_n(t) - x_{n-1}(t)\|) \\ &\le g(t, u_{n-1}(t)). \end{aligned}$$

Let $n \le m$. Then one can easily obtain

$$\|x'_n(t) - x'_m(t)\| \le g(t, u_{n-1}(t)) + g(t, u_{m-1}(t)) + g(t, \|x_n(t) - x_m(t)\|).$$

Since $u_{n+1}(t) \le u_n(t)$ for all n, it follows that

$$D^+\|x_n(t) - x_m(t)\| \le g(t, \|x_n(t) - x_m(t)\|) + 2g(t, u_{n-1}(t)),$$

where D^+ is the Dini derivative. An application of Theorem 1.4.1 in [11] gives

$$\|x_n(t) - x_m(t)\| \le r_n(t), \quad t_0 \le t \le t_0 + \alpha,$$

where $r_n(t)$ is the maximal solution of

$$v' = g(t, v) + 2g(t, u_{n-1}(t)), \quad v_n(t_0) = 0,$$

for each n. Since, as $n \to \infty$, $2g(t, u_{n-1}(t)) \to 0$ uniformly on $[t_0, t_0 + \alpha]$, it follows by Lemma 1.3.1 in [11] that $r_n(t) \to 0$ uniformly on $[t_0, t_0 + \alpha]$. This implies that $x_n(t)$ converges uniformly to $x(t)$ and it is now easy to show that $x(t)$ is a solution of (1) by standard arguments.

To show that this solution is unique; let $y(t)$ be another solution of (1) existing on $[t_0, t_0 + \alpha]$. Define

$$m(t) = \|x(t) - y(t)\|$$

and note that $m(t_0) = 0$. Then,

$$D^+ m(t) \le \|x'(t) - y'(t)\| = \|f(t, x(t)) - f(t, y(t))\| \le g(t, m(t)),$$

using the assumption (ii). Again applying [11, Theorem 1.4.1], we have

$$m(t) \le r(t), \quad t_0 \le t \le t_0 + \alpha,$$

where $r(t)$ is the maximal solution of (2). But, by assumption (i), $r(t) \equiv 0$ and this proves that $x(t) \equiv y(t)$. Hence, the limit of the successive approximations is the unique solution of (1). The proof is complete.   □

Isolating the ideas of the proof of Theorem 1, one can prove the following problem.

Problem 2.1 Let (E, ρ) be a complete metric space and let T be a mapping of E into itself such that

$$\rho(Tx, Ty) \le G(\rho(x, y)), \quad \text{when} \quad \rho(x, y) \le u_0.$$

Suppose that $G : [0, u_0] \to [0, u_0]$ is such that

(i) $Gu_0 < u_0$,

(ii) $G(u)$ is monotone nondecreasing in u, and

(iii) $Gu = u$ iff $u = 0$.

Then if $x_0 \in E$ is such that

$$\rho(Tx_0, x_0) \le u_0,$$

the sequence of iterates $\{T^n x_0\}$ converges to a unique fixed point of T.   □

It is important to note that the method of successive approximations does not work if the assumption (ii) is weakened in any form. For example if it is replaced by

$$(x - y, f(t, x) - f(t, y)) \le g(t, \|x - y\|)\|x - y\|.$$

3. Personal circumstances

Before we proceed further, it may be instructive to indicate the way I ended up with this type of research and my educational excursions so that the readers might appreciate the surrounding circumstances. Let me quote below that portion from the paper of R.P. Agarwal and S.G. Leela, "A brief biography and survey of collected works of V. Lakshmikantham", *Nonlinear Analysis*, **40** (2000), 1-20.

"Until he was thirteen years of age, Lakshmikantham did not have any formal schooling since his parents lived in a village where there was no school. He did have, however, a photographic memory and that helped him to absorb from his mother a good understanding of the native language, Telugu. He decided to get formal education at the age of thirteen and went to his uncle who worked in a city. By his own effort, he learned English and mathematics and joined high school. After completing two years of university eduction, he had to take up a job in a bank, where he served for five years. During that time, Lakshmikantham studied himself all the subjects needed for the bank service such as bookkeeping, banking, foreign exchange, economics, etc. By this time, India became independent and was divided into India and Pakistan. The different kingdoms that were in India were expected to join either with India or with Pakistan. One such large kingdom was Nizam State with the capital Hyderabad. Lakshmikantham learned that in that state, if one works for two years in any job, one gets study leave. Since he was anxious to study further, he left the job at the bank and took a low paying storekeeper's job in a biology department in the college of Osmania University in Hyderabad. Fortunately, in a few months, he got a position of Laboratory Assistant in the Chemistry Department of that university. He managed to get an approval which permits any employee of the university to independently study for any degree while working. Using this option, Lakshmikantham completed a Bachelor and Masters degrees successfully. He then proceeded to work for his Ph.D. degree utilizing the leave from his job that was available. Since there were not qualified experts to supervise, he chose to work independently again. He did not know in which area of mathematics to work and so Lakshmikantham went on perusing different mathematical journals in the library for a few days. He found that some papers of Wintner, Hartman, Bellman, Coddington and Levinson, he could follow, and therefore decided to work in differential equations. Concentrating on the papers that deal with uniqueness, convergence of successive approximations and boundedness of solutions of differential equations, Lakshmikantham soon realized that using integral equations corresponding to given differential equations resulted in integral inequalities, which needed an extra assumption of monotonicity. Since using differential inequalities requires no monotony assumption, Lakshmikantham's first attempt was to deduce differential inequality from the integral inequality, which meant utilizing Carathéodory conditions. He discovered that using integral equation is a mistake and went directly to utilize the given differential equation, to develop the comparison principle via differential inequality. At this stage, Lakshmikantham found that the convergence of successive approximations under any one of the uniqueness conditions required monotone assumption on the uniqueness function (comparison function). In *American Journal of Mathematics* (1946), 13-19, Wintner proved the convergence by Osgood's condition with monotony assumption and inserted a remark at the end of the proof, saying that

> "the restriction of monotony imposed is superfluous, follows if the consideration of the functions
> $$w_k(t) = |x_{k+1}(t) - x_k(t)|$$
> themselves is replaced by that of their best monotone majorant sequence, that is,
> $$w_k(t) = \max_{0 \le s \le t} |x_{k+1}(s) - x_k(s)|."$$

Lakshmikantham assumed the truth of this remark and proved the convergence under general Kamke's uniqueness condition using the theory of differential inequality without monotony assumption. Since he suspected the truth of the remark, Lakshmikantham wrote directly to Professor Aurel Wintner, incorporating it in the paper. Professor Wintner was kind to respond as follows:

> "Your criticism of the remark I made in 1946, concerning the omission of monotony, is fully justified and I don't see how I did make that remark. Unfortunately, this implies that theorem 4 of your manuscript will have to be omitted. I would suggest that you transfer your manuscript in a modified form to an appropriate periodical."

The letter of Wintner encouraged Lakshmikantham and provided much needed self-confidence, which does not exist due to one thousand years of subordination in India. His first paper on boundedness results appeared in 1957 in *Proceedings of the American Mathematical Society*. He also made some progress in the convergence theory, proved a non-uniqueness result and integral inequalities for systems.

In 1960, Lakshmikantham came to UCLA with the support of Earl Coddington where he pursued his work on differential inequalities and Lyapunov stability. He then spent one year each in MRC, University of Wisconsin, Madison, RIAS, Baltimore, and University of Calgary in Canada. Lakshmikantham was very active and published several papers in differential equations with delay, abstract differential equations, parabolic differential equations and popularized the method of vector Lyapunov functions. He then took the position of Head and Professor of Mathematics Department, in Marathwada University in India. He returned to the USA to assume the position of Head and Professor at the University of Rhode Island in 1966. He then went on similar status to the University of Texas at Arlington in 1973 and finally, to Florida Institute of Technology, in 1989."

Recall that in Theorem 1, the comparison function $g(t, u)$ is assumed to be monotone nondecreasing in u. The question whether this additional assumption of monotony is really needed has been open for many years starting from Wintner. A partial answer is given in [10], namely, it is sufficient if the function $g(t, u)$ of Theorem 1, without monotony assumption, dominates the function
$$g_0(t, u) = \max_{\|x-y\| \le u} \|f(t, x) - f(t, y)\|.$$

In Hartman [7] and Deimling [5], one finds results which claim that this problem is solved. Unfortunately, the proofs have a subtle error in both cases. In [12], the cause of the breakdown of the proof is isolated so that someone can concentrate on this open problem with an interesting history.

4. Monotone approximations

An interesting and fruitful technique for proving existence results for nonlinear problems is the method of lower and upper solutions. This method coupled with the monotone iterative technique manifests itself as an effective and flexible mechanism that offers theoretical as well as constructive existence results in a closed set, generated by the lower and upper solutions. The lower and upper solutions serve as rough bounds, which can be improved by monotone iterative procedures. Moreover, the iteration schemes can also be employed for the investigation of qualitative and quantitative properties of solutions. The ideas embedded in these techniques have proved to be of immense value and have played a crucial role in unifying a wide variety of nonlinear problems. This unification is well known and is related in the monographs [9], [14], where this technique is applied to ordinary and partial differential equations. Let us state a simple typical result that provides monotone approximations which are the solutions of corresponding linear problems. Consider the scalar initial value problem

$$u' = F(t, u), \quad u(0) = u_0, \quad t \in J = [0, T]. \tag{5}$$

One can prove the following result, see [9] for a proof.

Theorem 2. *Assume that*

(i) $\alpha_0, \beta_0 \in C^1[J, R], \alpha_0' \leq F(t, \alpha_0), \beta_0' \geq F(t, \beta_0)$ *and* $\alpha_0(t) \leq \beta_0(t)$ *on* J;

(ii) $F \in C[J \times R, R]$ *and for* $t \in J$,

$$F(t, u_1) - F(t, u_2) \geq -M(u_1 - u_2), \quad M \geq 0 \text{ and } \alpha_0(t) \leq u_2 \leq u_1 \leq \beta_0(t).$$

Then there exist monotone sequences $\{\alpha_n(t)\}, \{\beta_n(t)\}$ *such that* $\alpha_n \to \rho, \beta_n \to r$ *as* $n \to \infty$ *uniformly on* J *and* (ρ, r) *are the minimal and maximal solutions of* (5) *respectively.*

The monotone approximations are generated from the linear initial value problems

$$\alpha_{n+1}' + M\alpha_{n+1} = F(t, \alpha_n) + M\alpha_n, \qquad \alpha_{n+1}(0) = u_0,$$

$$\beta_{n+1}' + M\beta_{n+1} = F(t, \beta_n) + M\beta_n, \qquad \beta_{n+1}(0) = u_0, \quad n = 0, 1, 2, \ldots,$$

and α_0, β_0 are known as lower and upper solutions of (5).

We observe that the special case when $F(t, u)$ is monotone nondecreasing in u is covered in Theorem 2 with $M = 0$. However, the other case, when $F(t, u)$ is monotone nonincreasing in u is not included in Theorem 2 and is of particular interest. Under some special conditions, one can construct, when $F(t, u)$ is nonincreasing, a single iteration procedure yielding an alternative sequence which forms two monotone sequences bounding the solution from above and below. In this case, the iteration scheme is either

$$\alpha'_{n+1} = F(t, \alpha_n), \quad \alpha_{n+1}(0) = u_0,$$

or

$$\beta'_{n+1} = F(t, \beta_n), \quad \beta_{n+1}(0) = u_0.$$

Then we have the following result.

Theorem 3. *Assume that $F(t, u)$ is nonincreasing in u. Then $\alpha_n(t)$ and the unique solution $u(t)$ of (5) satisfy*

$$\alpha_0 \leq \alpha_2 \leq \ldots \leq \alpha_{2n} \leq u \leq \alpha_{2n+1} \leq \ldots \leq \alpha_3 \leq \alpha_1 \quad \text{on } J,$$

provided $\alpha_0 \leq \alpha_2$. Moreover, the alternating sequences $\{\alpha_{2n}\}$, $\{\alpha_{2n+1}\}$ converge uniformly as $n \to \infty$ to ρ, r on J and $\rho \leq u \leq r$ on J. Similarly, β_n and $u(t)$, satisfy

$$\beta_1 \leq \beta_3 \leq \ldots \leq \beta_{2n+1} \leq u \leq \beta_{2n} \leq \ldots \leq \beta_2 \leq \beta_0 \quad \text{on } J,$$

provided $\beta_0 \geq \beta_2$. Also, $\{\beta_{2n+1}\}$, $\{\beta_{2n}\}$ converge uniformly as $n \to \infty$ to ρ^*, r^* and $\rho^* \leq u \leq r^*$ on J.

Note that there is no assumption of the existence of lower and upper solutions α_0, β_0 in Theorem 3, since one can construct them when F is nonincreasing. Also, uniqueness of solutions of (5) easily follows. See [9].

This fruitful technique, however, is no longer applicable when the nonlinear terms involved are discontinuous in the dependent variable. By means of a generalized iteration principle, this technique has been extended in [6] to deal with discontinuous differential equations. Also, in [4], differential equations which can be represented as

$$Lu = Nu,$$

where L denotes some differential operator and N stands for the lower order terms, and may also depend on Lu in the implicit case are discussed. Moreover, differential inclusions of hemivariational type which can be considered as the multivalued versions of some hemivariational inequalities are also investigated in [3], [4].

A natural question that arises is whether it is possible to obtain the monotone sequences $\{\alpha_n\}$, $\{\beta_n\}$, when $F(t, u)$ is nonincreasing, without additional restrictions. The answer is positive and we shall illustrate it by finding a unified framework which offers several cases of interest. Moreover, this leads to the definition of different types of lower and upper solutions, each of which generates different situations. We shall discuss this new approach for elliptic equation utilizing the variational technique which is not well known.

Consider the BVP

$$\begin{cases} Lu = F(x, u) \text{ in } \Omega, \\ u = 0 \qquad \text{ on } \partial\Omega \text{ (in the sense of trace)}; \end{cases} \tag{6}$$

where L denotes the second order partial differential operator in the divergence form

$$Lu = -\sum_{i,j}^{n} (a_{ij}(x)u_{x_i})_{x_j} + c(x)u \tag{7}$$

and Ω is an open, bounded subset of R^n. We assume that

$$a_{ij}, c \in L^\infty(\Omega), \quad a_{ij} = a_{ji}, \quad i, j = 1, 2, \ldots, n,$$

$$\sum_{i,j=1}^{n} a_{ij}(x)\xi_i\xi_j \geq \theta \mid \xi \mid^2 \quad \text{for } x \in \Omega, \text{ a.e. } \xi \in R^n$$

with $\theta > 0$, (uniform elliptic condition), and $c(x) \geq 0$. We shall always mean that the boundary condition is in the sense of trace and hence we shall not repeat it to avoid monotony. Also,

$$F : \bar{\Omega} \times R \to R, \qquad F(x, u),$$

is a Carathéodory function that is $F(\cdot, u)$ measurable for all $u \in R$ and $F(x, \cdot)$ is continuous a.e. $x \in \Omega$. The bilinear form $B[\ ,\ ]$ associated with the operator L is

$$B[u, v] = \int_\Omega \left[\sum_{i,j=1}^{n} a_{ij}(x)u_{x_i}v_{x_j} + c(x)uv \right] dx, \tag{8}$$

for $u, v \in H_0^1(\Omega)$.

Definition 4. The function $u \in H_0^1(\Omega)$ is said to be a weak solution of (6) if

$$F(x, u) \in L^1(\Omega), \qquad F(x, u)u \in L^1(\Omega),$$

and

$$B[u, v] = (F, v) \tag{9}$$

for all $v \in H_0^1(\Omega)$ where $(\ ,\ )$ denotes inner product in $L_2(\Omega)$.

Definition 5. The function $\alpha_0 \in H^1(\Omega)$ is said to be a weak lower solution of (6) if, $\alpha_0 \leq 0$ on $\partial\Omega$ and

$$\int_\Omega \left[\sum_{i,j=1}^{n} a_{ij}(x)(\alpha_0)_{x_i}v_{x_j} + c(x)\alpha_0 v \right] dx \leq \int_\Omega F(x, \alpha_0)v dx, \tag{10}$$

for each $v \in H_0^1(\Omega)$, $v \geq 0$. If the inequalities are reversed, then α_0 is said to be a weak upper solution of (6).

The following result on existence can be proved employing Lax-Milgram Lemma, see [8].

Lemma 6. *Consider the linear BVP*

$$\begin{bmatrix} Lu = h(x) & \text{in } \Omega, \\ u = 0 & \text{on } \Omega \text{ (in the sense of trace).} \end{bmatrix} \tag{11}$$

Then there exists a unique solution $u \in H_0^1(\Omega)$ for the linear BVP (11) provided

$$0 < c^* \leq c(x) \quad \text{a.e. in } \Omega \quad \text{and} \quad h \in L^2(\Omega).$$

Let us next prove the basic comparison result in the variational setup.

Theorem 7. *Let α_0, β_0 be weak lower and upper solutions of* (6). *Suppose further that F satisfies*

$$F(x, u_1) - F(x, u_2) \leq K(u_1 - u_2) \tag{12}$$

whenever $u_1 \geq u_2$ a.e., $x \in \Omega$ and $K > 0$. Then, if $0 < c - K \in L^1(\Omega)$, we have

$$\alpha_0(x) \leq \beta_0(x) \text{ in } \Omega, \text{ a.e.} \tag{13}$$

Proof. From the definition of lower and upper solutions, we get

$$\int_\Omega [\sum_{i,j=1}^n a_{ij}(x)(\alpha_{0,x_i} - \beta_{0,x_i})v_{x_j} + c(x)(\alpha_0 - \beta_0)v]dx$$

$$\leq \int_\Omega [f(x, \alpha_0) - f(x, \beta_0)]vdx, \tag{14}$$

for each $v \in H_0^1(\Omega)$, $v \geq 0$, a.e. Choose

$$v = (\alpha_0 - \beta_0)^+ \in H_0^1(\Omega), \qquad v \geq 0, \text{ a.e.}$$

Then we have

$$\int_\Omega [\sum_{i,j=1}^n a_{ij}(x)(\alpha_{0,x_i} - \beta_{0,x_i})(\alpha_0 - \beta_0)_{x_j}^+ + c(x)(\alpha_0 - \beta_0)(\alpha_0 - \beta_0)^+]dx$$

$$\leq \int_\Omega [f(x, \alpha_0) - f(x, \beta_0)](\alpha_0 - \beta_0)^+ dx.$$

Since

$$(\alpha_0 - \beta_0)_{x_j}^+ = \begin{bmatrix} (\alpha_{0,x_j} - \beta_{0,x_j}) & \text{a.e. on } \{\alpha_0 > \beta_0\}, \\ 0 & \text{a.e. on } \{\alpha_0 \leq \beta_0\}, \end{bmatrix}$$

we obtain using the ellipticity condition and (12)

$$\int_{\alpha_0 > \beta_0} [\theta \mid (\alpha_{0,x_i} - \beta_{0,x_i}) \mid^2 + c(x) \mid \alpha_0 - \beta_0 \mid^2]dx \leq K \int_{\alpha_0 > \beta_0} \mid \alpha_0 - \beta_0 \mid^2 dx,$$

which implies $\alpha_0(x) \leq \beta_0(x)$ in Ω a.e. This completes the proof. $\qquad\square$

The following corollary is useful in our discussion.

Corollary 8. *For $p \in H^1(\Omega)$ satisfying*

$$\int_\Omega \left[\sum_{i,j=1}^n a_{ij}(x)p_{x_i}v_{x_j} + c_0(x)pv \right] dx \leq 0,$$

for each $v \in H_0^1(\Omega)$, $v \geq 0$, a.e. and $p \leq 0$ on $\partial\Omega$ we have $p(x) \leq 0$ in Ω, a.e., provided $c_0(x) > 0$.

Let us consider the following semilinear elliptic boundary value problem in the divergence form

$$\begin{bmatrix} Lu = f(x,u) + g(x,u) & \text{in } \Omega, \\ u = 0 & \text{on } \partial\Omega \text{ (in the sense of trace).} \end{bmatrix} \tag{15}$$

In order to develop monotone iterative technique for the BVP (15), we need to utilize appropriate coupled lower and upper solutions of (15). We shall only define the following two types.

Definition 9. Relative to the BVP (15), the functions $\alpha_0, \beta_0 \in H^1(\Omega)$ are said to be

(i) coupled weak lower and upper solutions of type I, if

$$B[\alpha_0, v] \le (f(x,\alpha_0) + g(x,\beta_0), v),$$

$$B[\beta_0, v] \ge (f(x,\beta_0) + g(x,\alpha_0), v),$$

for each $v \in H_0^1(\Omega)$, $v \ge 0$ a.e. in Ω;

(ii) coupled weak lower and upper solutions of type II, if

$$B[\alpha_0, v] \le (f(x,\beta_0) + g(x,\alpha_0), v),$$

$$B[\beta_0, v] \ge (f(x,\alpha_0) + g(x,\beta_0), v),$$

for each $v \in H_0^1(\Omega)$, $v \ge 0$, a.e. in Ω.

We are now in a position to prove the following result which contains several results of interest.

Theorem 10. *Assume that*

(A_1) $\alpha_0, \beta_0 \in H^1(\Omega)$ *are the weak coupled lower and upper solutions of type I with* $\alpha_0(x) \le \beta_0(x)$ *a.e. in* Ω;

(A_2) $f, g : \bar{\Omega} \times R \to R$ *are Carathéodory functions such that* $f(x,u)$ *is nondecreasing in* u, $g(x,u)$ *is nonincreasing in* u *for* $x \in \Omega$, *a.e.;*

(A_3) $c(x) \ge N > 0$ *in* Ω, *a.e. and for any* $\eta, \mu \in H^1(\Omega)$ *with* $\alpha_0 \le \eta$, $\mu \le \beta_0$, *the function* $h(x) = f(x,\eta) + g(x,\mu) \in L^2(\Omega)$.

Then there exist monotone sequences

$$\{\alpha_n(x)\}, \quad \{\beta_n(x)\} \in H_0^1(\Omega)$$

such that $\alpha_n \to \rho$, $\beta_n \to r$ *weakly in* $H_0^1(\Omega)$ *as* $n \to \infty$ *and* (ρ, r) *are weak coupled minimal and maximal solutions of* (15), *respectively, that is,*

$$L\rho = f(x,\rho) + g(x,r) \text{ in } \Omega, \qquad \rho = 0 \text{ on } \partial\Omega,$$

$$Lr = f(x,r) + g(x,\rho) \text{ in } \Omega, \qquad r = 0 \text{ on } \partial\Omega.$$

Proof. Consider the linear BVPs

$$L\alpha_{n+1} = f(x, \alpha_n) + g(x, \beta_n) \quad \text{in } \Omega, \qquad \alpha_{n+1} = 0 \quad \text{on } \partial\Omega, \tag{16}$$

and

$$L\beta_{n+1} = f(x, \beta_n) + g(x, \alpha_n) \quad \text{in } \Omega, \qquad \beta_{n+1} = 0 \quad \text{on } \partial\Omega. \tag{17}$$

The variational forms associated with (16) and (17) are

$$B[\alpha_{n+1}, v] = \int_\Omega [f(x, \alpha_n) + g(x, \beta_n)] v \, dx, \tag{18}$$

$$B[\beta_{n+1}, v] = \int_\Omega [f(x, \beta_n) + g(x, \alpha_n)] v \, dx, \tag{19}$$

for all $v \in H_0^1(\Omega)$, $v \geq 0$ a.e. in Ω.

We shall show that the weak solutions α_n, β_n of (16) and (17) are uniquely defined and satisfy

$$\alpha_0 \leq \alpha_1 \leq \alpha_2 \leq \dots \leq \alpha_n \leq \beta_n \leq \dots \leq \beta_2 \leq \beta_1 \leq \beta_0 \quad \text{a.e. in } \Omega. \tag{20}$$

For each $n \geq 1$, if we have $\alpha_0 \leq \alpha_n \leq \beta_n \leq \beta_0$, then by assumption (A_3),

$$h_1(x) \equiv f(x, \alpha_n) + g(x, \beta_n) \in L_2(\Omega)$$

and

$$h_2(x) \equiv f(x, \beta_n) + g(x, \alpha_n) \in L_2(\Omega).$$

Therefore, Lemma 6 implies that BVPs (16) and (17) have unique solutions α_n and β_n in view of $c(x) \geq N > 0$. Hence, we need to show (20) holds.

For this, we first claim that $\alpha_1 \geq \alpha_0$ a.e. in Ω. Now, let $p = \alpha_0 - \alpha_1$ so that $p \leq 0$ on $\partial\Omega$ and for $v \in H_0^1(\Omega)$, $v \geq 0$ a.e. in Ω,

$$B[p, v] = \int_\Omega \left[\sum_{i,j=1}^n a_{ij}(x) p_{x_i} v_{x_j} + c(x) p v \right] dx$$

$$\leq \int_\Omega [f(x, \alpha_0) + g(x, \beta_0)] v \, dx - \int_\Omega [f(x, \alpha_0) + g(x, \beta_0)] v \, dx = 0.$$

Hence by Corollary 8, $p(x) \leq 0$, that is, $\alpha_0 \leq \alpha_1$ a.e. in Ω. Similarly, we can show that $\beta_1 \leq \beta_0$ a.e. in Ω.

Assume that, for some fixed $n > 1$,

$$\alpha_n \leq \alpha_{n+1} \quad \text{and} \quad \beta_n \geq \beta_{n+1} \quad \text{a.e. in } \Omega.$$

Now consider

$$p = \alpha_{n+1} - \alpha_{n+2}.$$

Note that $p = 0$ on $\partial\Omega$, and using the monotone nature of f, g, we get

$$B[p, v] = \int_\Omega [f(x, \alpha_n) - f(x, \alpha_{n+1}) + g(x, \beta_n) - g(x, \beta_{n+1})] v \, dx \leq 0.$$

This implies by Corollary 8 that

$$\alpha_{n+1} \leq \alpha_{n+2} \quad \text{a.e. in} \quad \Omega.$$

Also, by setting $p = \beta_{n+2} - \beta_{n+1}$, one can easily show in the same way, that $\beta_{n+2} \leq \beta_{n+1}$ a.e. in Ω. Hence, using the induction argument, we have $\alpha_n \geq \alpha_{n+1}$ and $\beta_n \geq \beta_{n+1}$ a.e. in Ω for all $n \geq 1$.

We now show that $\alpha_1 \leq \beta_1$ a.e. in Ω. Consider $p = \alpha_1 - \beta_1$ and note that $p = 0$ on $\partial\Omega$, and

$$B[p, v] = \int_{\Omega} [f(x, \alpha_0) + g(x, \beta_0) - f(x, \beta_0) - g(x, \alpha_0)]v dx \leq 0,$$

because of the fact f, g are monotone by assumption.

Hence Corollary 8 yields $\alpha_1 \leq \beta_1$ a.e. in Ω. Employing similar arguments, it is easy to show that if we assume, for some fixed $n > 1$, we have $\alpha_n \leq \beta_n$ a.e. in Ω, then it follows that $\alpha_{n+1} \leq \beta_{n+1}$ a.e. in Ω. Consequently, using the induction argument, it is clear that (20) holds for all $n \geq 1$.

By the monotone character of $\{\alpha_n\}, \{\beta_n\}$ there exist pointwise limits

$$\lim_{n \to \infty} \alpha_n(x) = \rho(x) \text{ a.e. in } \Omega \text{ and } \lim_{n \to \infty} \beta_n(x) = r(x) \text{ a.e. in } \Omega.$$

Moreover, since

$$\alpha_0 \leq \alpha_n \leq \beta_n \leq \beta_0 \quad \text{a.e. in} \quad \Omega,$$

it follows by Lebesgue's dominated convergence theorem that

$$\alpha_n \to \rho \quad \text{and} \quad \beta_n \to r \quad \text{in } L^2(\Omega).$$

For each $n \geq 1$, we note that α_n satisfies for each $v \in H_0^1(\Omega)$, $v \geq 0$, a.e. in Ω

$$\int_{\Omega} \left[\sum_{i,j=1}^{n} a_{ij}(x)(\alpha_n)_{x_i} v_{x_j} + c(x)\alpha_n v \right] dx = \int_{\Omega} h_{n-1}(x)v dx,$$

where

$$h_{n-1}(x) = f(x, \alpha_{n-1} + g(x, \beta_{n-1}).$$

We now use the ellipticity condition and the fact that $c(x) \geq N > 0$ with $v = \alpha_n$ to get

$$\int_{\Omega} [\theta \mid (\alpha_n)_x \mid^2 + N \mid \alpha_n \mid^2] dx \leq \int_{\Omega} h_{n-1}(x)v dx.$$

Since the integrand on the right-hand side belongs to $L^2(\Omega)$, we obtain the estimate

$$\sup_n \| \alpha_n \|_{H_0^1(\Omega)} < \infty.$$

Hence there exists a subsequence $\{\alpha_{n_k}\}$ which converges weakly to $\rho(x)$ in $H_0^1(\Omega)$.

A similar argument implies that

$$\sup_n \| \beta_n \|_{H_0^1(\Omega)} < \infty.$$

Hence there exist subsequences $\{\alpha_{n_k}\}, \{\beta_{n_k}\}$ which converge weakly in $H_0^1(\Omega)$ to (ρ, r) $\in H_0^1(\Omega)$ respectively.

Since α_n and β_n satisfy (18) and (19) for a fixed $v \in H_0^1(\Omega)$, $v \geq 0$ a.e. in Ω by taking the limit as $n \to \infty$, we see that

$$B[\rho, v] = \int_\Omega]f(x, \rho) + g(x, r)]v dx$$

and

$$B[r, v] = \int_\Omega [f(x, r) + g(x, \rho)]v dx.$$

Finally, we claim that ρ and r are the weak coupled minimal and maximal solutions of (15) that is, if u is any weak solution of 15 such that $\alpha_0(x) \leq u(x) \leq \beta_0(x)$ a.e. in Ω, then

$$\alpha_0(x) \leq \rho(x) \leq u(x) \leq r(x) \leq \beta_0 \quad \text{a.e. in} \quad \Omega. \tag{21}$$

Suppose that for some fixed $n \geq 1$, $\alpha_n \leq u \leq \beta_n$ a.e. in Ω. Setting $p = \alpha_{n+1} - u$, we have $p = 0$ on $\partial\Omega$ and employing the monotone character of f and g, it follows that

$$B[p, v] = \int_\Omega [f(x, \alpha_n) + g(x, \beta_n) - f(x, u) - g(x, u)]v dx \leq 0,$$

which implies by Corollary 8 $\alpha_{n+1} \leq u$ a.e. in Ω. In a similar way, we obtain $u \leq \beta_{n+1}$ a.e. in Ω so that $\alpha_{n+1} \leq u \leq \beta_{n+1}$ a.e. in Ω. By induction, $\alpha_n \leq u \leq \beta_n$ a.e. in Ω for all $n \geq 1$. Now taking the limit as $n \to \infty$, we get (21), completing the proof. □

To show the uniqueness, we have the following corollary.

Corollary 11. *Assume, in addition to the conditions of Theorem 10, that f and g satisfy*

$$f(x, u_1) - f(x, u_2) \leq N_1(u_1 - u_2)$$

and

$$g(x, u_1) - g(x, u_2) \geq -N_2(u_1 - u_2)$$

where $u_1 \geq u_2$, $N_1 > 0$, $N_2 > 0$ and

$$c(x) - (N_1 + N_2) > 0 \quad \text{a.e. in} \quad \Omega.$$

Then, $\rho = u = r$ is the unique weak solution of (15).

Proof. Since we have $\rho \leq r$, it is enough to show that $r \leq \rho$. Setting

$$p = r - \rho,$$

we have $p = 0$ on $\partial\Omega$ and, using the assumptions of f, g, it follows that

$$\begin{aligned} B[p, v] &= \int_\Omega [f(x, r) + g(x, \rho) - f(x, \rho) - g(x, r)]v dx \\ &\leq \int_\Omega [N_1(r - \rho) + N_2(r - \rho)]v dx \\ &= ((N_1 + N_2)p, v). \end{aligned}$$

This implies by Corollary 8, $p \leq 0$ a.e. in Ω. Hence

$$u = \rho = r$$

is the weak solution of (15).

Now consider the linear iteration scheme

$$
\begin{aligned}
L\alpha_{n+1} &= f(x, \beta_n) + g(x, \alpha_n) \quad \text{in } \Omega, \quad \alpha_{n+1} = 0 \quad \text{on } \partial\Omega, \\
L\beta_{n+1} &= f(x, \alpha_n^*) + g(x, \beta_n) \quad \text{in } \Omega, \quad \beta_{n+1} = 0 \quad \text{on } \partial\Omega
\end{aligned}
\tag{22}
$$

and the associated variational forms

$$B[\alpha_{n+1}, v] = \int_\Omega [f(x, \beta_n) + g(x, \alpha_n)] v \, dx \tag{23}$$

$$B[\beta_{n+1}, v] = \int_\Omega [f(x, \alpha_n) + g(x, \beta_n)] v \, dx \tag{24}$$

for all $v \in H_0^1(\Omega)$, $v \geq 0$ a.e. in Ω. Then, for the weak solutions of (15), we have the following theorem which gives alternative sequences. $\qquad\square$

Theorem 12. *Assume that the conditions (A_2) and (A_3) of Theorem 10 hold. Then for any weak solution $u(x)$ of (15) such that*

$$\alpha_0(x) \leq u(x) \leq \beta_0(x) \quad \text{a.e. in } \Omega,$$

we have the iterates $\{\alpha_n(x)\}$, $\{\beta_n(x)\}$ satisfying

$$\alpha_0 \leq \alpha_2 \leq \ldots \leq \alpha_{2n} \leq u \leq \alpha_{2n+1} \leq \ldots \leq \alpha_3 \leq \alpha_1, \ a.e. \ in \ \Omega, \tag{25}$$

$$\beta_1 \leq \beta_3 \leq \ldots \leq \beta_{2n+1} \leq u \leq \beta_{2n} \leq \ldots \leq \beta_2 \leq \beta_0, \ a.e. \ on \ \Omega, \tag{26}$$

provided

$$\alpha_0 \leq \alpha_2 \quad and \quad \beta_2 \leq \beta_0 \quad a.e. \ in \ \Omega$$

where the iterative schemes are given by (22). Moreover, the monotone sequences

$$\{\alpha_{2n}\}, \ \alpha_{2n+1}\}, \ \{\beta_{2n}\}, \ \{\beta_{2n+1}\} \in H_0^1(\Omega)$$

converge weakly in $H_0^1(\Omega)$ to ρ, r, ρ^*, r^*, respectively, and they satisfy the relations

$$
\begin{aligned}
Lr &= f(x, \rho^*) + g(x, \rho) \ in \ \Omega, & r &= 0 \ on \ \partial\Omega, \\
L\rho &= f(x, r^*) + g(x, r) \ in \ \Omega, & \rho &= 0 \ on \ \partial\Omega, \\
Lr^* &= f(x, \rho) + g(x, \rho^*) \ in \ \Omega, & r^* &= 0 \ on \ \partial\Omega, \\
L\rho^* &= f(x, r) + g(x, r^*) \ in \ \Omega, & \rho^* &= 0 \ on \ \partial\Omega.
\end{aligned}
$$

Also,

$$\rho \leq u \leq r \quad and \quad r^* \leq u \leq \rho^*, \quad a.e. \ in \ \Omega.$$

For a proof, see [8]. We do not list the various special cases of known and new results which are included in Theorems 2 and 7. We shall provide such information in the next section, for the result on rapid convergence.

5. Rapid convergence

The method of quasilinearization developed by Bellman and Kalaba [1], [2] uses convexity assumption and provides lower bounding monotone sequence that converges to the assumed unique solution, once the initial approximation is chosen in an adroit fashion. If we utilize the technique of lower and upper solutions combined with the method of quasilinearization and employ the idea of Newton-Fourier, it is possible to construct concurrently lower and upper bounding monotone sequences whose elements are the solutions of the corresponding linear problems. Of course, both sequences converge rapidly to the solution. Furthermore, this unification provides a framework to enlarge the class of nonlinear problems considerably to which the method is applicable. For example, it is not necessary to impose usual convexity assumption on the nonlinear function involved, since one can allow much weaker assumptions. In fact, several possibilities can be investigated with this unified methodology and consequently, this technique is known as generalized quasilinearization [13]. Moreover, these ideas are extended, refined and generalized to various other types of nonlinear problems.

We shall discuss the method of generalized quasilinearization relative to the BVP (15) using a unified approach so that several important cases are included in the results proved.

We shall prove the following result for the boundary value problem (15), employing Corollary 8 and Lemma 6.

Theorem 13. *Assume that*

$(B1)$ $\alpha_0, \beta_0 \in H^1(\Omega)$ *are lower and upper solutions of* (15) *such that*

$$\alpha_0 \leq \beta_0 \quad in \quad \Omega, \ a.e.;$$

$(B2)$ $f, g : \bar{\Omega} \times R \to R$ *are Carathéodory functions,* $f_u(x, u)$, $g_u(x, u)$, $f_{uu}(x, u)$, $g_{uu}(x, u)$ *exist and are Carathéodory functions, and*

$$f_{uu}(x, u) \geq 0, \qquad g_{uu}(x, u) \leq 0,$$

for $\alpha_0 \leq u \leq \beta_0$ *a.e. in* Ω;

$(B3)$ $0 < N \leq c(x) - f_u(x, \beta_0) - g_u(x, \alpha_0)$ *a.e. in* Ω, *and for any* $\mu, \eta \in H^1(\Omega)$ *satisfying* $\alpha_0 \leq \mu \leq \eta \leq \beta_0$, *the function* $h \in L^2(\Omega)$, *where*

$$h(x) = f(x, \eta) + g(x, \eta) - f_u(x, \mu)\eta - g_u(x, \eta)\eta.$$

Then there exist monotone sequences $\{\alpha_k\}$, $\{\beta_k\} \in H_0^1(\Omega)$ *such that* $\alpha_k \to \rho$, $\beta_k \to r$ *weakly in* $H_0^1(\Omega)$ *as* $k \to \infty$, *with* $\rho = r = u$ *is the unique weak solution of* (15) *satisfying* $\alpha_0 \leq u \leq \beta_0$, *a.e. in* Ω *and the convergence is quadratic.*

Proof. We prove the conclusion in several steps.

(a) **Iterative schemes and generalized quasilinearization.** Let us introduce the linearization of $f + g$ in the form

$$G(x, u; \alpha, \beta) = f(x, \beta) + g(x, \beta) + f_u(x, \alpha)(u - \beta) + g_u(x, \beta)(u - \beta), \quad (27)$$

and

$$F(x, u; \alpha, \beta) = f(x, \alpha) + g(x, \alpha) + f_u(x, \alpha)(u - \alpha) + g_u(x, \beta)(u - \alpha), \quad (28)$$

and consider the following related iterative schemes for $k = 0, 1, 2, \ldots$

$$L\beta_{k+1} = G(x, \beta_{k+1}; \alpha_k, \beta_k) \text{ in } \Omega, \quad \beta_{k+1} = 0 \text{ on } \partial\Omega, \quad (29)$$

and

$$L\alpha_{k+1} = F(x, \alpha_{k+1}; \alpha_k, \beta_k) \text{ in } \Omega, \quad \alpha_{k+1} = 0 \text{ on } \partial\Omega. \quad (30)$$

The variational forms associated with (29) and (30) are given by

$$B[\beta_{k+1}, v] = \int_\Omega G(x, \beta_{k+1}; \alpha_k, \beta_k) v dx, \quad (31)$$

and

$$B[\alpha_{k+1}, v] = \int_\Omega F(x, \alpha_{k+1}; \alpha_k, \beta_k) v dx, \quad (32)$$

for all $v \in H_0^1(\Omega)$, $v \geq 0$ a.e. We shall show that the weak solutions β_k, α_k of (29) and (30) respectively, are uniquely defined and satisfy

$$\alpha_0 \leq \alpha_1 \leq \alpha_2 \leq \ldots \leq \alpha_k \leq \beta_k \leq \ldots \leq \beta_2 \leq \beta_1 \leq \beta_0 \text{ a.e. in } \Omega. \quad (33)$$

Let us consider $k = 0$. Then according to (27), $G(x, u; \alpha_0, \beta_0)$ is of the form

$$G(x, u; \alpha_0, \beta_0) = d(x)u + h(x),$$

where

$$d(x) = f_u(x, \alpha_0) + g_u(x, \beta_0)$$

and

$$h(x) = f(x, \beta_0) + g(x, \beta_0) - f_u(x, \alpha_0)\beta - g_u(x, \beta_0)\beta_0.$$

Since $f_u(x, u)$ is nondecreasing in u and $g_u(x, u)$ is nonincreasing in u for each $x \in \Omega$, a.e., we find that

$$\begin{aligned} c(x) - d(x) &= c(x) - f_u(x, \alpha_0) - g_u(x, \beta_0) \\ &\geq c(x) - f_u(x, \beta_0) - g_u(x, \alpha_0) \\ &\geq N > 0 \end{aligned}$$

and $h \in L^2(\Omega)$, by $(B3)$. Thus by Lemma 6 there is a unique weak solution $\beta_1 \in H_0^1(\Omega)$ of 29). Similarly, one gets the existence of a weak solution $\alpha_1 \in H_0^1(\Omega)$ of (30). We shall next show that

$$\alpha_0 \leq \alpha_1 \leq \beta_1 \leq \beta_0 \quad \text{a.e. in } \Omega. \quad (34)$$

Since α_0 is a lower solution of (15), we see that

$$B[\alpha_0, v] \leq \int_\Omega F(x, \alpha_0; \alpha_0, \beta_0) v dx$$

for all $v \in H_0^1(\Omega)$, $v \geq 0$ a.e. By the definition of the iterates, α_1 is the unique weak solution of (30) for $k = 0$ and thus α_1 satisfies the variational form (32) with $k = 0$. Let

$$p = \alpha_0 - \alpha_1$$

so that $p(0) \leq 0$ on $\partial\Omega$ and for $v \in H_0^1(\Omega)$, $v \geq 0$ a.e.

$$\int_\Omega \sum_{i,j=1}^n a_{ij}(x)p_{x_j}v_{x_j} + \int_\Omega c_0(x)pv dx \leq 0$$

where

$$c_0(x) = c(x) - d(x) \geq 0.$$

Hence Corollary 8 yields $p \leq 0$, that is, $\alpha_0 \leq \alpha_1$ a.e. in Ω. Similarly, we can show that $\beta_1 \leq \beta_0$ a.e. in Ω. Next we shall show that

$$\alpha_1 \leq \beta_0 \quad \text{and} \quad \alpha_0 \leq \beta_1 \quad \text{a.e. in } \Omega.$$

To show that $\alpha_1 \leq \beta_0$ we employ the following inequalities which are consequences of $f_{uu}(x, u) \geq 0$ and $g_{uu}(x, u) \leq 0$ imposed by $(B2)$,

$$f(x, u) \geq f(x, v) + f_u(x, v)(u - v), \tag{35}$$

$$g(x, u) \geq g(x, v) + g_u(x, u)(u - v), \tag{36}$$

for all $\alpha_0 \leq v \leq u \leq \beta_0$. By (35) and (36), we obtain for all $v \in H_0^1(\Omega)$, $v \geq 0$ a.e.,

$$
\begin{aligned}
B[\alpha_1, v] &= \int_\Omega F(x, \alpha_1; \alpha_0, \beta_0)v dx \\
&\leq \int_\Omega [f(x, \beta_0) - f_u(x, \alpha_0)(\beta_0 - \alpha_0) + g(x, \beta_0) \\
&\quad\quad - g_u(x, \beta_0)(\beta_0 - \alpha_0) + f_u(x, \alpha_0)(\alpha_1 - \alpha_0) \\
&\quad\quad + g_u(x, \beta_0)(\alpha_1 - \alpha_0)]v dx \\
&\leq \int_\Omega [f(x, \beta_0) + g(x, \beta_0) + f_u(x, \alpha_0)(\alpha_1 - \alpha_0 + \alpha_0 - \beta_0) \\
&\quad\quad + g_u(x, \beta_0)(\alpha_0 - \beta_0 + \alpha_1 - \alpha_0)]v dx \\
&= \int_\Omega G(x, \alpha_1; \alpha_0, \beta_0)v dx.
\end{aligned}
\tag{37}
$$

Since β_0 is an upper solution of (15), it satisfies for all $v \in H_0^1(\Omega)$, $v \geq 0$ a.e.,

$$
\begin{aligned}
B[\beta_0, v] &\geq \int_\Omega [f(x, \beta_0) + g(x, \beta_0)]v dx \\
&= \int_\Omega G(x, \beta_0; \alpha_0, \beta_0)v dx.
\end{aligned}
$$

Hence by Corollary 8, $\alpha_1 \leq \beta_0$ a.e. in Ω. A similar argument proves that $\alpha_0 \leq \beta_1$ a.e. in Ω. To prove $\alpha_1 \leq \beta_1$, we use (35) and (36) and the fact $g_u(x, u)$ is nonincreasing in u and $\alpha_1 \leq \beta_0$ to get

$$B[\alpha_1, v] = \int_\Omega F(x, \alpha_1; \alpha_0, \beta_0) v dx$$

$$\leq \int_\Omega [f(x, \alpha_1) + g(x, \alpha_1) + g_u(x, \alpha_1)(\alpha_0 - \alpha_1)$$

$$+ g_u(x, \beta_0)(\alpha_1 - \alpha_0)] v dx \tag{38}$$

$$\leq \int_\Omega [f(x, \alpha_1) + g(x, \alpha_1)] v dx$$

$$= \int_\Omega F(x, \alpha_1; \alpha_1, \beta_1) v dx$$

for all $v \in H_0^1(\Omega)$, $v \geq 0$ a.e. Similarly, for all $v \in H_0^1(\Omega)$, $v \geq 0$ a.e.,

$$B[\beta_1, v] = \int_\Omega G(x, \beta_1; \alpha_0, \beta_0) v dx$$

$$\geq \int_\Omega [f(x, \beta_1) + f_u(x, \beta_1)(\beta_0 - \beta_1)$$

$$+ f_u(x, \alpha_0)(\beta_1 - \beta_0) + g(x, \beta_1)] v dx$$

$$= \int_\Omega [f(x, \beta_1) + (-f_u(x, \beta_1) + f_u(x, \alpha_0))(\beta_1 - \beta_0) \tag{39}$$

$$+ g(x, \beta_1)] v dx$$

$$\geq \int_\Omega [f(x, \beta_1) + g(x, \beta_1)] v dx$$

$$= \int_\Omega F(x, \beta_1; \alpha_1, \beta_1) v dx,$$

because of the fact that $f_u(x, u)$ is nondecreasing in u and $\alpha_0 \leq \beta_0$. The inequalities (38) and (39) imply that α_1 is a lower solution and β_1 is an upper solution of (15). Since $f + g$ is Lipschitz continuous in $\alpha_0 \leq u \leq \beta_0$, it follows from Theorem 7 that $\alpha_1 \leq \beta_1$ in Ω proving (34).

We shall next prove that if

$$\alpha_{k+1} \leq \alpha_k \leq \beta_k \leq \beta_{k+1} \text{ a.e. in } \Omega, \tag{40}$$

for some $k > 1$, then it follows that

$$\alpha_k \leq \alpha_{k+1} \leq \beta_{k+1} \leq \beta_k \text{ a.e. in } \Omega. \tag{41}$$

Since α_k satisfies for $v \in H_0^1(\Omega)$, $v \geq 0$ a.e.,

$$B[\alpha_k, v] = \int_\Omega F(x, \alpha_k; \alpha_{k-1}, \beta_{k-1}) v dx$$

by utilizing the arguments employed to obtain (38) and (39), we can show because of (40) that

$$B[\alpha_k, v] \leq \int_\Omega F(x, \alpha_k; \alpha_k, \beta_k) v dx \tag{42}$$

and

$$B[\beta_k, v] \geq \int_\Omega G(x, \beta_k; \alpha_k, \beta_k) v dx, \tag{43}$$

for each $v \in H_0^1(\Omega)$, $v \geq 0$ a.e. Furthermore, $\alpha_{k+1}, \beta_{k+1}$ satisfy (30) and (29), respectively. Hence we conclude from Corollary 8 that

$$\alpha_k \leq \alpha_{k+1} \quad \text{and} \quad \beta_{k+1} \leq \beta_k \quad \text{a.e. in } \Omega.$$

Next we show that

$$\alpha_{k+1} \leq \beta_k \quad \text{and} \quad \alpha_k \leq \beta_{k+1}.$$

We find using (35) and (36) that

$$
\begin{aligned}
B[\alpha_{k+1}, v] &= \int_\Omega F(x, \alpha_{k+1}; \alpha_k, \beta_k) v \, dx \\
&\leq \int_\Omega [f(x, \beta_k) - f_u(x, \alpha_k)(\beta_k - \alpha_k) + g_u(x, \beta_k)(\beta_k - \alpha_k) \\
&\qquad + g(x, \beta_k) + f_u(x, \alpha_k)(\alpha_{k+1} - \alpha_k) \\
&\qquad + g_u(x, \beta_k)(\alpha_{k+1} - \alpha_k)] v \, dx \\
&\leq \int_\Omega [f(x, \beta_k) + g(x, \beta_k) \\
&\qquad + f_u(x, \alpha_k)(-\beta_k + \alpha_k + \alpha_{k-1} - \alpha_k) \\
&\qquad + g_u(x, \beta_k)(\alpha_k - \beta_k + \alpha_{k+1} - \alpha_k)] v \, dx \\
&= \int_\Omega G(x, \alpha_{k+1}; \alpha_k, \beta_k) v \, dx.
\end{aligned}
$$

This together with (43) gives by Corollary 8,

$$\alpha_{k+1} \leq \beta_k.$$

Similarly, we can show that

$$\alpha_k \leq \beta_{k+1},$$

using (42) and for each $v \in H_0^1(\Omega)$, $v \geq 0$ a.e.

$$
\begin{aligned}
B[\beta_{k+1}, v] &= \int_\Omega G(x, \beta_{k+1}; \alpha_k, \beta_k) v \, dx \\
&\geq \int_\Omega [f(x, \alpha_k) + f_u(x, \alpha_k)(\beta_k - \alpha_k) + g(x, \alpha_k) + f_u(x, \beta_k)(\beta_k - \alpha_k) \\
&\qquad + f_u(x, \alpha_k)(\beta_{k+1} - \beta_k) + g_u(x, \beta_k)(\beta_{k+1} - \beta_k)] v \, dx \\
&= \int_\Omega [f(x, \alpha_k) + g(x, \alpha_k) + f_u(x, \alpha_k)(\beta_k - \alpha_k + \beta_{k+1} - \beta_k) \\
&\qquad + g_u(x, \beta_k)(\beta_k - \alpha_k + \beta_{k+1} - \beta_k)] v \, dx \\
&= \int_\Omega F(x, \beta_{k+1}; \alpha_k, \beta_k) v \, dx.
\end{aligned}
$$

Finally, to prove

$$\alpha_{k+1} \leq \beta_{k+1} \quad \text{a.e. in } \Omega,$$

we need to show that α_{k+1} and β_{k+1} satisfy

$$B[\alpha_{k+1}, v] \leq \int_\Omega F(x, \alpha_{k+1}; \alpha_k, \beta_k) v \, dx$$

and

$$B[\beta_{k+1}, v] \geq \int_{\Omega} G(x, \beta_{k+1}; \alpha_k, \beta_k)v dx,$$

for each $v \in H_0^1(\Omega)$, $v \geq 0$ a.e. This precisely employs the same arguments as we have utilized in proving (38) and (39) replacing $\alpha_0, \beta_0, \alpha_a, \beta_1$ by $\alpha_k, \beta_k, \alpha_{k+1}, \beta_{k+1}$ and using (35) and (36) as well as monotone character of f_u and g_u. Consequently, we get from Theorem 7 that

$$\alpha_{k+1} \leq \beta_{k+1},$$

proving 41). Thus by induction, it follows that (33) is true for each $k = 1, 2, \ldots$.

(b) Convergence of $\{\alpha_k\}, \{\beta_k\}$ to the unique solution of (15). By the monotone character of the iterates $\{\alpha_k\}, \{\beta_k\}$, according to (33) there exist pointwise limits

$$\rho(x) = \lim_{k \to \infty} \alpha_k(x), \quad r(x) = \lim_{k \to \infty} \beta_k(x), \quad \text{a.e. in } \Omega.$$

Moreover, since

$$\alpha_0 \leq \alpha_k \leq \beta_k \leq \beta_0 \quad \text{a.e. in } \Omega,$$

it follows by Lebesgue's dominated convergence theorem that

$$\alpha_k \to \rho \quad \text{and} \quad \beta_k \to r \quad \text{in } L^2(\Omega).$$

We note that α_k satisfies for each $v \in H_0^1(\Omega)$, $v \geq 0$ a.e. in Ω,

$$\int_{\Omega} \left[\sum_{i,j}^{n} a_{ij}(x)\alpha_{k,x_i} v_{x_j} + c_0(x)\alpha_k v \right] dx$$

$$= \int_{\Omega} [f(x, \alpha_{k-1}) - f_u(x, \alpha_{k-1})\alpha_{k-1} + g(x, \alpha_{k-1}) - g_u(x, \beta_{k-1})\alpha_{k-1}]v dx,$$

where

$$c_0(x) = c(x) - f_u(x, \alpha_{k-1}) - g_u(x, \beta_{k-1}).$$

We now use the ellipticity condition and $(B3)$ with $v = \alpha_k$ to get

$$\int_{\Omega} [\theta \mid \alpha_{k,x} \mid^2 + N \mid \alpha_k \mid^2] dx \leq \int_{\Omega} [f(x, \alpha_{k-1}) - f_u(x, \alpha_{k-1})\alpha_{k-1}$$
$$+ g(x, \alpha_{k-1}) - g_u(x, \beta_{k-1})\alpha_{k-1}]\alpha_k dx.$$

We then get, since by $(B3)$ the integrand on the right-hand side belongs to $L^2(\Omega)$, in view of the fact that $\alpha_{k-1}, \beta_{k-1} \in H_0^1(\Omega)$, and the estimate

$$\sup_k \| \alpha_k \|_{H_0^1(\Omega)} < \infty.$$

A similar argument implies that

$$\sup_k \| \beta_k \|_{H_0^1(\Omega)} < \infty.$$

Hence there exist subsequences $\{\alpha_{k_j}\}, \{\beta_{k_j}\}$ which converge weakly in $H_0^1(\Omega)$ to $\rho, r \in H_0^1(\Omega)$, respectively. To verify that ρ, r are weak solutions of (15), we fix

$v \in H_0^1(\Omega)$, $v \geq 0$ a.e. and find that α_{k+1}, β_{k+1} satisfy (31) and (32) with F and G defined by (27) and (28). Taking limits as $k \to \infty$, we obtain

$$B[\rho, v] = \int_\Omega [f(x, \rho) + g(x, \rho)] v \, dx,$$

and

$$B[r, v] = \int_\Omega [f(x, r) + g(x, r)] v \, dx,$$

showing that ρ, r are weak solutions of (15).

To prove that $\rho = r = u$ is the unique solution of (15), it is enough to prove that $r \leq \rho$ a.e. in Ω, since we know that $\rho \leq r$ a.e. in Ω. Taking $\alpha = r$, $\beta = \rho$, and applying Theorem 7, we find that $r \leq \rho$ a.e. in Ω proving the claim.

(c) **Quadratic convergence of $\{\alpha_k\}$, $\{\beta_k\}$.** To prove the quadratic convergence of sequences $\{\alpha_k\}$, $\{\beta_k\}$ to the unique solution u respectively, we set

$$p_{k-1} = u - \alpha_{k+1}, \qquad q_{k+1} = \beta_{k+1} - u,$$

so that $p_{k+1}(x) = 0$ on Ω and $q_{k+1}(x) = 0$ on Ω. We then have for $v \in H_0^1(\Omega)$, $v \geq 0$ a.e. in Ω, using the fact that f_u is nondecreasing in u and g_u is nonincreasing in u,

$$
\begin{aligned}
B[p_{k+1}, v] &= \int_\Omega [f(x, u) + g(x, u) - f(x, \alpha_k) - f_u(x, \alpha_k)(\alpha_{k+1} - \alpha_k) \\
&\qquad - g(x, \alpha_k) - g_u(x, \beta_k)(\alpha_{k+1} - \alpha_k)] v \, dx \\
&\leq \int_\Omega [(f_u(x, u) - f_u(x, \alpha_k)) p_k - (g_u(x, \beta_k) - g_u(x, \alpha_k)) p_k \\
&\qquad + (f_u(x, \alpha_k) + g_u(x, \beta_k) p_{k+1}] v \, dx \\
&= \int_\Omega [f_{uu}(x, \xi) p_k^2 - g_{uu}(x, \sigma)(\beta_k - \alpha_k) \\
&\qquad + (f_u(x, \alpha_k) + g_u(x, \beta_k)) p_{k+1}] v \, dx,
\end{aligned}
$$

where

$$\alpha_k \leq \xi \leq u, \qquad \alpha_k \leq \sigma \leq \beta_k.$$

But

$$
\begin{aligned}
-g_{uu}(x, \sigma)(\beta_k - \alpha_k) p_k &\leq N_2(q_k + p_k) p_k \\
&\leq N_2(p_k^2 + p_k q_k) \\
&\leq \frac{3}{2} N_2 p_k^2 + \frac{1}{2} N_2 q_k^2,
\end{aligned}
$$

where

$$| g_{uu}(x, u) |_{L^\infty(\Omega)} \leq N_2.$$

Thus we get

$$\int_{\Omega} \left[\sum_{i,j=1}^{n} a_{ij}(x)(p_{k+1})_{x_i} v_{x_j} + c_0(x)p_{k+1}v \right] dx$$

$$\le \int_{\Omega} [(N_1 + \frac{3}{2}N_2)p_k^2 + \frac{1}{2}N_2 q_k^2]v dx,$$

where

$$| f_{uu}(x,u) |_{L^{\infty}(\Omega)} \le N_1.$$

Taking

$$v = p_{k+1}$$

and using the ellipticity condition, we arrive at

$$\int_{\Omega} [\theta \mid (p_{k+1})_x \mid^2 + N \mid p_{k+1} \mid^2]dx \le N_0 \int_{\Omega} (\mid p_k \mid^2 + \mid q_k \mid^2)p_{k+1} dx,$$

where

$$N_0 = N_1 + \frac{3}{2}N_2.$$

Let

$$\theta_0 = \min(\theta, N).$$

Then

$$\theta_0 \parallel p_{k+1} \parallel^2_{H_0^1(\Omega)} \le N_0 \left(\frac{2}{\theta_0} \right)^{1/2} [\parallel p_k^2 \parallel_{L^2(\Omega)}$$

$$+ \parallel q_k^2 \parallel_{L^2(\Omega)}] \left(\frac{\theta_0}{2} \right)^{1/2} \parallel p_{k+1} \parallel_{L^2(\Omega)}$$

$$\le \frac{1}{2} \left[\frac{2N_0^2}{\theta_0} \parallel p_k^2 \parallel^2_{L^2(\Omega)} + \frac{\theta_0}{2} \parallel p_{k+1} \parallel^2_{L^2(\Omega)} \right]$$

$$+ \frac{1}{2} \left[\frac{2N_0^2}{\theta_0} \parallel q_k^2 \parallel^2_{L^2(\Omega)} + \frac{\theta_0}{2} \parallel p_{k+1} \parallel^2_{L^2(\Omega)} \right]$$

$$= \frac{N_0^2}{\theta_0}[\parallel p_k^2 \parallel^2_{L^2(\Omega)} + \parallel q_k^2 \parallel^2_{L^2(\Omega)}] + \frac{\theta_0}{2} \parallel p_{k+1} \parallel^2_{L^2(\Omega)} .$$

We then get

$$\frac{\theta_0}{2} \parallel p_{k+1} \parallel^2_{H_0^1(\Omega)} \le \frac{N_0^2}{\theta_0}[\parallel p_k^2 \parallel^2_{H_0^1(\Omega)} + \parallel q_k^2 \parallel^2_{H_0^1(\Omega)}]$$

or, equivalently,

$$\parallel p_{k+1} \parallel^2_{H_0^1(\Omega)} \le \frac{2N_0^2}{\theta_0^2}[\parallel p_k^2 \parallel^2_{H_0^1(\Omega)} + \parallel q_k^2 \parallel^2_{H_0^1(\Omega)}].$$

One can get a similar estimate for v_{k+1}. We omit the details. $\qquad\square$

Several remarks are now in order.

Remarks 14.

(1) If $g(x, u) \equiv 0$ in (15), we get a result when $f(x, u)$ is convex.

(2) If $f(x, u) \equiv 0$ in (15), we obtain a result when $g(x, u)$ is concave.

(3) Consider the case when $g(x, u) \equiv 0$ and $f(x, u)$ is not convex. Assume that

$$\tilde{f}(x, u) = f(x, u) + G(x, u)$$

is convex, where $G(x, u)$ is convex. Then it would be a special case of Theorem 13 since we can write (15) with f replaced by $\tilde{f}$ and $\tilde{g} = -G(x, u)$ and note that conditions of Theorem 13 are satisfied. Consequently, we obtain the same conclusion even when f is not convex.

(4) A dual situation of (3) arises when $f(x, u) \equiv 0$ and $g(x, u)$ is not concave but

$$\tilde{g}(x, u) = g(x, u) + F(x, u)$$

is concave with $F(x, u)$ concave. We then rewrite (15) with f, g replaced by

$$\tilde{f} = -F, \quad \tilde{g} = g + F,$$

respectively. Since the assumptions of Theorem 13 hold, we get the same conclusion even when $g(x, u)$ is not concave.

(5) Suppose that $g(x, u)$ is concave and $f(x, u)$ is not convex but

$$\tilde{f}(x, u) = f(x, u) + G(x, u)$$

is convex with $G(x, u)$ convex. Then Theorem 13 gives the same conclusion by replacing f, g by $\tilde{f}$ and $\tilde{g} = g - G$.

(6) A dual case of (5) is also valid, that is, $f(x, u)$ is convex but $g(x, u)$ is not concave and

$$\tilde{g}(x, u) = g(x, u) + F(x, u)$$

is concave with $F(x, u)$ concave. We then consider Theorem 13 with f, g replaced by $\tilde{f} = f - F$ and $\tilde{g}$ to obtain the same conclusion of Theorem 13.

(7) Suppose now that both f, g do not enjoy the required convex and concave properties in Theorem 13 for (15). However, $f + G$, $g + F$ are convex, concave respectively with F, G being convex and concave. Then we replace in Theorem 13, f, g by

$$\tilde{f} = f + G - F, \quad \tilde{g} = g + F - G,$$

so that the required condition of Theorem 13 are satisfied and Theorem 13 yields the same result even in this case.

Such results can be extended to various nonlinear problems with some technical difficulty, which can be of interest to applied scientists. Extending to multivalued differential inclusions poses some challenge.

References

1. Bellman, R., *Methods of Nonlinear Analysis, Vol. II*, Academic Press, New York, 1973.

2. Bellman, R., and Kalaba, R., *Quasilinearization and Nonlinear Boundary Value Problems*, American Elsevier, New York, 1965.
3. Carl, S., A combined variational-monotone iterative method for elliptic boundary value problems with discontinuous nonlinearity, *Appl. Anal.* **43** (1992), 21–45.
4. Carl, S., and Heikkilä, S., *Nonlinear Differential Equations in Ordered Spaces*, Chapman and Hall/CRC, Boca Raton, 2000.
5. Deimling, K., On approximate solutions of differential equations in Banach spaces, *Math. Ann.* **212** (1974/75), 79–88.
6. Heikkilä, S., and Lakshmikantham, V., *Monotone Iterative Technique for Discontinuous Nonlinear Differential Equations*, Marcel Dekker Inc., New York, 1994.
7. Hartman, Ph., *Ordinary Differential Equations*, John Wiley, New York, 1964.
8. Köksal, S., and Lakshmikantham, V., *Monotone Flows, Rapid Convergence for Nonlinear Partial Differential Equations*, Taylor & Francis, England, 2003.
9. Ladde, G.S., Lakshmikantham, V., and Vatsala, A.S., *Monotone Iterative Technique for Nonlinear Differential Equations*, Pitman, Boston, 1985.
10. Lakshmikantham, V., On Kamke's function in the uniqueness theorem of ordinary differential equations, *Proc. Nat. Acad. Sci. India* **34** (1964), 11–14.
11. Lakshmikantham, V., and Leela, S.G., *Differential and Integral Inequalities, Vol. I*, Academic Press, New York, 1969.
12. Lakshmikantham, V., and Mitchell, R., On the redundancy of monotony assumption, *J. Math and Phys. Sci.* **10** (1976), 219–230.
13. Lakshmikantham, V., and Vatsala, A.S., *Generalized Quasilinearization for Nonlinear Problems*, Kluwer Academic Publishers, Dordrecht, 1998.
14. Pao, C.V., *Nonlinear Parabolic and Elliptic Equations*, Plenum Press, New York, 1992.

Ten Mathematical Essays on Approximation in Analysis and Topology
J. Ferrera, J. López-Gómez, F. R. Ruiz del Portal, Editors

Spectral Theory and Nonlinear Analysis

J. López-Gómez

Matemática Aplicada, Universidad Complutense de Madrid,
28040-Madrid, Spain

Abstract

The topological degree is a generalized counter of the number of zeros of a nonlinear map in an open set. Thus, it is one of the main legacies of 20th century mathematics, as finding out zeros of maps and counting them has been one of the main tasks of mathematicians of all times. Nonlinear analysis is the part of mathematics analyzing non-linear maps and equations. A general principle in nonlinear analysis is that any change of parity in the topological degree as some parameter crosses some critical value entails the existence of a global component of the solution set. It turns out that algebraic multiplicities in the context of spectral theory for linear operators provide us with a number of finite algorithms to compute any change of the local topological degree. Consequently, algebraic multiplicities are algebraic invariants providing us with a lot of topological information about the global structure of the components of the solution sets of nonlinear equations. In this review we give a rather complete summary of recent results in these areas, crossing edges from spectral theory to nonlinear analysis. The number of applications of the abstract theory revisited here might be certainly unlimited.

Key words: Nonlinear analysis, nonlinear eigenvalue, topological degree, algebraic multiplicities, spectral theory, bifurcation theory

1. Introduction

This paper surveys several central problems in spectral theory and nonlinear functional analysis related to the structure of the set of fixed points of a general class of compact

---

[1] This work has been supported by the Spanish Ministry of Education and Science under grant REN2003-00707.

nonlinear operators defined in a real Banach space for which a smooth curve of solutions is assumed to be known; by a change of variable, one can assume that the curve is a straight line. We work with real spaces and real curves since these are the most common situations arising in the applications of the abstract theory. Precisely, we will focus our attention into the following two issues:

1. Characterizing the points of the curve from which a continuum of solutions emanates for any perturbing nonlinearity
2. Analyzing the topological structure of the component of the solution set emanating from any of those *nonlinear eigenvalues*

Although both issues can be solved by means of the Brouwer–Leray–Schauder degree, we will study how several of the most important generalized algebraic multiplicities introduced independently in the literature provide us with optimal analytic-algebraic invariants for calculating any change of the local topological degree —or fixed point index— as one moves along the known curve of solutions. So, viewing all these algebraic multiplicities as finite algorithms for detecting any change of index. The topological degree is a generalized counter of the number of fixed points that a compact nonlinear operator possesses in a certain bounded open set; one of the most powerful analytical-topological tools developed during 20th century.

In Section 1 we fix our functional setting and introduce some of the main questions to be addressed later. As we know that mathematics are suffering an extremely high fragmentation, we will briefly introduce the axiomatic and fundamental properties of the topological degree, so that any potential reader with no previous background about degree theory can easily read this review. In Section 3 we give the topological characterization of nonlinear eigenvalues. In Section 4 we introduce and establish some very deep connections between some of the most important generalized algebraic multiplicities available in the literature. Although some of them were introduced to solve a number of problems originated in very separated fields, all of them have shown to be optimal algebraic-analytic invariants for ascertaining any change of sign of the local topological degree. The relevance of the algebraic multiplicities constructions coming from the fact that in some circumstances they provide us with very powerful, eventually optimal, constructive methods for solving nonlinear equations and systems, which cannot be carried out by merely using purely topological techniques like obstruction theory. The associated resolution schemes should share a certain amount of light in singularity theory and algebraic geometry. Finally, in Section 5 we collect some of the most important results available in global bifurcation theory, where purely topological methods have shown its greatest versatility.

Experts should be indulgent with the general tone and flavor of this review; its main objective being to convince specialists from a number of separated areas of the relevance of the general theory and techniques treated here. Mathematical advances are mainly promoted by establishing hidden bisociations connecting a priori separated fields.

2. General assumptions and basic concepts

Throughout this paper we consider a real Banach space, U, a nonlinear abstract operator

$$\mathfrak{F} : \mathbb{R} \times U \to U$$

of the form

$$\mathfrak{F}(\lambda, u) := \mathfrak{L}(\lambda)u - \mathfrak{N}(\lambda, u),$$

and the associated nonlinear abstract equation

$$\mathfrak{F}(\lambda, u) := 0, \tag{1}$$

where the following conditions are assumed to be satisfied:

Assumption L: $\mathfrak{L}(\lambda) \in \mathcal{L}(U)$ for each $\lambda \in \mathbb{R}$, where $\mathcal{L}(U)$ stands for the space of linear continuous operators of U, and the family of operators

$$\mathfrak{K}(\lambda) := I - \mathfrak{L}(\lambda), \qquad \lambda \in \mathbb{R}, \tag{2}$$

is compact (the closure of the image of any bounded set is a compact subset of U), where I stands for the identity of U. Moreover, the map

$$\mathbb{R} \longrightarrow \mathcal{L}(U)$$

$$\lambda \mapsto \mathfrak{L}(\lambda)$$

is of class C^r, for some $r \geq 0$.

Assumption N: $\mathfrak{N} : \mathbb{R} \times U \to U$ is a compact operator of class C^q, $q \geq 0$, such that

$$\lim_{u \to 0} \frac{\mathfrak{N}(\lambda, u)}{\|u\|} = 0 \quad \text{uniformly on compact subsets of } \lambda \in \mathbb{R}.$$

Thus, Equation (1) reduces to a fix point equation for a compact operator,

$$u = \mathfrak{K}(\lambda)u + \mathfrak{N}(\lambda, u). \tag{3}$$

Assumption N entails that

$$(\lambda, u) = (\lambda, 0)$$

is a solution of (1) for each $\lambda \in \mathbb{R}$ and that $\mathfrak{N}(\lambda, u)$ is a nonlinear perturbation for the linear equation

$$\mathfrak{L}(\lambda)u = 0.$$

Actually, if $r \geq 1$ and $q \geq 1$, then

$$\mathfrak{N}(\lambda, 0) = 0 \quad \text{and} \quad D_u \mathfrak{N}(\lambda, 0) = 0 \quad \text{for each} \quad \lambda \in \mathbb{R},$$

where D_u stands for Fréchet u–differentiation, and, hence,

$$\mathfrak{F}(\lambda, 0) = 0 \quad \text{and} \quad D_u \mathfrak{F}(\lambda, 0) = \mathfrak{L}(\lambda).$$

In order to emphasize that it is already known, the state $(\lambda, u) = (\lambda, 0)$ will be called the *trivial solution* of Equation (1).

Under Assumption L, $\mathfrak{L}(\lambda)$ is a Fredhom operator of index zero for each $\lambda \in \mathbb{R}$, i.e., $R[\mathfrak{L}(\lambda)]$ is a closed subspace of U and

$$\dim N[\mathfrak{L}(\lambda)] = \operatorname{codim} R[\mathfrak{L}(\lambda)] < \infty,$$

since it is a compact perturbation of the identity. Subsequently, for any $T \in \mathcal{L}(U)$, $N[T]$ and $R[T]$ will stand for the null space (or kernel) and the range (or image) of T, respectively. Thanks to the open mapping theorem, $\mathfrak{L}(\lambda)$ is an isomorphism if

$$\dim N[\mathfrak{L}(\lambda)] = 0.$$

Let Σ denote the *spectrum of the family* $\mathfrak{L}(\lambda)$ —its elements are called *eigenvalues*—

$$\Sigma := \{ \lambda \in \mathbb{R} \ : \ \dim N[\mathfrak{L}(\lambda)] \geq 1 \}, \tag{4}$$

and $\mathfrak{S}$ the set of *non-trivial solutions* of Equation (1),

$$\mathfrak{S} := \left[\mathfrak{F}^{-1}(0) \cap (\mathbb{R} \times (U \setminus \{0\})) \right] \cup \{ (\lambda, 0) \ : \ \lambda \in \Sigma \}. \tag{5}$$

The following concept plays a crucial role in nonlinear analysis.

Definition 1. Given $\lambda_0 \in \mathbb{R}$, $(\lambda_0, 0)$ is said to be a **bifurcation point** of Equation (1) from the curve of trivial solutions $(\lambda, 0)$ if there exists a sequence

$$(\lambda_n, u_n) \in \mathfrak{F}^{-1}(0) \cap (\mathbb{R} \times (U \setminus \{0\})), \qquad n \geq 1,$$

such that

$$\lim_{n \to \infty} (\lambda_n, u_n) = (\lambda_0, 0).$$

The following result collects some well known general properties of Σ and $\mathfrak{S}$ (cf. [32, Section 6.1]).

Proposition 2. *For each $\mu \in \mathbb{R} \setminus \Sigma$, there exists an open neighborhood of $(\mu, 0)$ in $\mathbb{R} \times U$, say B, such that $B \cap \mathfrak{S} = \varnothing$. Thus, Σ is a closed subset of $\mathbb{R}$ and $(\lambda_0, 0)$ cannot be a bifurcation point of Equation (1) from $(\lambda, u) = (\lambda_0, 0)$ if $\lambda_0 \notin \Sigma$. Moreover, $\mathfrak{S}$ is a closed subset of $\mathbb{R} \times U$.*

As a result, if $(\lambda_0, 0)$ is a bifurcation point from $(\lambda, 0)$, then

$$N := \dim N[\mathfrak{L}(\lambda_0)] \geq 1. \tag{6}$$

However, very simple algebraic examples show that (6) might not be sufficient for bifurcation. Indeed, for any $\lambda_0 \in \mathbb{R}$, the system

$$\begin{cases} (\lambda - \lambda_0)u_1 + u_2^3 = 0, \\ (\lambda - \lambda_0)u_2 - u_1^3 = 0, \end{cases} \tag{7}$$

fits into our abstract setting by choosing

$$U = \mathbb{R}^2, \qquad u = (u_1, u_2),$$

and

$$\mathfrak{L}(\lambda) = (\lambda - \lambda_0)I_{\mathbb{R}^2}, \qquad \mathfrak{N}(\lambda, u) = \begin{pmatrix} -u_2^3 \\ u_1^3 \end{pmatrix}.$$

Although $\mathfrak{L}(\lambda_0) = 0$, and, hence, (6) holds true with $N = 2$, multiplying the first equation of (7) by u_2, the second one by u_1, and subtracting the resulting identities, gives

$$u_1^4 + u_2^4 = 0$$

and, therefore,

$$u_1 = u_2 = 0.$$

Consequently, $(\lambda_0, 0)$ is not a bifurcation point from $(\lambda, 0)$. However this example shows that (6) is not sufficient for bifurcation, if we change the nonlinearity, then $(\lambda_0, 0)$ might be a bifurcation point for the new equation from $(\lambda, 0)$. Indeed, if instead of (7), we consider the following system

$$\begin{cases} (\lambda - \lambda_0)u_1 - u_2^3 = 0, \\ (\lambda - \lambda_0)u_2 - u_1^3 = 0, \end{cases} \tag{8}$$

then, for each $\lambda > \lambda_0$,

$$(\lambda, u_1, u_2) := \left(\lambda, \sqrt{\lambda - \lambda_0}, \sqrt{\lambda - \lambda_0}\right)$$

provides us with a nontrivial solution of (8), and therefore $(\lambda_0, 0)$ is a bifurcation point from $(\lambda, 0)$. These examples actually show that λ_0 is not a *nonlinear eigenvalue* of the family

$$\mathfrak{L}(\lambda) = (\lambda - \lambda_0)I_{\mathbb{R}^2}.$$

Definition 3. λ_0 is said to be a **nonlinear eigenvalue** of $\mathfrak{L}(\lambda)$ if $(\lambda_0, 0)$ is a bifurcation point of (1) from $(\lambda, 0)$ for any $\mathfrak{N}(\lambda, u)$ satisfying Assumption N.

In other words, λ_0 is a nonlinear eigenvalue of $\mathfrak{L}(\lambda)$ if the fact that bifurcation occurs is exclusively based on the linear part of (1). It should be noted that even in the case when λ_0 is a nonlinear eigenvalue of $\mathfrak{L}(\lambda)$, the local structure of the set of solutions of (1) around $(\lambda_0, 0)$ will depend on the nature of the nonlinearity $\mathfrak{N}(\lambda, u)$. Actually, the problem of the algebraic classification of all possible solution varieties of (1) around $(\lambda_0, 0)$, according to the nature of the nonlinearity $\mathfrak{N}(\lambda, u)$, is one of the central problems in singularity theory and real algebraic geometry (e.g., [21]), but that analysis is outside the scope of this review.

3. A brief introduction to the topological degree

The topological degree is a generalized counter of the number of zeros that a compact perturbation of the identity has within an open and bounded set. The faster way to introduce it is by means of the **uniqueness theorem** of H. Amann and S. A. Weiss [1].

Theorem 4. *Let $\mathcal{O}$ be the set of bounded open subsets of U, and, for each $\Omega \in \mathcal{O}$, denote by $K(\bar{\Omega})$ the set of compact perturbations of the identity on $\bar{\Omega}$ with the topology of the uniform convergence. Consider the set of pairs*

$$\mathcal{P} := \{ (f, \Omega) \in K(\bar{\Omega}) \times \mathcal{O} \; : \; f^{-1}(0) \cap \partial\Omega = \varnothing \}.$$

Then, there exists a unique map

$$\mathrm{Deg} \; : \; \mathcal{P} \longrightarrow \mathbb{Z}$$

satisfying the following properties:

D1 $\mathrm{Deg}\,(I, \Omega) = 1$ *if* $0 \in \Omega$.

D2 *If $\Omega \in \mathcal{O}$, Ω_1 and Ω_2 are two disjoint open subsets of Ω and $f \in K(\bar{\Omega})$ with*

$$f^{-1}(0) \cap [\bar{\Omega} \setminus (\Omega_1 \cup \Omega_2)] = \varnothing,$$

   then,

$$\mathrm{Deg}\,(f, \Omega) = \mathrm{Deg}\,(f, \Omega_1) + \mathrm{Deg}\,(f, \Omega_2).$$

D3 *If $\Omega \in \mathcal{O}$, $h : [0, 1] \to K(\bar{\Omega})$ is continuous, and*

$$(h(t), \Omega) \in \mathcal{P} \quad \text{for each} \quad t \in [0, 1],$$

   then

$$\mathrm{Deg}\,(h(t), \Omega) = \mathrm{Deg}\,(h(0), \Omega) \qquad \text{for each} \quad t \in [0, 1].$$

The integer $\mathrm{Deg}\,(f, \Omega)$ is called the topological degree of f in Ω.

Property **D1** is usually refereed to as the *normalization* property; it establishes that the identity map has one zero in any bounded open set containing the origin. Property **D2** entails three basic properties that any counter of zeros must satisfy. Indeed, by choosing

$$\Omega = \Omega_1 = \Omega_2 = \varnothing,$$

gives

$$\mathrm{Deg}\,(f, \varnothing) = 0. \tag{9}$$

Secondly, by choosing

$$\Omega = \Omega_1 \cup \Omega_2,$$

shows that

$$\mathrm{Deg}\,(f, \Omega_1 \cup \Omega_2) = \mathrm{Deg}\,(f, \Omega_1) + \mathrm{Deg}\,(f, \Omega_2), \tag{10}$$

so establishing the *additivity* of the counter. Thirdly, by choosing $\Omega_2 = \varnothing$, provides us with the identity

$$\mathrm{Deg}\,(f, \Omega) = \mathrm{Deg}\,(f, \Omega_1), \tag{11}$$

so establishing the *excision* property of the counter. As an easy consequence, from these properties it is apparent that

$$f^{-1}(0) \cap \bar{\Omega} = \varnothing \quad \Longrightarrow \quad \mathrm{Deg}\,(f, \Omega) = 0. \tag{12}$$

In other words, f must have a zero in Ω if

$$\mathrm{Deg}\,(f, \Omega) \neq 0.$$

Property **D3** establishes the *homotopy invariance* of the counter, i.e., the fact that the number of zeros remains unchanged if the map is continuously deformed in such a way that it does not loose nor win zeros through the boundary. In other words, it establishes the continuity of the counter in the quotient space defined by the relation *being homotopic*; as it is an integer, it must be constant.

In the special case when $U = \mathbb{C}$, γ is a Jordan curve, Ω_γ is the component enclosed by γ and f is an holomorphic function in $\bar{\Omega}_\gamma$ such that

$$f^{-1}(0) \cap \mathrm{Tray}\,\gamma = \varnothing,$$

then, $\mathrm{Deg}\,(f, \Omega)$, equals the total number of zeros of f in Ω_γ counted according to their orders. Therefore, the topological degree is indeed a generalized counter of the number of zeros that f possesses within Ω (cf. [33, Chapter 11] for an elementary introduction at undergraduate level).

The first construction of the topological degree in $\mathbb{R}^N$ was carried over by L. E. J. Brouwer [5] to prove his celebrated fixed point theorem. Brouwer's degree was later extended by J. Leray and J. Schauder [29] to cover the infinite dimensional setting dealt with here. Naturally, it yields to the simplest proof of Schauder's fixed point theorem [51] (cf. A. N. Tychonoff [55]).

Now, we will describe the standard procedure to calculate the topological degree in applications. Subsequently, we will consider the subset of $\mathcal{P}$ consisting of all pairs (f, Ω) with $f \in C^1(\bar{\Omega})$, such that $f^{-1}(0) \cap \Omega$ is finite, possibly empty, and

$$Df(u) \quad \text{is an isomorphism for each} \quad u \in f^{-1}(0) \cap \Omega.$$

Any of those pairs will be called *regular*. The set of regular pairs will be denoted by $\mathcal{R}$. The next result can be obtained from the infinite dimensional version of S. Smale [54] of Sard's theorem [49].

Theorem 5. *For any $(f, \Omega) \in \mathcal{P}$ there exists a continuous map*

$$h \; : \; [0, 1] \to K(\bar{\Omega})$$

such that

$$h(0) = f, \qquad (h(1), \Omega) \in \mathcal{R}, \qquad (h(t), \Omega) \in \mathcal{P} \qquad \forall\, t \in [0, 1].$$

In other words, any compact perturbation of the identity is homotopic, within $\mathcal{P}$, to a regular function.

Consequently, thanks to the homotopy invariance of the degree,

$$\mathrm{Deg}\,(f, \Omega) = \mathrm{Deg}\,(h(1), \Omega).$$

Moreover, it is independent of the *smoothing homotopy*. Thus, calculating the topological degree is reduced to calculate the degree of a regular pair $(f, \Omega) \in \mathcal{R}$. In calculating the degree of a regular pair the following scheme should be followed. Suppose

$$(f, \Omega) \in \mathcal{R} \quad \text{and} \quad f^{-1}(0) \cap \Omega = \{u_1, ..., u_N\}. \tag{13}$$

Note that, thanks to (12),

$$\text{Deg}(f, \Omega) = 0 \quad \text{if} \quad f^{-1}(0) \cap \Omega = \varnothing.$$

Due to Property **D2**, (13) implies

$$\text{Deg}(f, \Omega) = \sum_{j=1}^{N} \text{Deg}(f, B_\epsilon(u_j)) \tag{14}$$

for any sufficiently small $\epsilon > 0$ satisfying

$$B_\epsilon(u_j) \subset \Omega, \quad 1 \le j \le n,$$

and

$$\bar{B}_\epsilon(u_i) \cap \bar{B}_\epsilon(u_j) = \varnothing \quad \text{if} \quad i \ne j.$$

Actually, (14) is still valid for any $(f, \Omega) \in \mathcal{P}$ having a finite number of zeros in Ω. Throughout the rest of this paper $B_R(u)$ stands for the open ball of radius $R > 0$ centered at u. Note that, for each $1 \le j \le N$,

$$\text{Deg}(f, B_\epsilon(u_j)) \quad \text{is independent of} \quad \epsilon$$

as soon as

$$B_\epsilon(u_j) \subset \Omega \quad \text{and} \quad f^{-1}(0) \cap \bar{B}_\epsilon(u_j) = \{u_j\}.$$

More generally, if $(f, \Omega) \in \mathcal{P}$ and $u_0 \in \Omega$ is an isolated zero of f, then $\text{Deg}(f, B_\epsilon(u_0))$ is well defined and independent of $\epsilon > 0$ as soon as

$$f^{-1}(0) \cap \bar{B}_\epsilon(u_0) = \{u_0\};$$

this value being called the *index* of f at u_0, and denoted by $\text{Ind}(f, u_0)$. Using this concept, (14) can be rewritten in the form

$$\text{Deg}(f, \Omega) = \sum_{j=1}^{N} \text{Ind}(f, u_j). \tag{15}$$

As the index is a local concept and, thanks to the inverse function theorem, one can establish a local homotopy between f and its linearization at u_j, $Df(u_j)$ (by neglecting $f(u) - Df(u_j)(u - u_j)$), because $Df(u_j)$ is an isomorphism for each $1 \le j \le N$, it turns out that (15) gives rise to

$$\text{Deg}(f, \Omega) = \sum_{j=1}^{N} \text{Ind}(Df(u_j), 0). \tag{16}$$

Finally, to complete the calculation of the degree the following result should be used. It is usually refereed to as the *Leray–Schauder formula*.

Theorem 6. *Suppose K is a linear compact operator and $T := I - K$ is a linear isomorphism. Then,*

$$\mathrm{Ind}\,(T,0) = (-1)^{n(T)} \tag{17}$$

where $n(T)$ is the sum of the algebraic multiplicities of all negative eigenvalues of T. In particular, if $U = \mathbb{R}^N$, then (17) becomes into

$$\mathrm{Ind}\,(T,0) = \mathrm{sign}\,\det T$$

in any basis of $\mathbb{R}^N$.

Thanks to Theorem 6, (16) implies

$$\mathrm{Deg}\,(f,\Omega) = \sum_{j=1}^{N}(-1)^{n(Df(u_j))}. \tag{18}$$

Conversely, defining the degree of a pair $(f,\Omega) \in \mathcal{R}$ through (18) and extending it to the class $\mathcal{P}$ by means of Theorem 5, one is naturally driven to the standard analytic construction of the topological degree. In applications, to calculate the topological degree of a pair $(f,\Omega) \in \mathcal{P}$, one should follow the next steps:

- Constructing the smoothing homotopy.
- Once deformed f into a regular operator in $\bar{\Omega}$, say F, finding out the zeros of F.
- Calculating $n(DF(u_0))$ for each $u_0 \in F^{-1}(0) \cap \Omega$.

More general developments of the Leray–Schauder degree for wider classes of fixed point equations were introduced by F. E. Browder, R. D. Nussbaum and W. V. Petryshyn in the late sixties (see N. G. Lloyd [30] and K. Deimling [9]) and by J. Mawhin [42]. More recently, P. M. Fitzpatrick and J. Pejsachowitz [16] developed a degree theory for quasilinear Fredholm mappings, which has been substantially tided up in a series of very recent papers by P. Benevieri and M. Furi, [2], [3], [4]. These theories remain aside the scope of this introductory review.

4. Topological characterization of nonlinear eigenvalues

In this section, the general assumptions and notations introduced in Section 1 are maintained. The following result characterizes the nonlinear eigenvalues of $\mathfrak{L}(\lambda)$ by means of the topological degree.

Theorem 7. *Suppose $\lambda_0 \in \mathbb{R}$ is an isolated point of Σ. Then, the following assertions are true:*

1.– *If $\mathrm{Ind}\,(\mathfrak{L}(\lambda),0)$ changes as λ crosses λ_0, then λ_0 is a nonlinear eigenvalue of $\mathfrak{L}(\lambda)$.*

2.– *If $\mathfrak{L}(\lambda)$ is of class C^1 and λ_0 is a nonlinear eigenvalue of $\mathfrak{L}(\lambda)$, then*

$$\mathrm{Ind}\,(\mathfrak{L}(\lambda),0))\ \ \textit{changes as}\ \ \lambda\ \ \textit{crosses}\ \ \lambda_0.$$

Consequently, if $\mathfrak{L} \in C^1$, then $\lambda_0 \in \Sigma$ is a nonlinear eigenvalue of $\mathfrak{L}(\lambda)$ if and only if the parity of the integer number

$$n(\mathfrak{L}(\lambda)) = \sum_{\mu_\lambda \in \sigma(\mathfrak{L}(\lambda)) \cap (-\infty, 0)} m_a[\mathfrak{L}(\lambda); \mu_\lambda]$$

changes as λ crosses λ_0. We have denoted by $m_a[T; \mu]$ the algebraic multiplicity of μ as an eigenvalue of T, i.e.,

$$m_a[T; \mu] = \dim \bigcup_{j=0}^{\infty} N[(T - \mu I)^j] = \dim N[(T - \mu I)^{\nu(T, \mu)}]$$

where $\nu(T, \mu)$ is the *algebraic ascent* of μ as an eigenvalue of T. Since under our assumptions the complex, non real, eigenvalues appear by pairs, λ_0 is a nonlinear eigenvalue of $\mathfrak{L}(\lambda)$ if and only if

$$d(\lambda) := \sum_{\mu_\lambda \in \sigma(\mathfrak{L}(\lambda)) \cap [\mathrm{Re}\,(z) < 0]} m_a[\mathfrak{L}(\lambda); \mu_\lambda] = \sum_{\rho_\lambda \in \sigma(\mathfrak{K}(\lambda)) \cap [|z| > 1]} m_a[\mathfrak{K}(\lambda); \rho_\lambda]$$

changes as λ crosses λ_0 (cf. (2)). Therefore, λ_0 is a nonlinear eigenvalue of $\mathfrak{L}(\lambda)$ if and only if the parity of the dimension of the *unstable manifold* of zero as a fixed point of the compact family $\mathfrak{K}(\lambda)$ changes as λ crosses λ_0. This feature has a huge relevance from the point of view of the dynamics of the discrete dynamical system defined by the fixed point equation (3).

Basically, Part 1 of Theorem 7 goes back to M. A. Krasnosel'skii [27] and J. Ize [22], and Part 2 is attributable to J. Ize [23] and P. M. Fitzpatrick & J. Pejsachowicz [14], though the most pioneering classifications of nonlinear eigenvalues were given in terms of the parity of the generalized algebraic multiplicity of J. Esquinas and J. López-Gómez [31], [11], [12] and [10] (cf. [32] for further details), rather than in terms of the fixed point index.

Part 1 is based upon a very classical principle in bifurcation theory going back to H. Poincaré, establishing that associated with any change of index there is a bifurcation phenomenon. It actually entails that a continuum of solutions emanate from the trivial state, as a result of the *hyperbolic structure* of the set of zeros of (1). It turns out that the change of index implies the set of zeros of any regular approximation of (1) around $(\lambda_0, 0)$ to be of hyperbolic type and, therefore, it must contain something else than the trivial state (cf. the beautiful geometric proof of this nonlinear analysis principle given in [32, Chapter 3], which is based upon some ideas coming from S. N. Chow and J. K. Hale [6] and the references therein). Although they cannot reach the generality of the further topological characterizations found in J. Ize [23] and P. M. Fitzpatrick & J. Pejsachowicz [14], the pioneering proofs of Part 2 given by J. Esquinas and J. López-Gómez enjoy the tremendous advantage, over the subsequent purely topological proofs, of being entirely constructive. The proofs of [23] and [14] used rather sophisticated topological non-constructive tools —obstruction theory techniques.

5. Algebraic characterizations of nonlinear eigenvalues.

This section gives the construction scheme of a finite algorithm to calculate the change of $\mathrm{Ind}\,(\mathfrak{L}(\lambda), 0)$ as λ crosses an isolated eigenvalue $\lambda_0 \in \Sigma$. This algorithm is based upon the construction of a generalized algebraic multiplicity of $\mathfrak{L}(\lambda)$ at λ_0 which extends all classical concepts of generalized algebraic multiplicities available in the specialized literature. Besides its intrinsic interest, it connects two areas that have temporarily evolved through rather separated paths. Namely, spectral theory and bifurcation theory.

Subsequently, we denote by $\Phi_0(U)$ the space of Fredholm operators of index zero in U, not necessarily of the form (2). By definition, $T \in \Phi_0(U)$ if $R[T]$ is closed and

$$\dim N[T] = \mathrm{codim}\,R[T] < \infty\,.$$

The general assumption of this section is that

$$\mathfrak{L} \in C^r(\mathbb{R}; \Phi_0(U))\,, \qquad r \in \mathbb{N} \cup \{\infty\}\,, \quad r \geq 1\,,$$

and that

$$\lambda_0 \in \Sigma \quad \text{is an isolated eigenvalue of the family } \mathfrak{L}(\lambda)\,.$$

Given a family $\mathfrak{M}(\lambda)$ of class C^r, we will denote

$$\mathfrak{M}_j := \frac{1}{j!} \frac{d^j \mathfrak{M}}{d\lambda^j}(\lambda_0)\,, \qquad 0 \leq j < r + 1\,.$$

The following concepts go back to J. Esquinas and J. López-Gómez [11].

Definition 8. Given an integer $1 \leq k \leq r$, it is said that λ_0 is a k-**transversal eigenvalue** of $\mathfrak{L}(\lambda)$ if

$$\bigoplus_{j=1}^{k} \mathfrak{L}_j(N[\mathfrak{L}_0] \cap \cdots \cap N[\mathfrak{L}_{j-1}]) \oplus R[\mathfrak{L}_0] = U$$

and

$$\mathfrak{L}_k(N[\mathfrak{L}_0] \cap \cdots \cap N[\mathfrak{L}_{k-1}]) \neq \mathrm{span}\,[0]\,. \tag{19}$$

The integer $k = k(\lambda_0)$ is called the order of transversality of $\mathfrak{L}(\lambda)$ at λ_0.

Note that $\lambda_0 = 0$ is not a transversal eigenvalue in $U = \mathbb{R}$ of the family

$$\mathfrak{L}(\lambda) := \begin{cases} e^{-\lambda^{-2}} & \text{if } \lambda \in \mathbb{R} \setminus \{0\}\,, \\ 0 & \text{if } \lambda = 0\,. \end{cases} \tag{20}$$

Definition 9. Suppose λ_0 is a k-transversal eigenvalue of $\mathfrak{L}(\lambda)$. Then, the **algebraic multiplicity** of $\mathfrak{L}(\lambda)$ at λ_0 is defined by

$$\chi[\mathfrak{L}(\lambda); \lambda_0] := \sum_{j=1}^{k} j \, \dim \mathfrak{L}_j(N[\mathfrak{L}_0] \cap \cdots \cap N[\mathfrak{L}_{j-1}])\,.$$

The celebrated transversality condition of M. G. Crandall and P. H. Rabinowitz [7], as well considered by D. Westreich [56], equals the one given by Definition 8 with $k = 1$. Within that context,

$$\chi[\mathfrak{L}(\lambda); \lambda_0] := \dim N[\mathfrak{L}_0]. \tag{21}$$

Although transversal eigenvalues very rarely appear in applications, a broad class of families $\mathfrak{L}(\lambda)$ can be *transversalized* by means of a polynomial pencil. The following concept provides us with all admissible families for the validity of the subsequent transversalization theorem; it goes back to [32].

Definition 10. λ_0 is said to be an **algebraic eigenvalue** of $\mathfrak{L}(\lambda)$ if there exist $\epsilon > 0, C > 0$ and an integer $\nu \geq 1$ such that for each λ satisfying $0 < |\lambda - \lambda_0| < \epsilon$ the operator $\mathfrak{L}(\lambda)$ is an isomorphism and

$$\left\|\mathfrak{L}^{-1}(\lambda)\right\| \leq \frac{C}{|\lambda - \lambda_0|^\nu}. \tag{22}$$

The least integer $\nu \geq 1$ for which this estimate holds true is called the order of λ_0.

Note that $\lambda_0 = 0$ is not an algebraic eigenvalue of the family $\mathfrak{L}(\lambda)$ defined by (20). The following result allows us extending the concept of multiplicity introduced by Definition 9 to cover the case of general families of operators (cf. [32, Chapters 4, 5]). It characterizes the families that can be transversalized by means of a family of isomorphism.

Theorem 11. *For each integer $1 \leq k \leq r$ there exists a polynomial family of operators $P^k : \mathbb{R} \to \Phi_0(U)$,*

$$P^k(\lambda) = I + \sum_{i=1}^{\deg P^k} (\lambda - \lambda_0)^i P_i^k,$$

for which the new family of operators

$$\mathfrak{L}^{P^k}(\lambda) := \mathfrak{L}(\lambda)P^k(\lambda), \qquad \lambda \sim \lambda_0,$$

satisfies

$$U_k := \bigoplus_{j=1}^{k} \mathfrak{L}_j^{P^k} \left(N[\mathfrak{L}_0^{P^k}] \cap \cdots \cap N[\mathfrak{L}_{j-1}^{P^k}]\right) \oplus R[\mathfrak{L}_0^{P^k}] \subset U.$$

Moreover, the following assertions are true:

1. λ_0 *is a ν-transversal eigenvalue of $\mathfrak{L}^{P^\nu}(\lambda)$ if λ_0 is an algebraic eigenvalue of $\mathfrak{L}(\lambda)$ of order $1 \leq \nu \leq r$.*

2. U_k *is a proper subspace of U for all integers $1 \leq k \leq r$ if λ_0 is not an algebraic eigenvalue of $\mathfrak{L}(\lambda)$, or if it is an algebraic eigenvalue of order $\nu > r$.*

3. *Suppose*

$$P, Q : (\lambda_0 - \epsilon, \lambda_0 + \epsilon) \to \mathcal{L}(U)$$

are two C^r operator families satisfying the following:

(a) $P(\lambda_0)$ *and $Q(\lambda_0)$ are isomorphisms,*

(b) λ_0 is a k_1-transversal eigenvalue of $\mathfrak{L}^P(\lambda) := \mathfrak{L}(\lambda)P(\lambda)$ for some $1 \leq k_1 \leq r$,
(c) λ_0 is a k_2-transversal eigenvalue of $\mathfrak{L}^Q(\lambda) := \mathfrak{L}(\lambda)Q(\lambda)$ for some $1 \leq k_2 \leq r$.

Then, $k_1 = k_2$ and, for each $1 \leq j \leq k_1 = k_2$,

$$\dim \mathfrak{L}_j^P(\bigcap_{i=0}^{j-1} N[\mathfrak{L}_i^P]) = \dim \mathfrak{L}_j^Q(\bigcap_{i=0}^{j-1} N[\mathfrak{L}_i^Q]) .$$

In particular,

$$\chi[\mathfrak{L}^P(\lambda); \lambda_0] = \chi[\mathfrak{L}^Q(\lambda); \lambda_0] .$$

Theorem 11 shows the consistency of the following concept of multiplicity which extends the one introduced by Definition 9.

Definition 12. Suppose λ_0 is an algebraic eigenvalue of $\mathfrak{L}(\lambda)$ of order $1 \leq \nu \leq r$. Then, the algebraic multiplicity of $\mathfrak{L}(\lambda)$ at λ_0 is defined by

$$\chi[\mathfrak{L}(\lambda); \lambda_0] := \chi[\mathfrak{L}^{P^\nu}(\lambda); \lambda_0]$$

where $\mathfrak{L}^{P^\nu}(\lambda)$ is any of the families whose existence is guaranteed by Theorem 11.

The relevance of this generalized concept of algebraic multiplicity within the context of nonlinear analysis relies on the following result (cf. [32, Theorem 5.6.2]).

Theorem 13. *Suppose* (2) *and* λ_0 *is an algebraic eigenvalue of* $\mathfrak{L}(\lambda)$ *of order* $1 \leq \nu \leq r$. *Then, there exists* $\eta \in \{-1, 1\}$ *such that*

$$\mathrm{Ind}\,(\mathfrak{L}(\lambda), 0) = \eta \operatorname{sign}(\lambda - \lambda_0)^{\chi[\mathfrak{L}; \lambda_0]} \qquad if \quad 0 < |\lambda - \lambda_0| < \epsilon. \qquad (23)$$

Thus, λ_0 *is a nonlinear eigenvalue of* $\mathfrak{L}(\lambda)$ *if, and only if,*

$$\chi[\mathfrak{L}; \lambda_0] \in 2\mathbb{N} + 1 .$$

When $\mathfrak{L}$ is real analytic and $\mathfrak{L}(\lambda_i)$ is invertible for some $\lambda_i \in \mathbb{R}$, then the spectrum of the family $\mathfrak{L}(\lambda)$, Σ, is discrete and any $\lambda_0 \in \Sigma$ is an algebraic eigenvalue of $\mathfrak{L}(\lambda)$ (cf. [32, Theorem 4.4.4]). Therefore, $\chi[\mathfrak{L}; \lambda_0]$ is well defined and, thanks to Theorem 13, the multiplicity $\chi[\mathfrak{L}; \lambda_0]$ characterizes the nonlinear eigenvalues of $\mathfrak{L}$.

The characterization of nonlinear eigenvalues given by Theorem 13 in terms of the oddity of $\chi[\mathfrak{L}; \lambda_0]$ is valid for general families of operators in $\Phi_0(U)$, not necessarily compact perturbations of the identity. In this general context, if

$$\chi[\mathfrak{L}; \lambda_0] \in 2\mathbb{N},$$

then Theorem 11 enables us to construct a nonlinearity $\mathfrak{N}(\lambda, u)$ for which the unique solution of (1) in a neighborhood of $(\lambda_0, 0)$ is $(\lambda, 0)$ (cf. the proof of [32, Theorem 4.2.4]). Actually, using these constructive techniques it is easy to see that if $(\lambda_0, 0)$ is not a bifurcation point from $(\lambda, 0)$ of Equation (1), then the nonlinearity $\mathfrak{N}(\lambda, u)$ must exhibit some hidden strong symmetries around λ_0 (cf. the proof of [32, Theorem 4.2.4]). This fact might deserve a certain attention in algebraic geometry and singularity theory.

Besides its relevance in nonlinear analysis, Theorem 11 has shown to be a pivotal technical tool in connecting the generalized algebraic multiplicity $\chi[\mathfrak{L}; \lambda_0]$ with other concepts of multiplicity introduced independently in the literature and, in particular, provides us with a finite algorithm for constructing local Smith forms (cf. [53]), besides it allows extending many classical results in spectral theory (e.g., I. C. Göhberg and E. I. Sigal [20] and the references therein) to our C^r setting framework. We will clarify these points later.

The proof of Theorem 13 is based upon the algebraic multiplicity introduced by R. J. Magnus [40]. In the special case when

$$\mathfrak{L}(\lambda) = I - \lambda K$$

for some compact operator K, then $\chi[\mathfrak{L}(\lambda); \lambda_0]$ equals the classical algebraic multiplicity of λ_0 as a characteristic value of the compact operator K. Therefore, Theorem 13 provides us with a substantial extremely sharp extension of the celebrated Krasnosel'skii's bifurcation theorem —the detonating result—. Also, thanks to Theorem 13, under the transversality condition of M. G. Crandall and P. H. Rabinowitz we have (21), and hence λ_0 is a nonlinear eigenvalue of $\mathfrak{L}(\lambda)$ if and only if

$$\dim N[\mathfrak{L}_0] \text{ is odd.}$$

It seems this was the first characterization of nonlinear eigenvalues given in the literature (cf. [31]).

There are many other concepts of generalized algebraic multiplicities for families of Fredholm operators of index zero in nonlinear analysis, among them, those given by P. Sarreither [50], B. Laloux and J. Mawhin [28], J. Ize [22], H. Kielhöfer [25], [26], and P. J. Rabier [46]. We will focus our attention into the concept of P. J. Rabier, because it uses generalized Jordan chains and Smith's canonical forms.

Generalized Jordan chains were introduced to develop a spectral theory in the context of matrix polynomial pencils (e.g., I. C. Göhberg et al. [19] and the references therein), and have shown to be a powerful technical tool in a number of areas; among them, in studying the structure of the solution set of broad classes of linear systems of differential equations (cf. J. T. Wloka et al. [58]) and in analyzing nonlinear eigenvalue problems in bifurcation theory (cf. P. J. Rabier [46]). Closely related to generalized Jordan chains is the so-called *Smith form*. In dealing with polynomial families $\mathfrak{L}(\lambda)$ the existence of a *local Smith form* can be easily deduced from the *global Smith form* of I. C. Göhberg et al. [19, Chapter S1]. P. J. Rabier [46, Section 4] extended the available theory for polynomial pencils to build up a *local Smith form* for families of class C^∞ regularity at any eigenvalue λ_0 where no generalized Jordan chain can be continued indefinitely; this is what occurs, for example, when $\mathfrak{L}(\lambda)$ is of polynomial type and it is invertible at some value of λ. To gain precision, we briefly give the most basic concepts of the theory.

Any ordered finite set of $s + 1$ vectors

$$(x_0, \dots, x_s) \in U^{s+1}$$

with $0 \leq s < r + 1$, $x_0 \neq 0$, and

$$\sum_{i=0}^{j} \mathcal{L}_i x_{j-i} = 0, \qquad 0 \le j \le s, \tag{24}$$

is said to be a *Jordan chain of length* $s+1$ of the family $\mathcal{L}(\lambda)$ at λ_0 —originating at x_0—. When $j = 0$, (24) reduces to

$$\mathcal{L}_0 x_0 = 0$$

and, hence,

$$x_0 \in N[\mathcal{L}_0] \setminus \{0\}.$$

This is why λ_0 is said to be an *eigenvalue* of the family of operators $\mathcal{L}(\lambda)$ and x_0 is said to be an *eigenvector*. In the very special case when

$$\mathcal{L}(\lambda) = \lambda I - T,$$

where $\lambda \in \mathbb{R}$ and $T \in \mathcal{L}(U)$, the family $\mathcal{L}(\lambda)$ is analytic and

$$\mathcal{L}_0 = \lambda_0 I - T, \quad \mathcal{L}_1 = I, \quad \text{and } \mathcal{L}_j = 0 \quad \text{for any } j \ge 2.$$

Thus, (24) reduces to

$$\begin{cases} \mathcal{L}_0 x_0 = 0, \\ \mathcal{L}_1 x_0 + \mathcal{L}_0 x_1 = 0, \\ \cdots \\ \mathcal{L}_1 x_{s-1} + \mathcal{L}_0 x_s = 0. \end{cases}$$

Equivalently,

$$\begin{cases} T x_0 = \lambda_0 x_0, \\ T x_1 = \lambda_0 x_1 + x_0, \\ \cdots \\ T x_s = \lambda_0 x_s + x_{s-1}, \end{cases} \tag{25}$$

and, therefore, $(x_0, \ldots, x_s)$ is a Jordan chain of T in the classical sense. Note that (25) implies

$$x_j \in N[(T - \lambda_0 I)^{j+1}], \qquad 0 \le j \le s.$$

Also, if $(x_0, \ldots, x_s)$ is a Jordan chain, then so is $(x_0, \ldots, x_p)$ for any $0 \le p \le s$.

A Jordan chain $(x_0, \ldots, x_s)$ is said to be *maximal* if it does not exists $x \in U$ for which $(x_0, \ldots, x_s, x)$ is a Jordan chain of length $s + 2$. The *rank* of an eigenvector

$$x_0 \in N[\mathcal{L}_0] \setminus \{0\},$$

denoted in the sequel by $\mathrm{rank}(x_0)$, is defined as the maximum length of all Jordan chains originating at x_0 if such lengths are bounded above, while

$$\mathrm{rank}(x_0) := \infty$$

when the set of lengths of all Jordan chains originating at x_0 is unbounded.

Suppose the length of all Jordan chains of $\mathfrak{L}(\lambda)$ at λ_0 is uniformly bounded above by some natural number $k_1 \in \mathbb{N}$ such that

$$1 \leq k_1 < r+1 .$$

Without loss of generality, k_1 can be assumed to be optimal, i.e., there exists

$$x_{0,1} \in N[\mathfrak{L}_0] \setminus \{0\}$$

such that

$$\mathrm{rank}(x_{0,1}) = k_1 .$$

Proceeding inductively, having chosen $i-1$ linearly independent vectors,

$$x_{0,1} , \ldots , x_{0,i-1} , \qquad 2 \leq i \leq n = \dim N[\mathfrak{L}_0] ,$$

pick up

$$x_{0,i} \in C_i := N[\mathfrak{L}_0] \setminus \mathrm{span}\,[x_{0,1}, \ldots, x_{0,i-1}]$$

such that

$$\mathrm{rank}(x_{0,i}) = k_i := \max\,\{\mathrm{rank}(x) \,:\, x \in C_i\,\} \leq k_{i-1} .$$

This process can be repeated until n elements

$$x_{0,1}, \ldots, x_{0,n}$$

have been selected. Now, for each $i \in \{1, \ldots, n\}$, let

$$(x_{0,i},\, x_{1,i}, \ldots, x_{k_i-1,i})$$

be any Jordan chain of length k_i originating at $x_{0,i}$ of $\mathfrak{L}(\lambda)$ at λ_0. The ordered family

$$[x_{0,1}, \ldots, x_{k_1-1,1}; \cdots ; x_{0,i}, \ldots, x_{k_i-1,i}; \cdots ; x_{0,n}, \ldots, x_{k_n-1,n}]$$

is said to be a *canonical set of Jordan chains* of $\mathfrak{L}(\lambda)$ at λ_0 (cf. P. J. Rabier [46, pp. 906]). By construction,

$$k_1 \geq \cdots \geq k_n \geq 1 .$$

Moreover, due to [46, pp. 907], the integer numbers $k_1, \ldots, k_n$ are independent of the canonical set of Jordan chains of $\mathfrak{L}(\lambda)$ at λ_0. Thus, the following concept of multiplicity, attributable to P. J. Rabier [46], is consistent: The integers $k_1, \ldots, k_n$ are called the *partial multiplicities* of $\mathfrak{L}(\lambda)$ at λ_0. The number

$$R[\mathfrak{L}(\lambda); \lambda_0] := \sum_{j=1}^{n} k_j$$

will be referred to as Rabier's multiplicity of $\mathfrak{L}(\lambda)$ at λ_0. Moreover, if $\mathfrak{E}$ and $\mathfrak{F}$ are two families of operators of class C^r around λ_0 such that $\mathfrak{E}(\lambda_0)$ and $\mathfrak{F}(\lambda_0)$ are isomorphisms, then Rabier's partial multiplicities of $\mathfrak{L}(\lambda)$ at λ_0 equal those of the new family

$$\mathfrak{L}^{\mathfrak{E},\mathfrak{F}}(\lambda) := \mathfrak{E}(\lambda)\mathfrak{L}(\lambda)\mathfrak{F}(\lambda) , \qquad \lambda \sim \lambda_0 ,$$

and, in particular,

$$R[\mathfrak{L}(\lambda); \lambda_0] = R[\mathfrak{L}^{\mathfrak{E},\mathfrak{F}}(\lambda); \lambda_0] .$$

Moreover, the following result, generalizing I. C. Göhberg et al. [19, Chapter S1] and P. J. Rabier [46, Section 4] is satisfied (cf. J. López-Gómez & C. Mora-Corral [34], [35]). It establishes the existence of a *local Smith form* in finite dimension; the precise concept of local Smith form being incorporated in the statement of the theorem.

Theorem 14. *Suppose $U = \mathbb{R}^N$ and the length of all Jordan chains of $\mathfrak{L}(\lambda)$ at λ_0 is uniformly bounded above by some natural number $k_1 \in \mathbb{N}$ such that $1 \leq k_1 < r + 1$. Then, $\mathfrak{L}(\lambda)$ has a local Smith form at λ_0, i.e., $\mathfrak{L}(\lambda)$ admits a decomposition of the form*

$$\mathfrak{L}(\lambda) = \mathfrak{E}(\lambda)\mathfrak{D}(\lambda)\mathfrak{F}(\lambda), \qquad \lambda \sim \lambda_0, \tag{26}$$

where $\mathfrak{E}(\lambda)$ are $\mathfrak{F}(\lambda)$ are C^{r-k_1} matrices of order m such that $\mathfrak{E}(\lambda_0)$ and $\mathfrak{F}(\lambda_0)$ are isomorphisms, and $\mathfrak{D}(\lambda)$ is a diagonal matrix of the form

$$\mathfrak{D}(\lambda) = \mathrm{diag}\left\{(\lambda - \lambda_0)^{k_1}, \ldots, (\lambda - \lambda_0)^{k_n}, 1, \ldots, 1\right\},$$

where $k_1 \geq \cdots \geq k_n \geq 1$ are Rabier's partial multiplicities of $\mathfrak{L}(\lambda)$ at λ_0.

The following very recent result coming from J. López-Gómez & C. Mora-Corral [34], [35] characterizes the existence of the local Smith form in terms of the concept of algebraic eigenvalue introduced by Definition 10, helping to reveal the importance that it might deserve in spectral theory.

Theorem 15. *Suppose $U = \mathbb{R}^N$. Then, the following conditions are equivalent:*

- $\mathfrak{L}(\lambda)$ *possesses a local Smith form at λ_0.*
- $\det \mathfrak{L}(\lambda)$ *has a zero of finite order at $\lambda = \lambda_0$.*
- λ_0 *is an algebraic eigenvalue of $\mathfrak{L}(\lambda)$ of order $k \leq r$.*
- *The length of any Jordan chain of $\mathfrak{L}(\lambda)$ at λ_0 is bounded above by r.*

In the infinite dimensional case, the following conditions are equivalent:

- λ_0 *is an algebraic eigenvalue of $\mathfrak{L}(\lambda)$ of order $k \leq r$.*
- *The length of any Jordan chain of $\mathfrak{L}(\lambda)$ at λ_0 is bounded above by r.*

Actually, constructing canonical sets of Jordan chains of $\mathfrak{L}(\lambda)$ at transversal eigenvalues, i.e., constructing the Smith canonical form, can be accomplished in a rather simple and extremely systematic way, [34], [35]. Undoubtedly, the best algorithm to construct the Smith form consists in applying Theorem 11 and then constructing the form for the transversalized family.

Besides being extremely useful in constructing the Smith form, Theorem 11 gave the key to prove the following result (cf. López-Gómez & C. Mora-Corral [36]).

Theorem 16. *Suppose $\mathfrak{L} : \mathbb{C} \to \Phi_0(U)$ is an holomorphic family. Then, the generalized algebraic multiplicity $\chi[\mathfrak{L}; \lambda_0]$ is given by the trace of the logarithmic residue of the family $\mathfrak{L}$ at the eigenvalue λ_0, i.e.,*

$$\chi[\mathfrak{L}; \lambda_0] = \mathrm{tr}\, \frac{1}{2\pi i} \int_\gamma \mathfrak{L}'(\lambda)\mathfrak{L}(\lambda)^{-1}\, d\lambda \tag{27}$$

where γ is any rectifiable Jordan positively oriented curve surrounding λ_0 in $\mathbb{C} \setminus \Sigma$ homotopic to λ_0 in $\mathbb{C} \setminus \Sigma$. Moreover, the traces of the remaining singular Laurent coefficients vanish.

As Theorem 16 substantially improve some previous results of A. G. Ramm [48, Th. 1], it might be of a certain relevance in theoretical physics. Besides the fact that, due to (27), $\chi[\mathfrak{L}; \lambda_0]$ equals the multiplicity introduced by I. C. Göhberg and E. I. Sigal [20], so enjoying all the pleasant properties of that algebraic multiplicity —and, reciprocally, showing that the multiplicity of [20] inherits all properties of $\chi[\mathfrak{L}; \lambda_0]$—, in the very special case when, for a fixed $L \in \mathcal{L}(U)$,

$$\mathfrak{L}(\lambda) = \lambda I - L,$$

identity (27) reduces to

$$\chi[\mathfrak{L}; \lambda_0] = \operatorname{tr} \frac{1}{2\pi i} \int_\gamma (\lambda I - L)^{-1} \, d\lambda$$

and, hence, $\chi[\mathfrak{L}; \lambda_0]$ equals the classical algebraic multiplicity of λ_0 as an eigenvalue of $\mathfrak{L}$, since the trace of a projection equals its rank. This provides us with a simple and direct proof of [32, Th. 5.4.1] —without passing through the algebraic multiplicity of R. J. Magnus [40]. Even more, it turns out that, thanks to Theorem 11, one can obtain the following extremely sharp version of Theorem 16. It basically makes apparent that (22) is the solely requirement from which all previous *holomorphic theory* can be rebuilt (it goes back to J. López-Gómez and C. Mora-Corral [36]).

Theorem 17. *Suppose λ_0 is an algebraic eigenvalue of $\mathfrak{L} \in C^r$ of order k, and $r \geq 2k-1$. Then,*

$$\mathfrak{L}(\lambda)^{-1} = \sum_{i=-k}^{-1} R_i(\lambda - \lambda_0)^i + o((\lambda - \lambda_0)^{-1}), \qquad \lambda \sim \lambda_0, \quad \lambda \neq \lambda_0, \qquad (28)$$

where all the operators R_i, $-k \leq i \leq -1$, have finite rank and

$$\chi[\mathfrak{L}; \lambda_0] = \operatorname{tr} R_{-1}, \qquad 0 = \operatorname{tr} R_i, \quad -k \leq i \leq -2. \qquad (29)$$

Actually, condition (22) is the unique important property of $\mathfrak{L}$ from which all crucial properties of holomorphic families $\mathfrak{L}(\lambda)$ follow. Among them, (28), (29) and the product formula.

As an immediate consequence, since the parity of $\chi[\mathfrak{L}; \cdot]$ characterizes the nonlinear eigenvalues of the family $\mathfrak{L}(\lambda)$, the multiplicity of I. C. Göhberg and E. I. Sigal also characterizes them. The corresponding result establishes some deeply hidden connections between a number of a priori independent concepts originated in rather separated areas. Nevertheless, our results in abstract spectral theory open a door for further developments of non-holomorphic operational calculus showing that (22) is the pivotal property under which such a calculus should be build up.

Very recently, C. Mora-Corral [43], [44], [45] has shown that, similarly to what happened with the uniqueness of the topological degree, the algebraic multiplicity is uniquely determined by a few among its properties, though we refrain of giving further details here.

6. Global behaviour of compact components

The first global bifurcation result is due to P. H. Rabinowitz [47] who showed that if

$$\mathfrak{K}(\lambda) = \lambda K \tag{30}$$

for all $\lambda \in \mathbb{R}$, and $\lambda_0 \neq 0$ is a characteristic value of K with odd algebraic multiplicity, then the component —maximal continuum (closed and connected set) for the inclusion— of the set of nontrivial solutions of Equation (1) emanating from $(\lambda, 0)$ at $\lambda = \lambda_0$, subsequently denoted by $\mathfrak{C}$, must satisfy some of the following alternatives. Either

1. $\mathfrak{C}$ is unbounded in $\mathbb{R} \times U$

2. There exists another characteristic value of K, $\lambda_1 \neq \lambda_0$, such that $(\lambda_1, 0) \in \bar{\mathfrak{C}}$

Since the celebrated Rabinowitz's paper was published, that alternative is called the *global alternative of Rabinowitz*. Its proof is based on the fact that, thanks to Theorem 13,

$$\text{Ind}\,(I - \lambda K, \lambda_0) \text{ changes sign as } \lambda \text{ crosses } \lambda_0 .$$

In most of the literature where a generalized algebraic multiplicity was constructed, it was shown that an odd multiplicity implies the global alternative of Rabinowitz, because an odd multiplicity entails a change of $\text{Ind}\,(I - \mathfrak{K}(\lambda), 0)$ as λ crosses λ_0.

P. H. Rabinowitz and E. N. Dancer pointed out that the proof of Rabinowitz's alternative can be easily adapted to show that if $\mathfrak{C}$ is bounded in $\mathbb{R} \times U$, then there exists another characteristic value $\lambda_1 \neq \lambda_0$ of K with an odd algebraic multiplicity such that $(\lambda_1, 0) \in \bar{\mathfrak{C}}$. Actually, they remarked that the number of characteristic values of K with an odd algebraic multiplicity where $\mathfrak{C}$ meets $(\lambda, 0)$ must be even, and that these characteristic values cannot be arbitrarily ordered. Some time later, J. Ize [22] and R. J. Magnus [40] developed these results by using their respective concepts of multiplicity. To state their main result, we must introduce the concept of parity.

Although far from necessary, we will subsequently assume that Σ is discrete and that $\mathfrak{L}(\lambda)$ is a compact perturbation of the identity. Then, a parity map is any map

$$P : \Sigma \to \{-1, 0, 1\}$$

satisfying the following requirements:

1. $P(\mu) = 0$ if $\text{Ind}\,(\mathfrak{L}(\lambda), 0)$ remains unchanged as λ crosses μ.

2. $P(\mu) \in \{-1, 1\}$ if $\text{Ind}\,(\mathfrak{L}(\lambda), 0)$ changes as λ crosses μ.

3. $P(\mu_1)P(\mu_2) = -1$ if $\mu_1, \mu_2 \in \Sigma$ are two consecutive eigenvalues of $\mathfrak{L}(\lambda)$ where the index changes.

Using this concept of parity for each of their respective algebraic multiplicities J. I. Ize [22] and R. J. Magnus [40] saw that if $\mathfrak{C}$ is a bounded component of the set of nontrivial solutions of Equation (1) with

$$\bar{\mathfrak{C}} \cap \{(\lambda, 0) \ : \ \lambda \in \Sigma\} = \{(\mu_1, 0), ..., (\mu_N, 0)\},$$

then

$$\sum_{j=1}^{N} P(\mu_j) = 0.$$

In particular, if $\mathfrak{C}$ is the component emanating from $(\lambda, 0)$ at a nonlinear eigenvalue, say λ_0, then there exists another nonlinear eigenvalue, $\lambda_1 \neq \lambda_0$, such that $(\lambda_1, 0) \in \bar{\mathfrak{C}}$, and, hence, Rabinowitz's alternative holds true.

The previous result provides us with very sharp information about the distribution of the nonlinear eigenvalues where $\mathfrak{C}$ meets $(\lambda, 0)$. For example, in the case when $\lambda_0 < \lambda_1$ are the unique nonlinear eigenvalues where $\mathfrak{C}$ meets $(\lambda, 0)$, it guarantees that $\mathfrak{L}(\lambda)$ must possess an even number of eigenvalues of odd multiplicity in the interval (λ_0, λ_1), possibly zero.

All previous results are immediate consequences from the main theorem of J. López-Gómez & C. Mora-Corral [37], further generalized in [38], which provides us with an optimal counter of the number of solutions of $\mathfrak{C}$ at each value of λ where it projects into the λ–axis of $\mathbb{R} \times U$. To state it we need to introduce the concept of *open isolating neighborhood*, and a very important property of those neighborhoods. Recall that $\mathfrak{S}$ stands for the set of nontrivial solutions of Equation (1).

Definition 18. Suppose Σ is discrete in $\mathbb{R}$, $\mathfrak{C}$ is a compact component of $\mathfrak{S}$, and set

$$\mathcal{B} := \mathfrak{C} \cap (\Sigma \times \{0\}).$$

An open set $\Omega \subset \mathbb{R} \times U$ is said to be an open isolating neighborhood of $\mathfrak{C}$ if

$$\mathfrak{C} \subset \Omega, \quad \partial\Omega \cap \mathfrak{S} = \varnothing \text{ and } \mathcal{B} := \Omega \cap (\Sigma \times \{0\}).$$

Thanks to a classical result by G. T. Whyburn [57], open isolating neighborhoods of a compact component $\mathfrak{C}$ always exist (cf. [32, pp. 174]). Moreover, they satisfy the following property (cf. [32, pp. 176]).

Proposition 19. *Suppose Σ is discrete in $\mathbb{R}$ and $\mathfrak{C}$ is a compact component of $\mathfrak{S}$, and set*

$$\mathcal{B} := \mathfrak{C} \cap (\Sigma \times \{0\}).$$

Let Ω be an open isolating neighborhood of $\mathfrak{C}$, denote

$$\Omega_\lambda := \{ u \in U : (\lambda, u) \in \Omega \}, \qquad \lambda \in \mathbb{R},$$

and consider $\delta > 0$ satisfying

$$\mathcal{B} + B_\delta(0, 0) \subset \Omega,$$

where $B_\delta(0, 0)$ is the ball of radius δ centered at $(0, 0) \in \mathbb{R} \times U$. Then, there exists $\rho = \rho(\delta) > 0$ such that for any

$$\lambda \in \mathbb{R} \setminus \bigcup_{(\mu, 0) \in \mathcal{B}} \left(\mu - \frac{\delta}{2}, \mu + \frac{\delta}{2} \right) \tag{31}$$

some of the following alternatives occurs. Either

1. $\bar{B}_\rho \cap \Omega_\lambda = \varnothing$, *or*

2. $u = 0$ *if* $(\lambda, u) \in \mathfrak{F}^{-1}(0) \cap (\{\lambda\} \times \bar{B}_\rho)$, *where B_ρ stands for the ball of U of radius ρ centered at zero.*

Now, we are ready to state the main result of [37].

Theorem 20. *Suppose Σ is discrete in $\mathbb{R}$, $\mathfrak{C}$ is a compact component of $\mathfrak{S}$, and set*

$$\mathcal{B} := \mathfrak{C} \cap (\Sigma \times \{0\}) .$$

Let Ω be an open isolating neighborhood of $\mathfrak{C}$. Then, for each

$$(\lambda^*, 0) \in (\mathbb{R} \times \{0\}) \setminus \mathcal{B}$$

there exists $\rho^* > 0$ such that

$$\mathrm{Deg}\,(\mathfrak{F}(\lambda^*, \cdot), \Omega_{\lambda^*} \setminus B_{\rho^*}) = \pm 2 \sum_{\substack{(\mu,0)\in\mathcal{B} \\ \mu > \lambda^*}} P(\mu) = \mp 2 \sum_{\substack{(\mu,0)\in\mathcal{B} \\ \mu < \lambda^*}} P(\mu) \qquad (32)$$

for any parity map P.

Suppose, in addition, that $\mathfrak{F}$ is of class C^1 and that there exists

$$(\lambda^*, 0) \in (\mathbb{R} \times \{0\}) \setminus \mathcal{B}$$

with $\Omega_{\lambda^*} \cap \mathfrak{C} \neq \emptyset$ such that

$$D_u \mathfrak{F}(\lambda^*, u) \ \text{is an isomorphism for each } (\lambda^*, u) \in \mathfrak{C} \cap \Omega_{\lambda^*} .$$

Then,

$$\mathrm{Card}\,\left([\mathfrak{F}(\lambda^*, \cdot)]^{-1}(0) \cap \mathfrak{C}\right) \geq 2 \max \left\{ 1, \Big| \sum_{\substack{(\mu,0)\in\mathcal{B} \\ \mu > \lambda^*}} P(\mu)\Big|, \Big| \sum_{\substack{(\mu,0)\in\mathcal{B} \\ \mu < \lambda^*}} P(\mu)\Big| \right\}. \qquad (33)$$

Proof. The details of the proof of (32) can be easily accomplished arguing as in the proof of [32, Theorem 6.3.1]. Indeed, let $\delta > 0$ be sufficiently small so that

$$\mathcal{B} + B_\delta(0,0) \subset \Omega, \qquad \lambda^* \in \mathbb{R} \setminus \bigcup_{(\mu,0)\in\mathcal{B}} \left(\mu - \frac{\delta}{2}, \mu + \frac{\delta}{2}\right),$$

and pick $\rho^* = \rho^*(\delta)$ satisfying the requirements of Proposition 19. Then, for each λ satisfying (31), the topological degree $\mathrm{Deg}\,(\mathfrak{F}(\lambda, \cdot), \Omega_\lambda \setminus \bar{B}_{\rho^*})$ is well defined, and, by homotopy invariance, it is constant for λ in between the first components of two consecutive points of $\mathcal{B}$. Moreover, it equals zero if

$$\lambda \in (-\infty, \min \mathcal{B} - \frac{\delta}{2}] \cup [\max \mathcal{B} + \frac{\delta}{2}, \infty) .$$

Suppose

$$\mathcal{B} \cap ([\lambda^*, \infty) \times \{0\}) = \{\mu_1, ..., \mu_M\} .$$

Then, using the homotopy invariance and the the additivity property of the degree, the following chain of identities is obtained

$$\mathrm{Deg}\,(\mathfrak{F}(\lambda^*,\cdot),\Omega_{\lambda^*}\setminus B_{\rho^*}) = \mathrm{Deg}\,(\mathfrak{F}(\lambda^*,\cdot),\Omega_{\lambda^*}) - \mathrm{Deg}\,(\mathfrak{F}(\lambda^*,\cdot),B_{\rho^*})$$

$$= \mathrm{Deg}\,(\mathfrak{F}(\mu_1 - \frac{\delta}{2},\cdot),\Omega_{\mu_1-\frac{\delta}{2}}) - \mathrm{Ind}\,(\mathfrak{L}(\mu_1 - \frac{\delta}{2}),0)$$

$$= \mathrm{Deg}\,(\mathfrak{F}(\mu_1 + \frac{\delta}{2},\cdot),\Omega_{\mu_1+\frac{\delta}{2}}) - \mathrm{Ind}\,(\mathfrak{L}(\mu_1 - \frac{\delta}{2}),0)$$

$$= \mathrm{Deg}\,(\mathfrak{F}(\mu_1 + \frac{\delta}{2},\cdot),\Omega_{\mu_1+\frac{\delta}{2}}\setminus B_{\rho^*}) + \mathrm{Ind}\,(\mathfrak{L}(\mu_1 + \frac{\delta}{2}),0) - \mathrm{Ind}\,(\mathfrak{L}(\mu_1 - \frac{\delta}{2}),0)\,.$$

Repeating this argument M times, it is apparent that

$$\mathrm{Deg}\,(\mathfrak{F}(\lambda^*,\cdot),\Omega_{\lambda^*}\setminus B_{\rho^*}) = \sum_{j=1}^{M}\left(\mathrm{Ind}\,(\mathfrak{L}(\mu_j + \frac{\delta}{2}),0) - \mathrm{Ind}\,(\mathfrak{L}(\mu_j - \frac{\delta}{2}),0)\right)$$

$$= \pm 2\sum_{j=1}^{M} P(\mu_j)\,, \tag{34}$$

since

$$\mathrm{Deg}\,(\mathfrak{F}(\mu_M + \frac{\delta}{2},\cdot),\Omega_{\mu_M+\frac{\delta}{2}}\setminus B_{\rho^*}) = 0\,.$$

Going backwards gives the second identity of (32). The proof of (33) is an easily consequence from (32). Suppose

$$\mathfrak{C}\cap\Omega_{\lambda^*} = \{\,(\lambda^*,u_1),...,(\lambda^*,u_h)\,\}\,.$$

Then, by the implicit function theorem, around any of these points $\mathfrak{C}$ consists of a differentiable curve and, hence, the isolating open neighboring Ω can be constructed in such a way that

$$\mathrm{Deg}\,(\mathfrak{F}(\lambda^*,\cdot),\Omega_{\lambda^*}\setminus B_{\rho^*}) = \sum_{j=1}^{h}\mathrm{Ind}\,(D_u\mathfrak{F}(\lambda^*,u_j),0)\,. \tag{35}$$

Therefore, since

$$\mathrm{Deg}\,(\mathfrak{F}(\lambda^*,\cdot),\Omega_{\lambda^*}\setminus B_{\rho^*}) \in 2\mathbb{N}$$

and

$$\mathrm{Ind}\,(D_u\mathfrak{F}(\lambda^*,u_j),0) \in \{-1,1\}\,, \qquad 1\le j\le h\,,$$

it is apparent that $h \ge 2$ and, thanks to (34) and (35), we find that

$$2|\sum_{j=1}^{M} P(\mu_j)| = |\mathrm{Deg}\,(\mathfrak{F}(\lambda^*,\cdot),\Omega_{\lambda^*}\setminus B_{\rho^*})|$$

$$= |\sum_{j=1}^{h}\mathrm{Ind}\,(D_u\mathfrak{F}(\lambda^*,u_j),0)|$$

$$\le h\,.$$

Repeating the argument with the points of $\mathcal{B}$ behind $(\lambda^*,0)$ concludes the proof. $\square$

Note that $2 \mid \sum_{j=1}^{M} P(\mu_j) \mid$ might be arbitrarily large.

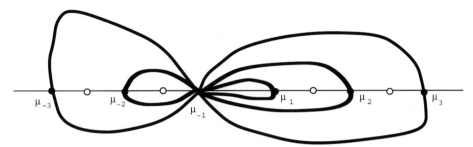

Figure 1. An admissible compact component

In Figure 1 we have represented a genuine situation where Theorem 20 predicts 6 solutions for each $\lambda \in (\mu_{-1}, \mu_1)$, 4 solutions for each $\lambda \in (\mu_{-2}, \mu_{-1}) \cup (\mu_1, \mu_2)$, and 2 solutions for each $\lambda \in (\mu_{-3}, \mu_{-2}) \cup (\mu_2, \mu_3)$. The Component $\mathfrak{C}$ is assumed to met the trivial state at $(\mu_j, 0)$, $j \in \{\pm 3. \pm 2, \pm 1\}$. We are assuming that $P(\mu_j) = 1$ if $j \in \{1, 2, 3\}$, and $P(\mu_j) = -1$ if $j \in \{-3, -2, -1\}$. The white filled points of Figure 1 represent all the nonlinear eigenvalues of $\mathfrak{L}(\lambda)$ in between the μ_j's, so that the values $P(\mu_j)$'s be the correct ones. In [37] it is shown how the multiplicity result of Theorem 20 is optimal.

For the validity of those results is far from necessary that Σ consists of isolated eigenvalues (cf. [38]). Moreover, using the topological degree introduced by P. Benevieri and M. Furi [2], [3], [4], our theory can be adapted to deal with general Fredholm operators of index zero, not necessarily compact perturbations of the identity.

By Sard-Smale's theorem, under the appropriate regularity assumptions, condition (33) is generically satisfied and, therefore, its right hand side provides us with the minimal number of zeros that such a component exhibits. Consequently, one is naturally driven to consider the problem of ascertaining the minimal topological complexity of the component in terms of shape theory by means of (33). But we refrain of giving more details here in, the interested reader being sent to [37], [38], [39] for a detailed discussion and further details.

The mathematical analysis of these and other related problems might certainly facilitate some further developments in shape theory and nonlinear analysis.

References

1. Amann, H., and Weiss, S.A., On the uniqueness of the topological degree, *Math. Z.* **130** (1973), 39–54.
2. Benevieri, P., and Furi, M., A simple notion of orientability for Fredholm maps of index zero between Banach manifolds and degree theory, *Ann. Sci. Math. Québec* **22** (1998), 131–148.
3. Benevieri, P., and Furi, M., On the concept of orientability for Fredholm maps between real Banach manifolds, *Topol. Meth. Nonl. Anal.* **16** (2000), 279–306.

4. Benevieri, P., and Furi, M., Bifurcation results for families of Fredholm maps of index zero between Banach spaces, in *Nonlinear Analysis and its Applications (St. John's, NF, 1999)*, pages 35–47, Nonlinear Anal. Forum 6, 2001.

5. Brouwer, L.E.J., Über Abbildung von Mannigfaltigkeiten, *Math. Ann.* **71** (1912), 97–115.

6. Chow, S.N., and Hale, J.K., *Methods of Bifurcation Theory*, Springer-Verlag, New York, 1982.

7. Crandall, M.G., and Rabinowitz, P.H., Bifurcation from simple eigenvalues, *J. Funct. Anal.* **8** (1971), 321–340.

8. Dancer, E.N., On the structure of solutions of non-linear eigenvalue problems, *Ind. Univ. Math. J.* **23** (1974), 1069–1076.

9. Deimling, K., *Nonlinear Functional Analysis*, Springer-Verlag, Berlin, 1985.

10. Esquinas, J., Optimal multiplicity in local bifurcation theory, II: General case, *J. Diff. Eqns.* **75** (1988), 206–215.

11. Esquinas, J., and López-Gómez, J., Resultados óptimos en teoría de bifurcación y aplicaciones, in *Actas del IX Congreso de Ecuaciones Diferenciales y Aplicaciones*, pages 159–162, Universidad de Valladolid, 1986.

12. Esquinas, J., and López-Gómez, J., Optimal results in local bifurcation theory, *Bull. Aust. Math. Soc.* **36** (1987), 25–37.

13. Esquinas, J., and López-Gómez, J., Optimal multiplicity in local bifurcation theory, I: Generalized generic eigenvalues, *J. Diff. Eqns.* **71** (1988), 72–92.

14. Fitzpatrick, P.M., and Pejsachowicz, J., A local bifurcation theorem for C^1-Fredholm maps, *Proc. Amer. Math. Soc.* **109** (1990), 995–1002.

15. Fitzpatrick, P.M., and Pejsachowicz, J., Parity and generalized multiplicity, *Trans. Amer. Math. Soc.* **326** (1991), 281–305.

16. Fitzpatrick, P.M., and Pejsachowicz, J., *Orientation and the Leray–Schauder Theory for Fully Nonlinear Elliptic Boundary Value Problems*, Memoirs of the American Mathematical Society 483, Providence, R.I., 1993.

17. Göhberg, I.C., Goldberg, S., and Kaashoek, M.A., *Classes of Linear Operators, Vol. 1.*, Operator Theory: Advances and Applications 49, Birkhäuser, Bassel, 1990.

18. Göhberg, I.C., and Krein, M.G., Fundamental aspects of defect numbers, root numbers and indices of linear operators, *Uspehi Mat. Nauk.* **12** (1957), 43–118.

19. Göhberg, I.C., Lancaster, P., and Rodman, L., *Matrix Polynomials*, Comp. Sci. Appl. Mathematics, Academic Press, New York, 1982.

20. Göhberg, I.C., and Sigal, E.I., An operator generalization of the logarithmic residue theorem and the theorem of Rouché, *Math. USSR Sbornik* **13** (1971), 603–625.

21. Golubitsky, M., and Schaeffer, D.G., *Singularities and Groups in Bifurcation Theory*, Appl. Math. Sci. 51, Springer-Verlag, New York, 1985.

22. Ize, J., *Bifurcation Theory for Fredholm Operators*, Memoirs of the American Mathematical Society 174, Providence, R.I., 1976.

23. Ize, J., Necessary and sufficient conditions for multiparameter bifurcation, *Rocky Mountain J. Math.* **18** (1988), 305–337.

24. Kato, T., *Perturbation Theory for Linear Operators*, Classics in Mathematics, Springer-Verlag, Berlin, 1995.

25. Kielhöfer, H., Degenerate bifurcation at simple eigenvalues and stability of bifurcating solutions, *J. Funct. Anal.* **38** (1980), 416–441.

26. Kielhöfer, H., Multiple eigenvalue bifurcation for Fredholm operators, *J. Reine Angew. Math.* **358** (1985), 104–124.

27. Krasnosel'skii, M.A., *Topological Methods in the Theory of Nonlinear Integral Equations*, Pergamon Press, New York, 1964.

28. Laloux, B., and Mawhin, J., Coincidence index and multiplicity, *Trans. Amer. Math. Soc.* **217** (1976), 143–162.

29. Leray, J., and Schauder, J., Topologie et équations functionelles, *Ann. Sci. École Norm. Sup. Sér. 3* **51** (1934), 45–78.

30. Lloyd, N.G., *Degree Theory*, Cambridge Tracts in Mathematics, Cambridge University Press, Cambridge, 1978.

31. López-Gómez, J., Autovalores genéricos y bifurcación, in *Actas X Jornadas Hispano-Lusas de Matemáticas, Vol. 4*, pages 40–44, Universidad de Murcia, 1985.

32. López-Gómez, J., *Spectral Theory and Nonlinear Functional Analysis*, Research Notes in Mathematics 426, Chapman & Hall/CRC, Boca Ratón, FL, 2001.

33. López-Gómez, J., *Ecuaciones Diferenciales y Variable Compleja*, Prentice-Hall, Madrid, 2001.

34. López-Gómez, J., and Mora-Corral, C., Characterizing the Existence of Local Smith forms for C^∞ families of matrix operators, in *Trends in Banach Spaces and Operator Theory* (Kaminska, A., Ed.), pages 139–151, Contemporary Mathematics 321, Amer. Math. Soc., Providence, 2003.

35. López-Gómez, J., and Mora-Corral, C., Characterizing the existence of local Smith forms for C^r families of operators, Preprint.

36. López-Gómez, J., and Mora-Corral, C., Finite Laurent developments and the logarithmic residue theorem in the real non-analytic case, *Int. Eqns. and Op. Th.* In Press.

37. López-Gómez, J., and Mora-Corral, C., Counting solutions of nonlinear abstract equations. *Top. Meth. in Nonl. Anal.* In Press.

38. López-Gómez, J., and Mora-Corral, C., Minimal complexity of semi-bounded components in bifurcation theory, *Nonl. Anal. TMA* **58** (2004), 749–777.

39. López-Gómez, J., and Mora-Corral, C., Counting zeros of C^1 Fredholm maps of index 1, *Bull London Math. Soc.*, to appear.

40. Magnus, R.J., A generalization of multiplicity and the problem of bifurcation, *Proc. London Math. Soc.* **32** (1976), 251–278.

41. Markus, A.S., and Sigal, E.I., The multiplicity of the characteristic number of an analytic operator function, *Math. Iss.* **3** (1970), 129–147.

42. Mawhin, J., Equivalence theorems for nonlinear operator equations and coincidence degree theory for some mappings in locally convex topological vector spaces, *J. Diff. Eqns.* **12** (1972), 610–636.

43. Mora-Corral, C., On the uniqueness of the algebraic multiplicity, *J. London Math. Soc. (2)* **69** (2004), 231–242.

44. Mora-Corral, C., Axiomatizing the algebraic multiplicity, in *The First 60 Years of Nonlinear Analysis of Jean Mawhin* (Delgado, M., López-Gómez, J., Ortega, R., and Suárez, A., Eds.), pages 175–187, World Scientific, Singapore, 2004.

45. Mora-Corral, C., *Algebraic Multiplicities and Bifurcation Theory*, Ph.D. Thesis, Universidad Complutense de Madrid, Madrid, 2004.
46. Rabier, P.J., Generalized Jordan chains and two bifurcation theorems of Krasnosel'skii, *Nonl. Anal. TMA* **13** (1989), 903–934.
47. Rabinowitz, P.H., Some global results for nonlinear eigenvalue problems, *J. Funct. Anal.* **7** (1971), 487–513.
48. Ramm, A.G., Singularities of the inverses of Fredholm operators, *Proc. Roy. Soc. Edinburgh* **102A** (1986), 117–121.
49. Sard, A., The measure of the critical values of differentiable maps, *Bull. Amer. Math. Soc.* **48** (1942), 883–890.
50. Sarreither, P., Transformationseigenschaften endlicher Ketten und allgemeine Verzweigungsaussagen, *Math. Scand.* **35** (1974), 115–128.
51. Schauder, J., Der Fixpunktsatz in Funktionalräumen, *Studia Math.* **2** (1930), 171–180.
52. Sigal, E.I., The multiplicity of a characteristic value of a product of operator functions, *Math. Iss.* **5**, (1970), 118–127.
53. Smith, H.J.S., On systems of linear indeterminate equations and congruences, *Phil. Trans. Roy. Soc. London* **151** (1861), 293–326.
54. Smale, S., An infinite dimensional generalisation of Sard's theorem, *Amer. J. Math.* **87** (1965), 861–866.
55. Tychonoff, A.N., Ein Fixpunksatz, *Math. Ann.* **11** (1935), 767–776.
56. Westreich, D., Bifurcation at eigenvalues of odd multiplicity, *Proc. Amer. Math. Soc.* **41** (1973), 609–614.
57. Whyburn, G.T., *Topological Analysis*, Princeton University Press, Princeton, 1958.
58. Wloka, J.T., Rowley, B., and Lawruk, B., *Boundary Value Problems for Elliptic Systems*, Cambridge University Press, New York, 1995.

Approximating Topological Spaces by Polyhedra

S. Mardešić

Department of Mathematics, University of Zagreb,
P.O.Box 335, 10 002 Zagreb, Croatia

Abstract

This paper discusses the problem of approximating topological spaces by polyhedra using various techniques based on inverse systems. The paper focuses on some of the author's contributions obtained over a period of forty five years. In particular, the paper lists properties preserved under inverse limits and states theorems concerning the representation of compact Hausdorff spaces as limits of inverse systems with desired additional properties. It also describes resolutions, a special kind of inverse limits, which are successfully used in non-compact situations. Furthermore, the paper discusses approximate inverse systems and approximate resolutions and describes situations where they show advantages over usual systems. Finally, the paper describes homotopy expansions, which are used in shape theory, i.e., in extending classical homotopy theory from the realm of polyhedra and CW-complexes to arbitrary topological spaces.

Key words: polyhedron, inverse system, inverse limit, resolution, approximate limit, approximate resolution, homotopy expansion, shape theory, strong expansion, strong shape theory

1. Introduction

One of the ultimate goals of topology is to understand the structure of topological spaces and (continuous) mappings, i.e., to obtain an insight in the category Top. Since in general this structure is rather complicated, it is natural to first study spaces having simpler structures, in particular, to study polyhedra, whose structure can be described in a combinatorial way. The next step consists in trying to extend results from polyhedra to spaces by approximating spaces by polyhedra. The approximation technique generally used is that of inverse systems. This program was inaugurated in 1926 by P.S. Aleksandrov [2], [3] and enabled him and other topologists to develop the foundations of algebraic topology for metric compacta [4] and more general spaces.

Recall that an n-dimensional simplex σ in a real vector space V is the convex hull of $n + 1$ points

$$v_0, \ldots, v_n \in V,$$

which are geometrically independent, i.e., the vectors

$$v_1 - v_0, \ldots, v_n - v_0$$

are linearly independent. A geometric simplicial complex is a set K (finite or infinite) of simplices contained in a real vector space V, such that every face of a simplex $\sigma \in K$ also belongs to K and the intersection $\sigma_1 \cap \sigma_2$ of two simplices of K is a face of both of them. Geometric simplicial complexes are completely determined by combinatorial objects known as abstract simplicial complexes $\mathcal{K}$. Such a complex consists of its vertices v and of its abstract simplices, i.e., finite non-empty sets of vertices such that

$$s \in \mathcal{K} \text{ and } \varnothing \neq t \subseteq s \text{ implies } t \in \mathcal{K}$$

and every singleton $\{v\}$ belongs to $\mathcal{K}$.

The carrier

$$|K| \subseteq V$$

of a geometric simplicial complex K is the union of all simplices from K. The (weak) topology on $|K|$ is obtained by defining open sets as subsets $U \subseteq |K|$, which have the property that, for every simplex $\sigma \in K$, endowed with the usual Euclidean topology, the intersection $U \cap \sigma$ is open in σ. A rectilinear polyhedron is the carrier $|K|$ of a geometric simplicial complex endowed with the described topology. More generally, a polyhedron is a topological space P, which admits a rectilinear polyhedron $|K|$ and a homeomorphism

$$\phi: |K| \to P.$$

One refers to ϕ as to a triangulation of P. Polyhedra are paracompact spaces (more precisely, they are stratifiable spaces), but they need not be metrizable. In fact, $|K|$ is metrizable if and only if K is locally finite, i.e., every vertex belongs to a finite number of simplices from K. The carrier $|K|$ is compact if and only if K is finite. In this case, $|K|$ is also metrizable.

The main technique used in the approximation of spaces by polyhedra is based on the notion of an inverse system of spaces

$$\mathbf{X} = (X_\lambda, p_{\lambda\lambda'}, \Lambda).$$

It consists of an index set Λ, endowed with a directed ordering $\leq$ (any two elements have an upper bound), of spaces X_λ, for $\lambda \in \Lambda$, and of mappings

$$p_{\lambda\lambda'}: X_{\lambda'} \to X_\lambda,$$

for $\lambda \leq \lambda'$. On the mappings $p_{\lambda\lambda'}$, called bonding mappings, one imposes the functorial requirements

$$p_{\lambda\lambda'}p_{\lambda'\lambda''} = p_{\lambda\lambda''}, \quad \text{for } \lambda \leq \lambda' \leq \lambda'',$$

and

$$p_{\lambda\lambda} = \text{id}, \quad \text{for } \lambda \in \Lambda.$$

A system $\mathbf{X}$ is cofinite if the ordering $\leq$ is antisymmetric and every element of Λ has only a finite number of predecessors. Inverse systems $\mathbf{X}$, whose index set

$$\Lambda = \mathbb{N},$$

are called inverse sequences and are usually denoted by (X_i, p_{ii+1}), because the mappings p_{ii+1} determine all other bonding mappings $p_{ii'}$. A mapping of systems

$$\mathbf{f} = (f, f_\lambda) \colon \mathbf{X} \to \mathbf{Y} = (Y_\mu, q_{\mu\mu'}, M)$$

consists of an increasing function

$$f \colon M \to \Lambda$$

and of mappings

$$f_\mu \colon X_{f(\mu)} \to Y_\mu, \quad \mu \in M,$$

such that

$$f_\mu p_{f(\mu)f(\mu')} = q_{\mu\mu'} f_{\mu'}, \quad \mu \leq \mu'. \tag{1}$$

In the special case when Λ is a singleton and thus, $\mathbf{X}$ is a single space X, $\mathbf{f} \colon X \to \mathbf{Y}$ consists of mappings

$$f_\mu \colon X \to Y_\mu, \quad \mu \in M,$$

such that

$$f_\mu = q_{\mu\mu'} f_{\mu'}, \quad \mu \leq \mu'. \tag{2}$$

With every inverse system of spaces $\mathbf{X}$ is associated its inverse limit. It consists of a space X and of a mapping

$$\mathbf{p} = (p_\lambda) \colon X \to \mathbf{X}$$

which has the following universal property. Whenever,

$$\mathbf{h} = (h_\lambda) \colon Z \to \mathbf{X}$$

is a mapping, then there exists a unique mapping

$$h \colon Z \to X$$

such that

$$\mathbf{p}h = \mathbf{h},$$

i.e., $p_\lambda h = h_\lambda$, for all $\lambda \in \Lambda$. It is easy to verify that the inverse limit is unique up to natural homeomorphism. To prove the existence of the inverse limit, one considers the set

$$X \subseteq \prod X_\lambda$$

of all points

$$x = (x_\lambda), \quad x_\lambda \in X_\lambda,$$

having the property that

$$p_{\lambda\lambda'}(x_{\lambda'}) = x_\lambda, \quad \text{for } \lambda \leq \lambda'.$$

Then the restrictions

$$p_\lambda = \pi_\lambda | X$$

of the canonical projections

$$\pi_\lambda \colon \prod X_\lambda \to X_\lambda$$

form a mapping $\mathbf{p} \colon X \to \mathbf{X}$ which has the desired universal property.

The following examples of inverse limits are especially useful.

Example 1. Let Λ be a directed set. Let S be a space and, for $\lambda \in \Lambda$, let X_λ be a subspace of S. For $\lambda \le \lambda'$, let $X_{\lambda'} \subseteq X_\lambda$ and let

$$p_{\lambda\lambda'} \colon X_{\lambda'} \to X_\lambda$$

be the inclusion map. Then

$$\mathbf{X} = (X_\lambda, p_{\lambda\lambda'}, \Lambda)$$

is an inverse system of spaces. If

$$X = \bigcap_{\lambda \in \Lambda} X_\lambda$$

and $p_\lambda \colon X \to X_\lambda$ is the inclusion mapping, then the mappings p_λ, for $\lambda \in \Lambda$, form the inverse limit $\mathbf{p} \colon X \to \mathbf{X}$ of $\mathbf{X}$.

Example 2. Let A be a set and let $(X_a, a \in A)$ be a collection of spaces. Let Λ consist of all non-empty finite subsets

$$\lambda = \{a_1, \ldots, a_n\} \subseteq A$$

ordered by inclusion $\subseteq$. Let

$$X_\lambda = X_{a_1} \times \ldots \times X_{a_n}$$

and for $\lambda \subseteq \lambda'$, let

$$p_{\lambda\lambda'} \colon X_{\lambda'} \to X_\lambda$$

be the canonical projection. Then

$$\mathbf{X} = (X_\lambda, p_{\lambda\lambda'}, \Lambda)$$

is an inverse system of spaces. Let

$$X = \prod_{a \in A} X_a$$

and let $p_\lambda \colon X \to X_\lambda$ be the canonical projection. Then

$$\mathbf{p} = (p_\lambda) \colon X \to \mathbf{X}$$

is the inverse limit of $\mathbf{X}$.

One can view the terms X_λ of a system $\mathbf{X}$ as approximations of the limit space X. Larger indices λ correspond to better approximations X_λ of X. The notion of inverse system and inverse limit was defined in its present form in 1931 by S. Lefschetz (see [18,19]). It was further developed by L. Pontryagin [57], H. Freudenthal [16] and others.

Compact Hausdorff spaces admit approximation by compact polyhedra, because of the following theorem (see [13]).

Theorem 3. *Every compact Hausdorff space* X *is the limit of an inverse system*

$$\mathbf{X} = (X_\lambda, p_{\lambda\lambda'}, \Lambda)$$

of compact polyhedra X_λ.

Remark 4. If the weight satisfies $w(X) \leq \kappa$, one can achieve that the cardinal $|\Lambda| \leq \kappa$. If X is a compact metric space, one can achieve that $\mathbf{X}$ be an inverse sequence.

The following generalization of Theorem 3 was first obtained by M. D. Alder in 1974 [1].

Theorem 5. *Every paracompact space* X *is the limit of an inverse system of polyhedra.*

This result was further generalized as follows (see [47] and [40]).

Theorem 6. *Every topologically complete space* X *is the limit of an inverse system of polyhedra.*

Recall that a space is topologically complete (or Dieudonné complete) if its topology is induced by a complete uniformity. It is well known that paracompact spaces, hence also metric spaces, are topologically complete [14].

We will consider in this paper results related to Theorems 3, 5 and 6. In particular, we will see that in various applications, for non-compact spaces, one needs a notion finer than that of an inverse limit. This is the notion of resolution, which can be viewed as a limit with additional desirable properties (see Section 4). Approximate inverse limits and approximate resolutions offer generalizations in a different direction (see Section 5 and 6). If one has in mind applications to shape theory, one needs other variants of the notion of inverse limit, called homotopy expansion and strong homotopy expansion (see Section 7).

2. Properties preserved under inverse limits

We say that a property C of spaces is inherited by the inverse limit, or equivalently, a class of spaces C is closed under inverse limits, provided $X_\lambda \in C$ implies $X \in C$. The following elementary theorem lists some of those properties (see e.g., [37], [14], [15]).

Theorem 7. *Any of the following classes of spaces are closed under inverse limits:*

  (i)   compact Hausdorff spaces,

  (ii) non-empty compact Hausdorff spaces,

  (iii) connected compact Hausdorff spaces,

  (iv) compact Hausdorff spaces of covering dimension $\leq n$.

Remark 8. In Theorem 7 compactness is an essential assumption. E.g., the limit of an inverse system of paracompact (metric) spaces need not be paracompact (metric). Even polyhedral counter-examples exist. Indeed, apply Example 2 to an uncountable collection of copies of the discrete space $\mathbb{N}$. One obtains an inverse system of copies of $\mathbb{N}$, whose limit is the product of an uncountable collection of copies of $\mathbb{N}$. However, it is well known that this product fails to be normal [60], hence, it cannot be paracompact. Consequently, the converse of Theorem 5 does not hold. The following is a positive result.

Theorem 9. *The limit of an inverse system of topologically complete spaces is topologically complete. In particular, the limit of an inverse system of polyhedra, metric spaces or paracompact spaces is topologically complete* [14].

Using Example 1, it is easy to find counter-examples for the non-compact analogues of statements (ii) and (iii) in Theorem 7. Concerning part (iv) of Theorem 7, there exist sophisticated examples showing that the compactness assumption cannot be omitted. In particular, M. G. Charalambous has exhibited an inverse sequence $\mathbf{X}$ of 0-dimensional Lindelöf spaces, whose limit X is normal, but its covering dimension $\dim X > 0$ [10]. Moreover, there exist examples of normal spaces X with $\dim X > 0$, which are $\mathbb{N}$-compact and thus, are limits of inverse systems consisting of copies of $\mathbb{N}$, [11]. The phenomenon that limits increase dimension is not possible for inverse sequences of metric spaces. This follows from the next result, due to K. Nagami [50].

Theorem 10. *If* $\mathbf{X}$ *is an inverse sequence of metric spaces* X_i *of dimension* $\dim X_i \leq n$, *then also the limit* X *has dimension* $\dim X \leq n$.

Nagami's theorem generalizes to the cohomological dimension $\dim_G$ and more generally, to the extension dimension ext-dim. For a fixed polyhedron P and a space X,

$$\mathrm{ext} - \dim X \leq P$$

means that every mapping $f \colon A \to P$, defined on a closed subset $A \subseteq X$ admits an extension $\tilde{f} \colon X \to P$. For paracompact spaces,

$$\dim_G X \leq n$$

is equivalent to

$$\mathrm{ext} - \dim X \leq P,$$

where P is an Eilenberg-MacLane complex of type $K(G, n)$.

Theorem 11. *([58])* *If* $\mathbf{X}$ *is an inverse sequence of metric spaces* X_i *of extension dimension*

$$\mathrm{ext} - \dim X_i \leq P,$$

where P *is an arbitrary polyhedron, then also the limit* X *has extension dimension*

$$\mathrm{ext} - \dim X \leq P.$$

Remark 12. Theorem 11 remains true if one replaces the assumption that the spaces X_i are metric by the weaker assumption that they are stratifiable [33]. In this case X is also stratifiable. Since polyhedra are stratifiable spaces and for normal spaces $\dim X \leq n$ is equivalent to ext-dim $X \leq P$, for $P = S^n$, one obtains the following result.

Corollary 13. *If* $\mathbf{X}$ *is an inverse sequence of polyhedra* X_i *of dimension* $\dim X_i \leq n$, *then the limit* X *is a stratifiable space of dimension* $\dim X \leq n$.

Corollary 13 also follows from the next theorem, which generalizes Nagami's Theorem 10 and is due to M.G. Charalambous [9].

Theorem 14. *If* $\mathbf{X}$ *is an inverse sequence of perfectly normal spaces* X_i *of dimension* $\dim X_i \leq n$, *then the limit* X *is a perfectly normal space of dimension* $\dim X \leq n$.

In some cases one can prove that a property is inherited by the limit, provided one imposes additional conditions on the bonding mappings $p_{\lambda\lambda'}$. Here are three examples of such theorems. The first one refers to $\mathcal{P}$-like spaces, where $\mathcal{P}$ is a class of polyhedra. A topological space X is said to be $\mathcal{P}$-like provided for every normal covering $\mathcal{U}$ of X there exist a polyhedron $P \in \mathcal{P}$, an open covering $\mathcal{V}$ of P and a surjective mapping $p \colon X \to P$ such that $p^{-1}(\mathcal{V})$ refines $\mathcal{U}$. Recall that normal coverings are open coverings which admit a partition of unity. In paracompact spaces every open covering is normal. In the case of compact Hausdorff spaces the definition assumes the following simpler form. For every open covering $\mathcal{U}$ of X there exist a polyhedron $P \in \mathcal{P}$ and a surjective mapping

$$p \colon X \to P$$

such that the fibers

$$p^{-1}(y), \quad y \in P,$$

refine $\mathcal{U}$. The following theorem is from [37] and [28].

Theorem 15. *Let* $\mathcal{P}$ *be a collection of compact polyhedra and let* $\mathbf{X}$ *be an inverse system of compact Hausdorff spaces* X_λ, *which are* $\mathcal{P}$-like. *If all the bonding mappings* $p_{\lambda\lambda'}$ *of* $\mathbf{X}$ *are surjective, then the limit* X *is* $\mathcal{P}$-like.

A much deeper result of this type is the following theorem [38], [39].

Theorem 16. *Let* $\mathcal{M}^n$ *be a class of closed connected compact triangulable n-manifolds. Let* $\mathbf{X}$ *be an inverse sequence such that* $X_i \in \mathcal{M}^n$ *and all the bonding mappings* $p_{ii'}$ *are surjective. If the limit* X *is an n-dimensional absolute neighborhood retract, then* X *is a generalized n-manifold over any principal ideal domain.*

The next theorem [27] involves monotone mappings. Recall that a mapping $f \colon X \to Y$ between compact Hausdorff spaces is monotone, provided all the fibers $f^{-1}(y)$, $y \in Y$, are connected.

Theorem 17. *Let $\mathbf{X}$ be an inverse system of locally connected Hausdorff continua. If the bonding mappings are monotone, then also the limit space X is a locally connected Hausdorff continuum. Moreover, the projections*

$$p_\lambda \colon X \to X_\lambda$$

are monotone mappings.

Inverse sequences make the ideal tool for constructing complicated metric compacta and were used or can be used in many classical examples, e.g.: the Cantor set, the solenoids, the Knaster continuum (pseudoarc), the Sierpiński carpet, etc. It is more difficult to use inverse systems to construct complicated compact Hausdorff spaces. An example of the use of this technique is my paper [25]. There I constructed a (non-metric) chainable continuum X, whose small inductive dimension ind $X = 2$. Note that the covering dimension dim of every chainable continuum is 1.

3. Spaces as limits of polyhedral systems with additional properties

The study of a space X having a certain property $\mathcal{C}$ is greatly facilitated if one can represent X as the limit of an inverse system of polyhedra X_λ having property $\mathcal{C}$. The first interesting result of this type is the following theorem, obtained in 1937 in H. Freudenthal's Ph.D. Thesis [16].

Theorem 18. *Every metric compact space X of dimension $\dim X \leq n$ is the limit of an inverse sequence $\mathbf{X} = (X_i, p_{ii+1})$ of compact polyhedra X_i of dimension $\dim X_i \leq n$.*

I used this result in a paper on the homology of the function space $(S^m)^X$, where S^m is the m-sphere and X is a metric compactum of dimension

$$\dim X = k < m.$$

The result establishes an isomorphism between the (singular) homology group

$$H_{m-k}((S^m)^X; \mathbb{Z})$$

and the Čech cohomology group

$$\check{H}^k(X; \mathbb{Z}),$$

[23]. It was natural to conjecture that an analogous result also holds for compact Hausdorff spaces X. My proof would work if I had the analogue of Theorem 18 for compact Hausdorff spaces. I was indeed surprised, when I discovered that this analogue is false. Indeed, I found a simple argument which shows that the limit X of an inverse system of compact polyhedra X_λ of dimension $\dim X_\lambda \leq 1$ always has small inductive dimension ind $X \leq 1$. Since in the literature there existed examples of compact Hausdorff spaces X with $\dim X = 1$, but ind $X = 2$ (see [22], [21]), it was clear that there exist 1-dimensional compact Hausdorff spaces, which are not limits of inverse systems of 1-dimensional polyhedra. I announced my result in 1958 [24]. Shortly after that, I found that in the Doklady

of the Soviet Academy that same year B. A. Pasynkov announced the same result [52]. I was not pleased by this discovery, but I did not get discouraged. To the contrary, it stimulated me to obtain the following positive result [26], which suffices to extend the result on function spaces $(S^m)^X$ to the desired case of compact Hausdorff spaces.

Theorem 19. *Every compact Hausdorff space X of dimension $\dim X \leq n$ is the limit of an inverse system $\mathbf{X} = (X_\lambda, p_{\lambda\lambda'}, \Lambda)$ of metric compacta X_λ of dimension $\dim X_\lambda \leq n$.*

The key step in the proof of Theorem 19 was the following factorization theorem [26].

Theorem 20. *Let $f \colon X \to Y$ be a mapping of an n-dimensional compact Hausdorff space X to a compact metric space Y. Then there exist an n-dimensional compact metric space Z and two mappings*

$$g \colon X \to Z \quad and \quad h \colon Z \to Y$$

such that

$$f = hg.$$

I liked the theorem because I have not seen anything similar before. At that time I was spending two academic years (1957/58 and 1958/59) at the Institute for Advanced Study in Princeton. Although Princeton was the world's capital of Topology, there was nobody there in dimension theory with whom I could discuss my result. Therefore, I mailed a copy of my paper to Moscow to professor P.S. Aleksandrov. There was no immediate reaction. However, in June of 1960 I attended in Zürich an international colloquium in honor of H. Hopf. There I saw Aleksandrov for the first time and I decided to tell him about my result. I introduced myself and was pleasantly surprised when he praised my paper and told me that they have studied it in details in his seminar in Moscow. Following my paper many different factorization theorems were discovered, especially by members of the Aleksandrov group in Moscow, and factorization theorems became a recognized technique in dimension theory of general spaces (see e.g., [6], [62], [5], [59], [8], [51], [17], [54], [55], [56]) etc. Factorization theorems are still of interest. E.g., M. Levin, L.R. Rubin and P.J. Schapiro have recently obtained a factorization theorem for extension dimension [20]. A simplified version reads as follows.

Theorem 21. *Let P be a polyhedron and let $f \colon X \to Y$ be a mapping of a compact Hausdorff space X of extension dimension*

$$\text{ext} - \dim X \leq P$$

to a compact metric space Y. Then there exist a compact metric space Z with

$$\text{ext} - \dim Z \leq P$$

and two mappings

$$g \colon X \to Z \quad and \quad h \colon Z \to Y$$

such that

$$f = hg.$$

In the case when P is separable, the assertion of Theorem 21 was obtained already in [56].

Freudenthal's theorem admits the following generalization to $\mathcal{P}$-like continua [37].

Theorem 22. *Let $\mathcal{P}$ be a class of compact connected polyhedra. Every metric $\mathcal{P}$-like continuum X is the limit of an inverse sequence $\mathbf{X}$, which consists of polyhedra belonging to $\mathcal{P}$ and of surjective bonding mappings.*

The following theorem from [27] is related to Theorem 17.

Theorem 23. *Every locally connected compact Hausdorff space is the limit of an inverse system, consisting of locally connected metric compacta and of monotone bonding mappings.*

4. Resolutions of spaces

The notion of an inverse limit looses most of its good properties when one abandons the realm of compact spaces. In the previous sections we saw examples supporting this claim. Here is yet another example to this effect. It is a sequence $\mathbf{X}$ of connected 1-dimensional polyhedra, whose limit fails to be connected in spite of the fact that all X_i are connected and all the bonding mappings are surjections.

Example 24. Let
$$X_i = (\mathbb{R} \times \{0\}) \cup (\{i\} \times \mathbb{R}^+) \subseteq \mathbb{R}^2,$$
where
$$\mathbb{R}^+ = \{t \in \mathbb{R} : t \geq 0\},$$
and let
$$p_{ii+1} \colon X_{i+1} \to X_i$$
be defined as follows: $p_{ii+1}|\mathbb{R} \times \{0\}$ is the identity mapping, p_{ii+1} maps the oriented segment $\{i+1\} \times [0,1]$ linearly onto the oriented segment $[i+1, i] \times \{0\}$ and maps the ray $\{i+1\} \times (1 + \mathbb{R}^+)$ linearly onto the ray $\{i\} \times \mathbb{R}^+$. It is readily seen that the limit of $\mathbf{X}$ is the disjoint union of two copies of $\mathbb{R}$.

It appears that the right modification of the notion of inverse limit is the notion of resolution. The following definition first appeared in 1981 in my paper [29].

A resolution of a space X is a mapping $\mathbf{p} \colon X \to \mathbf{X}$ of X to an inverse system
$$\mathbf{X} = (X_\lambda, p_{\lambda\lambda'}, \Lambda),$$
which has the property that, for every polyhedron P and every open covering $\mathcal{V}$ of P, the following conditions are satisfied:

(R1) For every mapping $f: X \to P$, there exist a $\lambda \in \Lambda$ and a mapping $h: X_\lambda \to P$ such that the mappings hp_λ and f are $\mathcal{V}$-near, i.e., every $x \in X$ admits a $V \in \mathcal{V}$ such that $hp_\lambda(x), f(x) \in V$.

(R2) There exists an open covering $\mathcal{V}'$ of P, such that whenever, for a $\lambda \in \Lambda$ and for two mappings $h, h': X_\lambda \to P$, the mappings $hp_\lambda, h'p_\lambda$ are $\mathcal{V}'$-near, then there exists a $\lambda' \geq \lambda$, such that the mappings $hp_{\lambda\lambda'}, h'p_{\lambda\lambda'}$ are $\mathcal{V}$-near.

The following example is the analogue of Example 1.

Example 25. Let M be a topological space, X a subset of M and $(X_\lambda, \lambda \in \Lambda)$ a basis of neighborhoods of X in M. Order Λ by putting $\lambda \leq \lambda'$, whenever $X_{\lambda'} \subseteq X_\lambda$, and let

$$p_{\lambda\lambda'}: X_{\lambda'} \to X_\lambda, \qquad p_\lambda: X \to X_\lambda,$$

be inclusion mappings. Then

$$\mathbf{X} = (X_\lambda, p_{\lambda\lambda'}, \Lambda)$$

is an inverse system and

$$\mathbf{p} = (p_\lambda): X \to \mathbf{X}$$

is a mapping. If all X_λ are paracompact spaces, $\mathbf{p}$ is a resolution of X.

Although the definition of resolution differs very much from the definition of inverse limit, the following theorem shows that, under very general conditions, resolutions can be viewed as special inverse limits (see [32]).

Theorem 26. *Let*

$$\mathbf{p}: X \to \mathbf{X} = (X_\lambda, p_{\lambda\lambda'}, \Lambda)$$

be a resolution. If all X_λ are Tychonoff spaces and X is a topologically complete space, then $\mathbf{p}$ is an inverse limit of $\mathbf{X}$.

The converse holds in the compact case (see [32]).

Theorem 27. *The limit $\mathbf{p}: X \to \mathbf{X}$ of an inverse system of compact Hausdorff spaces is always a resolution of X.*

Using resolutions instead of inverse limits one obtains the following generalization of Theorem 3, which shows that resolutions make a tool suitable for approximating arbitrary spaces by polyhedra.

Theorem 28. *Every topological space X admits a polyhedral resolution, i.e., a resolution $\mathbf{p}: X \to \mathbf{X} = (X_\lambda, p_{\lambda\lambda'}, \Lambda)$, where all X_λ are polyhedra.*

The following analogues of the respective parts of Theorem 7 hold for resolutions.

Theorem 29. *Let C be any of the following classes of spaces:*

  (i) *non-empty spaces,*

  (ii) *connected spaces,*

 (iii) *spaces having covering dimension (based on normal coverings) $\leq n$.*

If $\mathbf{p}\colon X \to \mathbf{X}$ *is a resolution of X and all terms X_λ of $\mathbf{X}$ belong to C, then so does X.*

Theorem 29 is easily proved if one uses the following characterization of resolutions (see [32]).

Theorem 30. *A mapping* $\mathbf{p}\colon X \to \mathbf{X}$ *is a resolution if and only if it has the following two properties:*

(B1) *For every normal covering $\mathcal{U}$ of X, there exist a $\lambda \in \Lambda$ and a normal covering $\mathcal{U}_\lambda$ of X_λ such that $p_\lambda^{-1}(\mathcal{U}_\lambda)$ refines $\mathcal{U}$.*

(B2) *For every $\lambda \in \Lambda$ and every normal covering $\mathcal{U}_\lambda$ of X_λ, there exists a $\lambda' \geq \lambda$ such that*

$$p_{\lambda\lambda'}(X_{\lambda'}) \subseteq \operatorname{St}(p_\lambda(X), \mathcal{U}_\lambda),$$

where $\operatorname{St}(A, \mathcal{V})$ denotes the star of A with respect to $\mathcal{V}$, i.e., the union of all sets $V \in \mathcal{V}$ with $A \cap V \neq \varnothing$.

The notion of resolution was developed in a series of papers by P. Bacon, K. Morita and myself. In 1975 Bacon considered mappings $\mathbf{p}\colon X \to \mathbf{X}$ having property (B1) and a stronger form of (B2) [7]. Also in 1975 Morita considered inverse limits having a certain property P. He called them proper limits [48]. He proved in 1984 that condition P is equivalent to the union of conditions (R1) and (R2) [49].

5. Approximate inverse systems

The fact that there exist 1-dimensional compact Hausdorff spaces, which are not limits of 1-dimensional polyhedra, and the fact that there exist Hausdorff chainable continua, which are not limits of arcs were among the reasons which led L. R. Rubin and myself to introduce in 1989 a more flexible kind of inverse systems of metric compact spaces, called approximate inverse systems [35]. The main idea was to abandon the rigid functorial requirement

$$p_{\lambda\lambda'}p_{\lambda'\lambda''} = p_{\lambda\lambda''}, \quad \text{for } \lambda \leq \lambda' \leq \lambda'',$$

and allow the mappings $p_{\lambda\lambda'}p_{\lambda'\lambda''}$ and $p_{\lambda\lambda''}$ to differ. However, the difference should be arbitrarily small when λ' is large enough. The notion was soon extended to arbitrary spaces by T. Watanabe and myself [44]. It was noticed by M. G. Charalambous that a simplified notion suffices if one is interested in approximating only spaces and not mappings [12] (also see [30] and [42]).

An approximate inverse system $\mathbf{X} = (X_\lambda, p_{\lambda\lambda'}, \Lambda)$ consists of the same data as an ordinary inverse systems. However, beside the requirement that $p_{\lambda\lambda}$ is the identity mapping, one imposes the following condition.

(A) For any $\lambda \in \Lambda$ and any normal covering $\mathcal{U}$ of X_λ, there exists an $\lambda' \geq \lambda$ such that the mappings $p_{\lambda\lambda_1}p_{\lambda_1\lambda_2}$ and $p_{\lambda\lambda_2}$ are $\mathcal{U}$ - near, for $\lambda_2 \geq \lambda_1 \geq \lambda'$.

Clearly, every ordinary inverse system is also an approximate system.

In the case of approximate systems $\mathbf{X}$, the role of mappings $\mathbf{h} \colon Z \to \mathbf{X}$ is taken up by approximate mappings $\mathbf{h}$. They consist of mappings $h_\lambda \colon Z \to X_\lambda$, for $\lambda \in \Lambda$. However, we do not require that $p_{\lambda\lambda'}h_{\lambda'}$ coincides with h_λ, for $\lambda \leq \lambda'$. Instead, we require that they satisfy the following condition.

(AM) For any $\lambda \in \Lambda$ and any normal covering $\mathcal{U}$ of X_λ, there exists a $\lambda' \geq \lambda$, such that the mappings $p_{\lambda\lambda''}h_{\lambda''}$ and h_λ are $\mathcal{U}$ - near, for every $\lambda'' \geq \lambda'$.

One defines the limit of an approximate inverse system $\mathbf{X}$ as an approximate mapping $\mathbf{p} \colon X \to \mathbf{X}$, which has the following universal property:

Whenever $\mathbf{h} \colon Z \to \mathbf{X}$ is an approximate mapping, then there exists a unique mapping $h \colon Z \to X$ such that

$$\mathbf{p}h = \mathbf{h},$$

i.e., $p_\lambda h = h_\lambda$, for all $\lambda \in \Lambda$.

Uniqueness of approximate limits (up to natural homeomorphism) follows immediately. If all X_λ are Tychonoff spaces, the approximate limit exists. It suffices to take for X the subspace of the product

$$\prod_\lambda X_\lambda,$$

which consists of all points $x = (x_\lambda)$, $x_\lambda \in X_\lambda$, satisfying the condition

$$\lim_{\lambda' \geq \lambda} p_{\lambda\lambda'}(x_{\lambda'}) = x_\lambda, \quad \text{for every } \lambda \in \Lambda.$$

For the mappings $p_\lambda \colon X \to X_\lambda$, one takes the restrictions of the canonical projections

$$\prod_\lambda X_\lambda \to X_\lambda.$$

The next theorem shows that good properties of limits of inverse systems of compact spaces are also properties of limits of approximate systems of compact spaces (see [35] and [41]). In particular, we have the following theorem.

Theorem 31. *Let C be any of the classes from Theorem 7. If in an approximate inverse system $\mathbf{X}$ all terms X_λ belong to C, then so does the limit X.*

Approximate systems of compact polyhedra can be used in constructing complicated compact Hausdorff spaces. E.g., in [36] L. R. Rubin and I proved the following non-metric version of a well-known result of R. D. Edwards and J. J. Walsh.

Theorem 32. *Let X be a compact Hausdorff space X, whose cohomological dimension*

$$\dim_{\mathbb{Z}} X \leq n, \qquad n \geq 1.$$

Then there exist a compact Hausdorff space Z, whose covering dimension $\dim Z \leq n$ and weight $w(Z) \leq w(X)$, and a cell-like mapping $f \colon Z \to X$.

In the proof we defined Z as the limit of an approximate system of compact polyhedra of dimension $\leq n$.

The usefulness of approximate systems of compact spaces is clearly seen from the next two theorems.

Theorem 33. **([35])** *Every compact Hausdorff space X of dimension $\dim X \leq n$ is the limit of an approximate system $\mathbf{X}$ of compact polyhedra X_λ of dimension $\dim X_\lambda \leq n$.*

Theorem 34. *Let $\mathcal{P}$ be a class of compact connected polyhedra. Every compact Hausdorff space X, which is $\mathcal{P}$-like, is the limit of an approximate inverse system $\mathbf{X}$, which consists of polyhedra from $\mathcal{P}$ and of surjective bonding mappings* [41].

Theorem 33 can be viewed as a generalization of the classical Freudenthal theorem (Theorem 18), while Theorem 34 can be viewed as a generalization of Theorem 22. In both cases the assumption that $\mathbf{X}$ is an approximate system is essential.

6. Approximate resolutions of spaces

Approximate resolutions were defined in my joint paper with T. Watanabe [44] by unifying the idea of resolution with the idea of approximate system (for a simplified version see [30]).

An approximate mapping $\mathbf{p} \colon X \to \mathbf{X}$ of a space X to an approximate system $\mathbf{X}$ is an approximate resolution provided the following two conditions (closely related to conditions (R1) and (R2) of Section 4) are satisfied, for any polyhedron P and any open covering $\mathcal{V}$ of P.

(R1)′ For every mapping $f \colon X \to P$, there exists a $\lambda \in \Lambda$ such that for any $\lambda' \geq \lambda$, there exists a mapping $h \colon X_{\lambda'} \to P$ such that the mappings $hp_{\lambda'}$ and f are $\mathcal{V}$-near.

(R2)′ There exists an open covering $\mathcal{V}'$ of P such that, for any $\lambda \in \Lambda$ and mappings $h, h' \colon X_\lambda \to P$ such that $hp_\lambda, h'p_\lambda$ are $\mathcal{V}'$-near mappings, there exists a $\lambda' \geq \lambda$, such that the mappings $hp_{\lambda\lambda''}, h'p_{\lambda\lambda''}$ are $\mathcal{V}$-near, for any $\lambda'' \geq \lambda'$.

An important feature of approximate resolutions is that they share all the good properties of resolutions. In particular, the analogue of Theorem 29 holds also for approximate resolutions. This is so because approximate resolutions can be characterized by properties (B1) and (B2) from Theorem 30 (see [44]). Similarly, the analogue of Theorem 26 holds also for approximate resolutions [44]. Theorems 33 and 34, which were valid only in the compact case, now have the following analogues.

Theorem 35. ([61]) *Every topological space X of dimension* $\dim X \leq n$ *admits an approximate resolution* $\mathbf{p} \colon X \to \mathbf{X}$, *where the approximate system* $\mathbf{X}$ *consists of polyhedra X_λ of dimension* $\dim X_\lambda \leq n$.

Theorem 36. ([34]) *Let $\mathcal{P}$ be a class of connected locally compact polyhedra. A topological space X is $\mathcal{P}$-like if and only if it admits an approximate resolution* $\mathbf{p} \colon X \to \mathbf{X}$ *such that all X_λ belong to $\mathcal{P}$ and all $p_{\lambda\lambda'}$ are surjective.*

Remark 37. In the case of normal spaces X, Vlasta Matijević [45] has improved Theorem 35 by showing that one can also achieve that all bonding mappings $p_{\lambda\lambda'}$ in $\mathbf{X}$ and all projection p_λ be irreducible. Recall that a mapping $f \colon X \to P$ to a polyhedron is irreducible provided P admits a triangulation K having the following property. There is no mapping

$$g \colon X \to |L|$$

to a proper subcomplex L of K such that, for every simplex $\sigma \in K$,

$$f(x) \in \sigma \quad \text{implies} \quad g(x) \in \sigma.$$

Note that irreducible mappings are always surjective.

Here is another result on approximate resolutions [46].

Theorem 38. *A topological space is finitistic (i.e., every normal covering admits a refinement having a finite-dimensional nerve) if and only if it admits an approximate resolution consisting of finite-dimensional polyhedra.*

7. Homotopy expansions of spaces

Inverse systems play an important role in the development of shape theory —an extension of homotopy theory from the realm of polyhedra and CW-complexes to arbitrary topological spaces. In ordinary shape theory one constructs a category $\mathrm{Sh}(\mathrm{Top})$, called the shape category, and a functor

$$S \colon \mathrm{H}(\mathrm{Top}) \to \mathrm{Sh}(\mathrm{Top}),$$

called the shape functor, defined on the homotopy category $\mathrm{H}(\mathrm{Top})$. Similarly, in strong shape theory one constructs a category $\mathrm{SSh}(\mathrm{Top})$, called the strong shape category, and a functor

$$\overline{S} \colon \mathrm{H}(\mathrm{Top}) \to \mathrm{SSh}(\mathrm{Top}),$$

called the strong shape functor. Moreover, one defines a forgetful functor

$$E \colon \mathrm{SSh}(\mathrm{Top}) \to \mathrm{Sh}(\mathrm{Top})$$

such that

$$E\overline{S} = S.$$

In all three categories H(Top), Sh(Top) and SSh(Top) objects are topological spaces and the functors $S, \overline{S}$ and E preserve objects. In the case when the target Y of a shape morphism is a polyhedron, shape morphisms

$$F\colon X \to Y$$

are simply morphisms of H(Top), i.e., homotopy classes of mappings

$$[f]\colon X \to Y.$$

To obtain shape morphisms in the general case, one approximates spaces by polyhedra. More precisely, one associates with X and Y suitable inverse systems of polyhedra **X**, **Y** and mappings

$$\mathbf{p}\colon X \to \mathbf{X}, \qquad \mathbf{q}\colon Y \to \mathbf{Y},$$

called expansions. In the case of ordinary shape these mappings are homotopy expansions and in the case of strong shape they are strong homotopy expansions (both defined below).

One also develops appropriate homotopy theories of inverse systems. More precisely, one constructs appropriate categories, whose objects are inverse systems of spaces. One usually requires that these systems be cofinite. In the case of ordinary shape, the appropriate category is the category $\mathrm{pro}-\mathrm{H}(\mathrm{Top})$, i.e., the procategory associated with the category Top. To define its morphisms $\mathbf{X} \to \mathbf{Y}$, one considers homotopy mappings

$$\mathbf{f}\colon \mathbf{X} \to \mathbf{Y}.$$

As in the case of mappings,

$$\mathbf{f} = (f, f_\lambda)\colon \mathbf{X} \to \mathbf{Y}$$

consists of an increasing function $f\colon \mathrm{M} \to \Lambda$ and of mappings

$$f_\mu\colon X_{f(\mu)} \to Y_\mu, \qquad \mu \in \mathrm{M}.$$

However, equality (1) from Section 1 is now replaced by the homotopy relation:

$$f_\mu p_{f(\mu)f(\mu')} \simeq q_{\mu\mu'} f_{\mu'}, \quad \mu \le \mu'. \tag{3}$$

Morphisms $[\mathbf{f}]\colon \mathbf{X} \to \mathbf{Y}$ are now defined as equivalence classes of homotopy mappings $\mathbf{f}\colon \mathbf{X} \to \mathbf{Y}$, where $\mathbf{f} \simeq \mathbf{f}'$ provided there exists an increasing function $g \ge f, f'$ such that

$$f_\mu p_{f(\mu)g(\mu)} \simeq f'_\mu p_{f'(\mu)g(\mu)}, \quad \text{for } \mu \le \mu'. \tag{4}$$

In the case of strong shape, the appropriate category is the coherent homotopy category CH(pro-Top). Its morphisms are equivalence classes of coherent mappings $\mathbf{f}\colon \mathbf{X} \to \mathbf{Y}$. The latter are given by increasing functions $f\colon \mathrm{M} \to \Lambda$, by mappings

$$f_\mu\colon X_{f(\mu)} \to Y_\mu$$

and also by n-homotopies $f_{\mu_0\mu_1\ldots\mu_n}$, for $n \ge 1$, where

$$f_{\mu_0\mu_1\ldots\mu_n}\colon X_{f(\mu_n)} \times \Delta^n \to Y_{\mu_0}$$

is a mapping, Δ^n is the standard n-simplex and

$$\mu_0 \leq \mu_1 \leq \dots \leq \mu_n$$

is any increasing sequence of indices in M. The 1-homotopies $f_{\mu\mu'}$ realize (3), the 2-homotopies

$$f_{\mu_0\mu_1\mu_2}: X_{f(\mu_2)} \times \Delta^2 \to Y_{\mu_0}$$

connect the 1-homotopies $q_{\mu_0\mu_1}f_{\mu_1\mu_2}$, $f_{\mu_0\mu_2}$ and $f_{\mu_0\mu_1}(p_{f(\mu_1)f(\mu_2)} \times 1)$, etc. In an alternative, but equivalent approach, one defines CH(pro-Top) (in this case usually denoted by Ho(pro-Top)) as the localization of the category pro-Top at homotopy level equivalences, i.e., at morphisms $[\mathbf{f}]$ of pro-Top, which have a representative $\mathbf{f}: \mathbf{X} \to \mathbf{Y}$, where f is the identity mapping and every $f_\lambda: X_\lambda \to Y_\lambda$ is a homotopy equivalence.

In both shape categories a morphism $F: X \to Y$ is given by two polyhedral expansions

$$\mathbf{p}: X \to \mathbf{X}, \qquad \mathbf{q}: Y \to \mathbf{Y}$$

and by a morphism $[\mathbf{f}]: \mathbf{X} \to \mathbf{Y}$. In the case of ordinary shape, $[\mathbf{f}]$ is from the category pro-H(Top) and in the case of strong shape, it is from CH(pro-Top). The conditions imposed on expansions insure that F does not depend on the particular choice of the expansions $\mathbf{p}$ and $\mathbf{q}$. More precisely, if $\mathbf{p}': X \to \mathbf{X}'$ and $\mathbf{q}': Y \to \mathbf{Y}'$ is another choice of polyhedral expansions, then there exist unique isomorphisms

$$[\mathbf{i}]: \mathbf{X} \to \mathbf{X}' \quad \text{and} \quad [\mathbf{j}]: \mathbf{Y} \to \mathbf{Y}'$$

such that

$$[\mathbf{i}][\mathbf{p}] = [\mathbf{p}'] \quad \text{and} \quad [\mathbf{j}][\mathbf{q}] = [\mathbf{q}'].$$

If $F': X' \to Y'$ is given by

$$\mathbf{p}': X \to \mathbf{X}', \quad \mathbf{q}': Y \to \mathbf{Y}'$$

and by a morphism

$$[\mathbf{f}']: \mathbf{X}' \to \mathbf{Y}',$$

then F' is identified with F whenever

$$[\mathbf{j}][\mathbf{f}] = [\mathbf{f}'][\mathbf{i}].$$

For a more detailed description of the above mentioned homotopy categories of inverse systems see [33]. Here we will describe precisely the two kinds of expansions.

We say that a mapping

$$\mathbf{p} = (p_\lambda): X \to \mathbf{X}$$

is a homotopy expansion provided the following two properties of Morita are fulfilled.

(M1) If P is a polyhedron and $f: X \to P$ is a mapping, then there exists a $\lambda \in \Lambda$ and there exists a mapping $h: X_\lambda \to P$ such that

$$hp_\lambda \simeq f.$$

(M2) If $\lambda \in \Lambda$ and $h_0, h_1 \colon X_\lambda \to P$ are mappings such that

$$h_0 p_\lambda \simeq h_1 p_\lambda,$$

then there exist a $\lambda' \geq \lambda$ such that

$$h_0 p_{\lambda\lambda'} \simeq h_1 p_{\lambda\lambda'}.$$

The following diagrams illustrate properties (M1) and (M2).

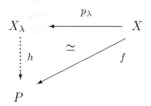 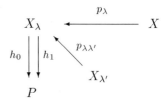

Resolutions and homotopy expansions are related by the following theorem (see [40]).

Theorem 39. *Every resolution* $\mathbf{p} \colon X \to \mathbf{X}$ *is a homotopy expansion.*

An immediate consequence of Theorems 28 and 39 is the following theorem, which plays an essential role in the construction of the category $\mathrm{Sh(Top)}$, because it insures the existence of desired polyhedral approximations of spaces.

Theorem 40. *Every topological space* X *admits a polyhedral homotopy expansion*

$$\mathbf{p} \colon X \to \mathbf{X}.$$

A mapping

$$\mathbf{p} = (p_\lambda) \colon X \to \mathbf{X}$$

is a strong homotopy expansion, for short a strong expansion, provided it has Morita's property (M1) as well as the following strong form of property (M2).

(S2) If $\lambda \in \Lambda$, $h_0, h_1 \colon X_\lambda \to P$ are mappings, and

$$F \colon X \times I \to P$$

is a homotopy which connects $h_0 p_\lambda$ and $h_1 p_\lambda$, then there exist a $\lambda' \geq \lambda$ and a homotopy

$$H \colon X_{\lambda'} \times I \to P,$$

which connects $h_0 p_{\lambda\lambda'}$ and $h_1 p_{\lambda\lambda'}$. Moreover, the homotopies

$$H(p_{\lambda'} \times 1), \quad F \colon X \times I \to P$$

are connected by a homotopy $(X \times I) \times I \to P$, which is fixed on $X \times \partial I$, i.e.,

$$H(p_{\lambda'} \times 1) \simeq F \, (\mathrm{rel} \, (X \times \partial I)).$$

The following diagram illustrates property (S2).

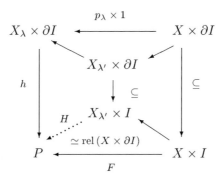

Clearly, every strong expansion is a homotopy expansion. Therefore, the following theorem is a strengthening of Theorem 39.

Theorem 41. *Every resolution* $p: X \to \mathbf{X}$ *is a strong expansion.*

An immediate consequence of Theorems 28 and 41 is the following theorem, which plays an essential role in the construction of the category SSh(Top), because it insures the existence of desired polyhedral approximations of spaces.

Theorem 42. *Every topological space* X *admits a strong expansion* $p: X \to \mathbf{X}$ *which consists of polyhedra.*

References

1. Alder, M.D., Inverse limits of simplicial complexes, *Compositio Math.* **29** (1974), 1–7.
2. Alexandroff, P.S., Simpliziale Approximationen in der allgemeinen Topologie, *Math. Ann.* **96** (1926), 489–511.
3. Alexandroff, P.S., Une définition des nombres de Betti pour un ensemble fermé quelconque, *C. R. Acad. Sci. Paris* **184** (1927), 317–319.
4. Alexandroff, P.S., Untersuchungen über Gestalt und Lage abgeschlossener Mengen beliebiger Dimension, *Ann. Math.* **30** (1928), 101–187.
5. Aleksandrov, P.S., and Pasynkov, B.A., *Introduction to Dimension Theory* (Russian), Nauka, Moscow, 1973.
6. Arhangel'skiĭ, A., The factorization of mappings by weight and dimension (Russian), *Dokl. Akad. Nauk SSSR* **174** (1967), 1243–1246.
7. Bacon, P., Continuous functors, *Gen. Top. Appl.* **5** (1975), 321–331.
8. Bogatyĭ, S., and Smirnov, Ju.M., Approximation by polyhedra and factorization theorems for ANR-bicompacta (Russian), *Fund. Math.* **87** (1975), 195–205.
9. Charalambous, M.G., The dimension of inverse limits, *Proc. Amer. Math. Soc.* **58** (1976), 289–296.

10. Charalambous, M.G., An example concerning inverse limit sequences of normal spaces, *Proc. Amer. Math. Soc.* **78** (1980), 605–608.

11. Charalambous, M.G., The dimension of inverse limit and N-compact spaces, *Proc. Amer. Math. Soc.* **85** (1982), 648–652.

12. Charalambous, M.G., Approximate inverse systems of uniform spaces and an application of inverse systems, *Comment. Math. Univ. Carolinae* **32** (1991), 551–565.

13. Eilenberg, S., and Steenrod, N.E., *Foundations of Algebraic Topology*, Princeton Univ. Press, Princeton, 1952.

14. Engelking, R., *General Topology*, Monografie Matematyczne 60, Polish Scientific Publishers, Warszawa, 1977.

15. Engelking, R., *Dimension Theory*, North-Holland, Amsterdam, 1978.

16. Freudenthal, H., Entwicklungen von Räumen und ihren Gruppen, *Compositio Math.* **4** (1937), 145–234.

17. Kulpa, W., *Factorization Theorems and Properties of the Covering Type*, Scientific Publications of the University of Silesia, Katowice, 1980.

18. Lefschetz, S., On compact spaces, *Ann. of Math.* **32** (1931), 521–538.

19. Lefschetz, S., *Algebraic Topology*, Amer. Math. Soc. Colloquium Publ. 27, New York, 1942.

20. Levin, M., Rubin, L.R., and Schapiro, P.J., The Mardešić factorization theorem for extension theory and C-separation, *Proc. Amer. Math. Soc.* **128** (2000), 3099–3106.

21. Lokucievskiĭ, O.V., On the dimension of bicompacta (Russian), *Dokl. Akad. Nauk SSSR* **67** (1949), 217–219.

22. Lunc, A.L., A bicompactum whose inductive dimension is greater than its dimension defined by means of coverings (Russian), *Dokl. Akad. Nauk SSSR (N.S.)* **66** (1949), 801–803.

23. Mardešić, S., Un théorème de dualité concernant les groupes d'homologie de l'espace fonctionnel $S_m{}^X$, *C.R. Acad. Sci. Paris* **242** (1956), 2214–2217.

24. Mardešić, S., Dimension and inverse limits of compact spaces, *Notices Amer. Math. Soc.* **5** (1958), 785.

25. Mardešić, S., Chainable continua and inverse limits, *Glasnik Mat. Fiz. Astron.* **14** (1959), 219–232.

26. Mardešić, S., On covering dimension and inverse limits for compact spaces, *Illinois J. Math.* **4** (1960), 278–291.

27. Mardešić, S., Locally connected, ordered and chainable continua (Croatian with English summary), *Rad Jugoslav. Akad. Znan. Umjetn.* **319** (1960), 147–166.

28. Mardešić, S., ε-mappings and inverse limits, *Glasnik Mat.-Fiz. Astron.* **18** (1963), 195–205.

29. Mardešić, S., Approximate polyhedra, resolutions of maps and shape fibrations, *Fund. Math.* **114** (1981), 53–78.

30. Mardešić, S., On approximate inverse systems and resolutions, *Fund. Math.* **142** (1993), 241–255.

31. Mardešić, S., Coherent and strong expansions of spaces coincide, *Fund. Math.* **158** (1998), 69–80.

32. Mardešić, S., *Strong Shape and Homology*, Springer Monographs in Mathematics, Springer, Berlin, 2000.

33. Mardešić, S., Extension dimension of inverse limits, *Glasnik Mat.* **35 (55)** (2000), 339–354.

34. Mardešić, S., and Matijević, V., $\mathcal{P}$-like spaces are limits of approximate $\mathcal{P}$-resolutions, *Top. and its Appl.* **45** (1992), 189–202.

35. Mardešić, S., and Rubin, L.R., Approximate inverse systems of compacta and covering dimension, *Pacific J. Math.* **138** (1989), 129–144.

36. Mardešić, S., and Rubin, L.R., Cell-like mappings and non-metrizable compacta of finite cohomological dimension, *Trans. Amer. Math. Soc.* **311** (1989), 53–79.

37. Mardešić, S., and Segal, J., ε-mappings onto polyhedra, *Trans. Amer. Math. Soc.* **109** (1963), 146–164.

38. Mardešić, S., and Segal, J., ε-mappings and generalized manifolds, *Michigan Math. J.* **14** (1967), 171–182.

39. Mardešić, S., and Segal, J., ε-mappings and generalized manifolds II, *Michigan Math. J.* **14** (1967), 423–426.

40. Mardešić, S., and Segal, J., *Shape Theory. The Inverse System Approach*, North-Holland, Amsterdam, 1982.

41. Mardešić, S., and Segal, J., $\mathcal{P}$-like continua and approximate inverse limits, *Math. Jap.* **33** (1988), 895–908.

42. Mardešić, S., and Uglešić, N., Morphisms of inverse systems require meshes, *Tsukuba J. Math.* **20** (1996), 357–363.

43. Mardešić, S., and Uglešić,U., On iterated inverse limits, *Top. and its Appl.* To appear.

44. Mardešić, S., and Watanabe, T., Approximate resolutions of spaces and mappings, *Glasnik Mat.* **24** (1989), 587–637.

45. Matijević, V., Approximate polyhedral resolutions with irreducible bonding mappings, *Rend. del'Istituto di Matem. Univ. Trieste* **25** (1993), 337–344.

46. Matijević, V., Spaces having approximate resolutions consisting of finite-dimensional polyhedra, *Publ. Math. Debrecen* **46** (1995), 301–314.

47. Morita, K., On expansions of Tychonoff spaces into inverse systems of polyhedra, *Sci. Rep. Tokyo Kyoiku Daigaku, A* **13** (1975), 66–74.

48. Morita, K., On shapes of topological spaces. *Fund. Math.* **86** (1975), 251–259.

49. Morita, K., Resolutions of spaces and proper inverse systems in shape theory, *Fund. Math.* **124** (1984), 263–270.

50. Nagami, K., Finite-to-one closed mappings and dimension, II, *Proc. Japan Acad.* **35** (1959), 437–439.

51. Nemec, A.G., and Pasynkov, B.A., Two general approaches to factorization theorems in dimension theory (Russian), *Dokl. Akad. Nauk SSSR* **233** (1977), no. 5, 788–791.

52. Pasynkov, B.A., On polyhedral spectra and dimension of bicompacta, in particular of bicompact groups (Russian), *Dokl. Akad. Nauk SSSR* **121** (1958), 45–48.

53. Pasynkov, B.A., Factorization of mappings onto metric spaces (Russian), *Dokl. Akad. Nauk SSSR* **182** (1968), 268–271.

54. Pasynkov, B.A., Factorization theorems in dimension theory (Russian), *Uspehi Mat. Nauk* **36** (1981), 147–175.

55. Pasynkov, B.A., Theorem on ω-mappings for mappings (Russian), *Uspehi Mat. Nauk* **39** (1984), 107–130.

56. Pasynkov, B.A., A factorization theorem for the cohomological dimensions of mappings (Russian), *Vestnik Moskov. Univ. Ser. I Mat. Mekh.* **4** (1991), 26–33.

57. Pontryagin, L., The theory of topological commutative groups, *Ann. of Math.* **35** (1934), 361–388.

58. Rubin, L.R., and Schapiro, P.J., Limit theorem for inverse sequences of metric spaces in extension theory, *Pacific J. Math.* **187** (1999), 177–186.

59. Skordev, G., and Smirnov, Ju.M., The factorization and approximation theorems for Aleksandrov-Čech cohomology in the class of bicompacta (Russian), *Dokl. Akad. Nauk SSSR* **220** (1975), 1031–1034.

60. Stone, A.H., Paracompactness and product spaces, *Bull. Amer. Math. Soc.* **54** (1948), 977–982.

61. Watanabe, T., Approximate resolutions and covering dimension, *Top. and its Appl.* **38** (1991), 147–154.

62. Zarelua, A.V., The method of the theory of rings of functions in the construction of bicompact extensions (Russian), in *Contributions to Extension Theory of Topological Structures (Proc. Sympos., Berlin, 1967)*, pages 249–256, Deutsch. Verlag Wissensch., Berlin, 1969.

Ten Mathematical Essays on Approximation in Analysis and Topology
J. Ferrera, J. López-Gómez, F. R. Ruiz del Portal, Editors

Periodic Solutions in the Golden Sixties: the Birth of a Continuation Theorem

J. Mawhin

Department of Mathematics, Université Catholique de Louvain,
Chemin de Cyclotron 2, B-1348 Louvain-la-Neuve, Belgium

Abstract

This paper describes the genesis of a continuation theorem introduced by the author in the late sixties, for proving the existence of periodic solutions for ordinary differential equations. After describing the state of the art in this domain both for weakly and strongly nonlinear differential equations, and in particular the work of Cesari, Hale and Cronin, one finds the description of the two stages in obtaining the theorem: first a combination of Cesari's method, a priori bounds and Brouwer degree, then a more direct approach based on Leray-Schauder's degree. A comparison is made with the evolution of Leray's approach to nonlinear functional equations in the thirties, as well as a connection with approximation theory.

Key words: periodic solutions, continuation theorems, alternative method, topological degree, integral equations

1. Introduction

In the early sixties, the study of T-periodic solutions of T-periodic ordinary differential equations or systems was still divided into two almost separated worlds: the case of *weakly nonlinear differential equations*

$$x'(t) = A(t)x + \varepsilon g(t, x, \varepsilon) \tag{1}$$

of arbitrary dimension n, where ε is a small parameter, and the case of *strongly nonlinear differential equations*

$$x'(t) = f(t, x(t)), \tag{2}$$

in dimension smaller or equal to 2. For example, in the Introduction of [12], Jack Hale writes:

> "Our knowledge of nonlinear systems is still far from being complete. For the case where the system of differential equations has order 2 (that is, one degree of freedom), much more is known than for higher-order systems. The reason for this is that analytical-topological methods may be applied very nicely for systems of order 2, whereas for higher dimensions, the techniques of topology are not sufficiently developed. For systems of order greater than 2, the differential equations are usually assumed to contain a given parameter, and some type of perturbation technique is used to discuss the behavior of solutions."

This remark is well documented by most fundamental treatises published at this time, like those of Lefschetz (1959) [18], Minorsky (1962) [25], Cesari (1963) [3], Sansone and Conti (1964) [32], Hale (1969) [13], which all separate the study of systems with small parameters and systems of dimension 2.

2. Weakly nonlinear systems

In the case of systems with a small parameter (1), it was necessary to distinguish between the *nonresonant* case, in which the associated linear system

$$x'(t) = A(t)x(t) \tag{3}$$

only has the trivial T-periodic solution, and the *resonant* one, in which (3) has non-trivial T-periodic solutions. Notice that the case where

$$A(t) \equiv 0$$

is always resonant.

In the nonresonant situation, *the system* (1) *always has at least one (small) T-periodic solution when* $|\varepsilon|$ *is sufficiently small* (a perturbation of the trivial solution when $\varepsilon = 0$). This is easily shown by applying the Schauder fixed point theorem if g is continuous, the Banach fixed point theorem if g is locally Lipschitzian in x or the implicit function theorem if g is of class C^1 with respect to x, to the equivalent integral equation

$$x(t) = \varepsilon \int_0^T G(t,s)g(s,x(s),\varepsilon)\,ds,$$

where $G(t,s)$ is the Green matrix associated to the periodic problem for (3).

In the resonant situation, the system (1) may or may not have, for small $|\varepsilon|$, T-periodic solutions, depending upon the nature of the nonlinear function g and upon the non-trivial T-periodic solutions of (3). Consider for example the resonant case where $A(t) \equiv 0$, for which the nontrivial T-periodic solutions of (3) are the constant functions. If $x(t;\varepsilon)$ is a T-periodic solution of (1) with $\varepsilon \neq 0$ which tend to a constant a when $\varepsilon \to 0$ (which is always the case if g is smooth enough), it is easy to see, by integrating both members of the identity

$$x'(t;\varepsilon) = \varepsilon g(t, x(t;\varepsilon), \varepsilon) \tag{4}$$

over $[0, T]$, using periodicity, and letting $\varepsilon \to 0$, that a must verify the system of equations

$$\int_0^T g(s, a, 0)\, ds = 0. \tag{5}$$

Thus, under mild regularity conditions upon g, *the existence of a solution to (5) is necessary for the existence of a T-periodic solution to (1) when $|\varepsilon|$ is small.*

To show that such a condition is sufficient when sufficiently strengthened, one uses an idea coming back to the work of A.M. Lyapunov and E. Schmidt on nonlinear integral equations, and referred as the *Lyapunov-Schmidt* or *alternative method*. We present it in the functional analytic version developed in the fifties by Cesari and Hale (see [3,12]). Assume that the Jacobian matrix $\partial_x g(t, x, \varepsilon)$ exist and is continuous. One can write the system (4) with $\varepsilon \neq 0$ in the equivalent form

$$\tilde{x} = \varepsilon H \left[g(\cdot, \overline{x} + \tilde{x}(\cdot), \varepsilon) - \frac{1}{T} \int_0^T g(s, \overline{x} + \tilde{x}(s), \varepsilon)\, ds \right], \tag{6}$$

$$\int_0^T g(s, \overline{x} + \tilde{x}(s), \varepsilon)\, ds = 0, \tag{7}$$

where

$$\overline{x} := \frac{1}{T} \int_0^T x(s)\, ds, \quad \tilde{x}(t) := x(t) - \overline{x},$$

and H is the linear mapping associating to any T-periodic function with mean value zero its (T-periodic) primitive with mean value zero. Using the implicit function theorem in the subspace $\tilde{C}_T$ of T-periodic functions having mean value zero, it is easy to see that, for each fixed $\overline{x}$ in a fixed ball, and $|\varepsilon|$ sufficiently small, the problem (6) has a unique T-periodic solution $\xi(\cdot; \overline{x}, \varepsilon) \in \tilde{C}_T$ such that

$$\xi(\cdot; \overline{x}, 0) = 0.$$

Thus $\overline{x} + \xi(t; \overline{x}, \varepsilon)$ will be a T-periodic solution of (4), if $\overline{x}$ satisfies the second equation (7) with $\tilde{x}$ replaced by ξ, namely

$$\Gamma(\overline{x}, \varepsilon) := \int_0^T g(s, \overline{x} + \xi(s; \overline{x}, \varepsilon), \varepsilon)\, ds = 0. \tag{8}$$

Using now the implicit function theorem in $\mathbb{R}^n$ to study equation (8), we easily see that *the system (4) has, for sufficiently small $|\varepsilon|$, a T-periodic solution, if the system*

$$G(\overline{x}) := \int_0^T g(s, \overline{x}, 0)\, ds = 0 \tag{9}$$

has a solution $\tilde{x}$ such that

$$\det \partial_x G(\overline{x}) = \det \left[\int_0^T \partial_x g(s, \overline{x}, 0)\, ds \right] \neq 0. \tag{10}$$

The problem has thus been reduced to the solution of the system (9) of n equations in n unknowns, the so-called *bifurcation equation*.

3. Cesari's method for strongly nonlinear systems

In the early sixties, Cesari proposed an extension of his approach to the case of systems of type (2), when f satisfies a Lipschitz condition in a suitable ball, namely, for some $L > 0$ and $R > 0$,

$$|f(t, x) - f(t, y)| \leq L|x - y| \tag{11}$$

for all $t \in \mathbb{R}$, $|x| \leq R$ and $|y| \leq R$, for some norm $|\cdot|$ in $\mathbb{R}^n$.

The fundamental idea (as soon improved by Knobloch [14]) is based upon the following simple properties of T-periodic functions. Let C_T denote the space of T-periodic continuous mappings x from $\mathbb{R}$ into $\mathbb{R}^n$, with uniform norm $\|\cdot\|$,

$$\omega := \frac{2\pi}{T},$$

and let

$$x(t) \sim a_0 + \sum_{s=1}^{\infty}(a_s \cos s\omega t + b_s \sin s\omega t),$$

be the Fourier series of x. If $m \geq 0$ is an integer, define the operator P_m by

$$P_m x(t) := a_0 + \sum_{s=1}^{m}(a_s \cos s\omega t + b_s \sin s\omega t).$$

Then, for each $x \in C_T$, one has

$$\|H(I - P_m)x\| \leq \sqrt{2}\omega^{-1}\sigma(m)\|x\| \tag{12}$$

where

$$\sigma^2(m) := \sum_{s=m+1}^{\infty} s^{-2},$$

so that

$$\frac{1}{m+1} < \sigma(m) < \frac{1}{\sqrt{m}}.$$

The proof of this inequality is a good exercise in the theory of Fourier series.

If $x = \overline{x}_m + \widetilde{x}_m$ with

$$\overline{x}_m = P_m x, \quad \widetilde{x}_m = (I - P_m)x,$$

denotes a T-periodic function, then, extending the idea of A. M. Lyapunov and E. Schmidt, the T-periodic solutions of system (2) are the solutions of the system of equations

$$\widetilde{x}_m = H(I - P_m)f(\cdot, \overline{x}_m(\cdot) + \widetilde{x}_m(\cdot)), \tag{13}$$

$$P_m x' = P_m f(\cdot, \overline{x}_m(\cdot) + \widetilde{x}_m(\cdot)). \tag{14}$$

Notice that the system (13) is nothing but the integrated form of

$$(I - P_m)x' = (I - P_m)f(\cdot, \overline{x}_m(\cdot) + \tilde{x}_m(\cdot)).$$

Now, if

$$r < R, \qquad K = \sup\{|f(t, x)| : t \in \mathbb{R}, |x| \leq R\}, \qquad \|\overline{x}_m\| \leq r,$$

and

$$\|\tilde{x}_m\| \leq R - r,$$

one has, using inequality (12),

$$\|H(I - P_m)f(\cdot, \overline{x}_m(\cdot) + \tilde{x}_m(\cdot))\| \leq \sqrt{2}\omega^{-1}\sigma(m)K,$$

and, if y is another element of C_T, one has

$$\|H(I - P_m)f(\cdot, \overline{x}_m(\cdot) + \tilde{x}_m(\cdot)) - H(I - P_m)f(\cdot, \overline{x}_m(\cdot) + \tilde{y}_m(\cdot))\|$$
$$\leq \sqrt{2}\omega^{-1}\sigma(m)L\|\tilde{x}_m - \tilde{y}_m\|.$$

Consequently, if m is sufficiently large so that

$$\sqrt{2}\omega^{-1}\sigma(m)K \leq R - r, \qquad \sqrt{2}\omega^{-1}\sigma(m)L < 1, \tag{15}$$

one can apply the Banach fixed point theorem in the space $(I - P_m)C_T$ to system (13), and obtain, for each $\overline{x}_m$ such that $\|\overline{x}_m\| \leq r$, a unique

$$\xi_m(\cdot, \overline{x}_m) \in (I - P_m)C_T$$

such that

$$\|\xi_m(\cdot, \overline{x}_m)\| \leq R - r$$

which solves (13). Thus $\overline{x}_m + \xi(\cdot; \overline{x}_m)$ will be a T-periodic solution of (2) if we can find some $\overline{x}_m \in P_m C_T$ such that $\|\overline{x}_m\| \leq r$, which satisfies the system

$$\overline{x}'_m = P_m f(\cdot, \overline{x}_m(\cdot) + \xi(\cdot, \overline{x}_m)). \tag{16}$$

Thus *the differential system (2) has a T-periodic solution if we can find some*

$$\overline{x}_m \in P_m C_T$$

satisfying the system (16).

Notice that (16) is a system of $(2m + 1)n$ equations in the $(2m + 1)n$ unknowns formed by the Fourier coefficients of $\overline{x}_m$. Like in the resonant case with a small parameter, the infinite-dimensional problem of finding a T-periodic solution has been reduced to a finite-dimensional one. However, the difficulties in applying Cesari's method lie in the two following facts:

1. The equations (16) are not explicitly known because the function ξ is not explicitly known.

2. One may have to take m quite large to verify conditions (15).

The fact that those difficulties are serious is reflected in the very small number of examples which have been treated in this way by Cesari and his coworkers. Notice that if we take $\xi = 0$ in (16), we obtain the equations for the *Galerkin approximation* of order m for the T-periodic solutions of (2).

When applied to system (1), Cesari's method does not make any *a priori* distinction between the fact that the system is non-resonant or resonant. I have used this observation and the implicit function theorem in [19] to show that the corresponding equations (16) has a solution for $|\varepsilon|$ small when the system (9) has a solution satisfying condition (10).

4. Topological degree and Cronin's monograph

One year after the publication of Cesari's paper [2], J. Cronin published her well known monograph on the use of topological degree in nonlinear analysis [4], calling the attention of analysts to more sophisticated topological methods. At this time, topological degree techniques were less than popular among experts in differential equations, with the exception of M.A. Krasnosel'skii and his school in Voronezh, whose important ideas had not yet really penetrated the Western world (having to wait 1968 and the English translation of the important monograph [15]). The use of topology was restricted to Brouwer or Schauder's fixed point theorem and to the most elementary version of Leray-Schauder fixed point theorem.

In Chapter I of [4], Cronin develops Brouwer degree, its properties and its computation using essentially the Alexandroff-Hopf approach based upon algebraic topology, and quite orthogonal to the habits of analysts. If

$$F : \overline{\Omega} \subset \mathbb{R}^n \to \mathbb{R}^n$$

is continuous, Ω open bounded, and if

$$F(x) \neq 0 \quad \text{on } \partial\Omega,$$

the *Brouwer degree*

$$\deg_B[F, \Omega, 0]$$

is some algebraic count of the number of zeros of F located in Ω, equal to 0 when G has no zero in Ω, and stable for sufficiently small perturbations of G. As a consequence, this degree has the important *homotopy invariance* property, namely

$$\deg_B[\Phi(\cdot, \varepsilon), \Omega, 0] \quad \text{remains constant for } \varepsilon \in [0, 1]$$

when Φ is continuous on $\overline{\Omega} \times [0, 1]$ and

$$\Phi(x, \varepsilon) \neq 0 \quad \text{on } \partial\Omega \times [0, 1].$$

In Chapter II, Cronin uses Brouwer degree to study the T-periodic solutions of weakly nonlinear systems (1). The main idea consists in replacing the use of implicit function theorem in studying equation (8) by the use of Brouwer degree. The existence of a solution of

(9) satisfying condition (10) is replaced by the more general condition that *the Brouwer degree* $\deg_B[G, B(r), 0]$ *is defined and different from zero for some* $r > 0$. This immediately implies that

$$\deg_B[\Gamma(\cdot, \varepsilon), B(r), 0] \neq 0$$

for sufficiently small $|\varepsilon|$, and hence that system (1) has at least one T-periodic solution. Notice that Cronin does not use a Lyapunov-Schmidt approach, but obtains equivalent results through *Poincaré's method.* In this approach, the Cauchy problem

$$\begin{cases} x'(t) = A(t)x(t) + \varepsilon g(t, x(t), \varepsilon), \\ x(0) = a, \end{cases} \tag{17}$$

is first solved, providing the solution $u(t; a, \varepsilon)$. The almost linear character of the system guarantees the global existence of $u(t; a, \varepsilon)$ on $[0, T]$. This solution is then T-periodic if a is such that

$$a = u(T; a, \varepsilon),$$

i.e. if a is a fixed point of *Poincaré's operator*

$$a \mapsto u(T; a, \varepsilon).$$

Chapter II of [4] ends with an application, due to Gomory [10], of the Brouwer degree on large balls for Poincaré's operator, to the study of some strongly nonlinear systems in dimension 2.

Chapter III of Cronin's book describes Leray-Schauder's extension of Brouwer degree to compact perturbations of identity in normed spaces [17]. Recall that a mapping

$$C : \overline{\Omega} \subset E \to E$$

is *compact* on the closure of an open bounded subset Ω of a normed space E, if it is continuous on $\overline{\Omega}$ and $C(\overline{\Omega})$ is relatively compact. In 1930, Schauder had shown that such *a compact mapping can be approximated by a sequence* (C_n) *of mappings defined on* $\overline{\Omega}$ *and taking values in finite-dimensional vector subspaces* E_n *of* E, and has used this fact to prove his famous *fixed point theorem* for compact mappings from a closed convex subset of E into itself. Four years later, Leray and Schauder have proved that, if

$$x - C(x) \neq 0 \quad \text{for } x \in \partial\Omega,$$

then for all sufficiently large n, the Brouwer degrees

$$\deg_B[(I - C_n)|_{E_n}, \Omega \cap E_n, 0]$$

stabilize, and their common value defines the *Leray-Schauder degree*

$$\deg_{LS}[I - C, \Omega, 0] := \lim_{n \to \infty} \deg_B[(I - C_n)|_{E_n}, \Omega \cap E_n, 0].$$

Leray and Schauder have extended to the more general situation the basic properties of Brouwer degree, in particular its *homotopy invariance,* and deduced from it new fixed point theorems, the simplest one stating that *if there exists* $R > 0$ *such that any possible solution of each equation*

$$x = \varepsilon C(x), \quad \varepsilon \in \,]0, 1[,$$

satisfies $\|x\| < R$, *then* C *has a fixed point in the closed ball* $B[R]$.

Chapter IV of Cronin's book gives some applications of Leray-Schauder degree, and in particular of the above *Leray-Schauder fixed point theorem,* to various problems for nonlinear integral and partial differential equations. The very last section of the book gives a short description of Cesari's method, without any application or any use of degree.

5. Injecting Brouwer degree in Cesari's method

After reading Cesari's paper and Cronin's interesting monograph, I realized in October 1967 that combining Brouwer degree with Cesari's method could provide a link between the small parameter results and the study of strongly nonlinear systems. The crucial observation was to show that, within the frame of Cesari's method, one had, for small nonzero $|\varepsilon|$, the equality,

$$\deg_B[G, B(r), 0] = \deg_B[\widehat{G}(\cdot, \varepsilon), B(r), 0],$$

where the map $\widehat{G}(\cdot, \varepsilon)$ is defined on the closed ball $B[r]$ of $\mathbb{R}^{(2m+1)n}$ by the left hand member of (16) with f remplaced by εg. By the homotopy invariance property, this equality would remain valid until $\varepsilon = 1$ under the assumption of the existence of an *a priori* bound r for the possible T-periodic solutions of the system (4). Hence, the following theorem could be proved.

Theorem 1. *Let*

$$g : \mathbb{R} \times \mathbb{R}^n \times [0, 1] \to \mathbb{R}^n, \qquad (t, x, \varepsilon) \to g(t, x, \varepsilon),$$

be T-periodic in t, locally Lipschitzian in x (uniformly in t and ε) and continuous. Assume there exists $r > 0$ such that the following conditions hold.

1. For each $\varepsilon \in {]0, 1]}$, any possible T-periodic solution of (4) is such that $\|x\| < r$.

2. Each possible zero a of G is such that $|a| < r$.

3. $\deg_B[G, B(r), 0] \neq 0$, where

$$G(a) := \frac{1}{T} \int_0^T g(t, a, 0) \, dt.$$

Then, for each $\varepsilon \in [0, T]$, the system

$$x'(t) = \varepsilon g(t, x(t), \varepsilon)$$

has at least one T-periodic solution $x \in B(r) \subset C_T$.

Such a result, connecting for the first time the conditions of the small parameter method in a resonant case with an existence result for strongly nonlinear systems of arbitrary dimension, was conceptually appealing. But it was necessary to test its practical applicability with some examples. In this time, the "paradigm" for nonlinear oscillations was the *forced Liénard equation*

$$y'' + f(y)y' + g(y) = e(t),$$

and its special case with f constant, the so-called *Duffing equation*. In [10], Gomory had obtained some results when f and g are polynomials, and Theorem 1 not only allowed new and simpler proofs of Gomory's results, but also several extensions. Another consequence was various improvements and corrections of results —quite difficult to read because of their sketchy style— obtained for dissipative Duffing equation and some special Liénard equation in 1964 and 1965 by Faure [7,8], using Leray-Schauder theory. The same approach had been used in 1960 by Ezeilo [6], in 1965 by Sedziwy [34], in 1965-66 by Villari [36–38] and in 1966 by Reissig [26] to study the T-periodic solutions of some *third order nonlinear differential equations*. With the exception of Villari, all those authors had deduced the *a priori bound* for the T-periodic solutions from the ultimate boundedness of all solutions. The application of Theorem 1 improved some of those existence conditions in several ways, showing that strongly nonlinear differential systems in dimension higher than two could be successfully treated.

My thesis advisor at the University of Liège was Paul Ledoux, a worldly renowned expert in the stability and vibrations of stars, whose (never realized) hope, in hiring me, had been the application of nonlinear methods to vibrating stars. Although not at all a specialist in applying topological methods to nonlinear ordinary differential equations, Ledoux was bold enough to propose me collecting the above results in a PhD Thesis. This was done with enthusiasm and the adviser received the complete text (about 250 pages) during the Spring 1968. It was only in the Fall that he called me to discuss the matter. The thesis [20] was (successfully) defended in February 1969, and a shortened version [21] was published the same year.

6. Applying Leray-Schauder's degree

We have seen in the previous section that the Leray-Schauder method had been successfully applied by various authors to problems of T-periodic solutions of some ordinary differential equations. The very first paper in this direction seems to be the one of Stoppelli [35] in 1952, dealing with the equation

$$y'' + |y'|y' + q(t)y' + y - p(t)y^3 = e(t),$$

where $p(t) > 0$. His results are described in [32] but, despite of this, Stoppelli's paper remained unnoticed untill some unfounded criticism in 1963 by Derwidué [5], who misused Leray-Schauder's method in confusing the existence of an *a priori* bound for all solutions with the requirement that each solution is bounded. He claimed as a consequence that each system (2) with a right-hand member globally Lipschitzian in x must have a T-periodic solution !

Between 1964 and 1966, Reissig [27–29] showed in successive steps, using Leray-Schauder theory, the following existence result for (1), published in book form in [30].

Theorem 2. *Let*

$$A : \mathbb{R} \to L(\mathbb{R}^n, \mathbb{R}^n) \quad and \quad g : [0, T] \times \mathbb{R}^n \times [0, 1] \to \mathbb{R}^n$$

be T-periodic in t and continuous. Assume that the following conditions hold.

1. *There exist $r > 0$ such that, for each $\varepsilon \in [0, 1]$, any possible T-periodic solution of each system*

$$x'(t) = A(t)x(t) + \varepsilon g(t, x(t), \varepsilon), \quad \varepsilon \in [0, 1],$$

   is such that $\|x\| < r$.

2. *The system*

$$x'(t) = A(t)x(t)$$

   has only the trivial T-periodic solution.

Then, for each $\varepsilon \in [0, 1]$, the system

$$x'(t) = A(t)x(t) + \varepsilon g(t, x(t), \varepsilon)$$

has at least one T-periodic solution x with $\|x\| < r$.

This result can be seen as a strongly nonlinear version of the existence theorem of the small parameter method in the non-resonant case. To prove it, Reissig wrote the problem in the equivalent fixed point form

$$x(t) = \varepsilon \int_0^T G(t, s)g(s, x(s), \varepsilon) \, ds,$$

with right-hand side compact on bounded subsets of C_T, where $G(t, s)$ is the Green matrix of the T-periodic problem associated to (3), and applied the elementary version of Leray-Schauder fixed point theorem mentioned at the end of Section 4. Notice that, in contrast to the case of Theorem 1, the Leray-Schauder degree associated to situations covered by Theorem 2 must be one.

By comparing Theorems 1 and 2, I was somewhat puzzled by the fact that Theorem 1 required for its proof the right-hand member to be locally Lipschitzian with respect to x, a condition absent in Theorem 2. I suspected that the assumption came from the method and not from "nature". Early in 1969, playing with the operators introduced by the Cesari-Hale method, I realized that the T-periodic solutions of system (4) for $\varepsilon \neq 0$ where the solutions of the *fixed point problem* in C_T

$$x = P_0 x + J P_0 g(\cdot, x(\cdot), \varepsilon) + \varepsilon H (I - P_0) g(\cdot, x(\cdot), \varepsilon), \tag{18}$$

where $J : \mathbb{R}^n \to \mathbb{R}^n$ is any automorphism, and that the right-hand member of (18) defined a compact nonlinear operator M on bounded subsets of C_T. Hence Leray-Schauder degree could be applied directly to $I - M$, leading to the following improved version of Theorem 1.

Theorem 3. *Let*

$$g : \mathbb{R} \times \mathbb{R}^n \times [0, 1] \to \mathbb{R}^n, \quad (t, x, \varepsilon) \to g(t, x, \varepsilon),$$

be T-periodic in t, and continuous. Assume there exists an open bounded set $\Omega \subset C_T$ such that the following conditions hold.

1. *For each $\varepsilon \in\,]0, 1]$, any possible T-periodic solution of (4) is such that $x \notin \partial \Omega$.*

2. *Each possible zero a of G is such that a $\notin \partial\Omega \cap \mathbb{R}^n$.*
3. $\deg_B[G, \Omega \cap \mathbb{R}^n, 0] \neq 0$, *where*

$$G(a) := \frac{1}{T} \int_0^T g(t, a, 0)\, dt.$$

Then, for each $\varepsilon \in [0, T]$, the system

$$x'(t) = \varepsilon g(t, x(t), \varepsilon)$$

has at least one T-periodic solution $x \in \Omega$.

By noticing that the T-periodic solutions of problem (1) were the solutions of the fixed point problem

$$x = P_0 x + JP_0[A(\cdot)x(\cdot) + \varepsilon g(\cdot, x(\cdot)), \varepsilon) + \varepsilon H(I - P_0)[A(\cdot)x(\cdot) + \varepsilon g(\cdot, x(\cdot), \varepsilon)],$$

it was also immediate to deduce from Leray-Schauder degree theory the following extension of Theorem 2.

Theorem 4. *Let*

$$A : \mathbb{R} \to L(\mathbb{R}^n, \mathbb{R}^n) \quad \text{and} \quad g : \mathbb{R} \times \mathbb{R}^n \times [0, 1] \to \mathbb{R}^n$$

be T-periodic in t and continuous. Assume that the following conditions hold.

1. *There exist an open bounded neighbourhood of zero $\Omega \subset C_T$ such that, for each $\varepsilon \in [0, 1]$, any possible T-periodic solution of each system*

$$x'(t) = A(t)x(t) + \varepsilon g(t, x(t), \varepsilon), \quad \varepsilon \in [0, 1]$$

is such that $x \notin \partial\Omega$.

2. *The system*

$$x'(t) = A(t)x(t)$$

has only the trivial T-periodic solution.

Then, for each $\varepsilon \in [0, 1]$, the system

$$x'(t) = A(t)x(t) + \varepsilon g(t, x(t), \varepsilon)$$

has at least one T-periodic solution $x \in \Omega$.

Another easy consequence was a result just proved by Güssefeldt [11] for systems odd with respect to x.

The whole thesis could be rewritten by replacing the use of Cesari's method and Brouwer degree by a direct application of Leray-Schauder degree to the compact operator M introduced above. As a consequence, some unnecessary local Lipschitz conditions on the nonlinear terms could be dropped in the main theorems and in the applications. But the thesis was already in the hands of the committee, and I could only keep the improvement for myself until the defense was performed. The new approach was published in the Fall

1969 [22], and appeared in book form in [31], including a version for the general resonant case with

$$A(t) \not\equiv 0.$$

Results of the type of Theorems 1 to 3 are generally referred as *continuation theorems,* in the sense that some solutions existing for $\varepsilon = 0$ are indeed "continued" until $\varepsilon = 1$. Given for example a problem of type (2), one imbeds it into a family of problems of type (4), with g chosen in such a way that

$$g(t, x, 1) \equiv f(t, x).$$

Three years later, the above continuation theorems were extended to L-compact perturbations N of a Fredholm linear operator L of index zero in a normed space, within the frame of *coincidence degree,* an extension of Leray-Schauder degree to mappings of the form

$$L - N$$

[23]. In the hands of many mathematicians, this theory, presented in book form in [9,24], has provided a large number of new existence and multiplicity theorems for various boundary value problems associated to ordinary, functional or partial differential equations.

7. Learning from history

Later, learning more about the history of mathematics, I have noticed two facts which are not unrelated to the above story, and which are connected with some aspect of approximation theory, like it has been the case several times in the previous sections.

First Fredholm's classical approach to study *linear integral equations*

$$x(t) - \int_0^T K(t, s)x(s)\, ds = h(t), \tag{19}$$

consists in approximating the integral by Riemann sums, solving the finite-dimension problem and going to the limit. Its abstract generalization in a Banach space given by F. Riesz for the equation

$$x - Kx = h \tag{20}$$

in a Banach space, with K linear and compact, corresponds to approximating of K by linear operators having finite-dimensional range.

In 1907, E. Schmidt [33] proposed the following alternative approach to study (19). We sketch it, for simplicity, in the case of solutions belonging to $L^2(0, T)$. Let (c_k) be a complete orthonormal systems in $L^2(0, T)$ such that

$$K(t, s) \sim \sum_{k=1}^{\infty} c_k(t)\varphi_k(s).$$

Then, there exists a positive integer m such that

$$\int_0^T \int_0^T |K(t, s) - \sum_{k=1}^{m} c_k(t)\varphi_k(s)|^2\, ds\, dt < 1. \tag{21}$$

Letting

$$\overline{K}_m(t,s) = \sum_{k=1}^{m} c_k(t)\varphi_k(s), \qquad \widetilde{K}_m(t,s) = K(t,s) - \overline{K}_m(t,s),$$

$$x_j = \int_0^T x(s)c_j(s)\,ds, \quad (j = 1,2,\ldots), \tag{22}$$

we notice that

$$\int_0^T \overline{K}_m(t,s)x(s)\,ds = \sum_{k=1}^{m} x_k c_k(t).$$

Thus we can write equation (19) in the form

$$x(t) - \int_0^T \widetilde{K}_m(t,s)x(s)\,ds = h(t) + \sum_{k=1}^{m} x_k c_k(t), \tag{23}$$

and, by Cauchy-Schwarz inequality,

$$\int_0^T \left| \int_0^T \widetilde{K}_m(t,s)x(s)\,ds \right|^2 dt \leq \int_0^T \left(\int_0^T |\widetilde{K}_m(t,s)|^2\,ds \right) \left(\int_0^T |x(s)|^2\,ds \right) dt$$

$$= \left(\int_0^T \int_0^T |\widetilde{K}_m(t,s)|^2\,ds\,dt \right) \|x\|^2.$$

Using (21), we see that the left-hand member of (23) has an inverse, which is easily shown to have the form of a linear integral operator with kernel say $L(t,s)$. Hence (19) is equivalent to

$$x(t) = \int_0^T L(t,s)h(s)\,ds + \sum_{k=1}^{m} x_k \int_0^T L(t,s)c_k(s)\,ds$$

$$:= h^*(t) + \sum_{k=1}^{m} x_k c_k^*(t).$$

Introducing this expression into (22) for $k = 1,2,\ldots,m$, we obtain the non-homogeneous system for $(x_1,\ldots,x_m)$,

$$x_j = h_j^* + \sum_{k=1}^{m} c_k^* x_k, \quad (j = 1,\ldots,m).$$

Thus equation (19) is equivalent to an linear non-homogeneous system of m scalar equations into m real unknowns x_j, for which Fredholm alternative follows from elementary linear algebra.

The reader will easily conclude that if Leray-Schauder's theory can be seen as a nonlinear version of Riesz approach based upon compactness, Cesari's method is a nonlinear version of Schmidt's method for solving a linear integral equation. In the linear situation, the two methods are equivalent, as continuous linear mappings are always Lipschitzian. The situation is different in nonlinear problems, as shown by the previous sections.

The second story deals with the PhD Thesis of Jean Leray of 1933 [16], devoted to the study of *nonlinear integral equations.* There Leray developed a method, that he modestly called the *Arzelá-Schmidt method,* to prove the existence of solutions of nonlinear integral equations. Consider for simplicity a problem the form

$$x(t) = \int_0^T F(t, s, x(s), \varepsilon)\, ds. \tag{24}$$

where F is sufficiently smooth. By combining the local Lyapunov-Schmidt theory with compactness arguments based upon Ascoli-Arzelá's theorem, Leray showed essentially that the existence of a solution to (24) for each $\varepsilon \in [0, 1]$ was insured if the following conditions were satisfied:

1. For each $\varepsilon \in [0, 1]$, and each possible solution $\xi(t)$ of (24), the variational equation

$$y(t) = \int_0^T \partial_x F(t, s, \xi(s), \varepsilon) y(s)\, ds$$

   has only the trivial solution (local uniqueness condition).

2. The set of possible solutions of (24) is *a priori* bounded by a constant r independent of ε.

3. For $\varepsilon = 0$, equation (24) has an odd number of solutions.

One year later, with Schauder [17], Leray injected the concept of topological degree into the study of such equations (and of more general ones), and their new continuation theorem allowed them to assume F only continuous and to drop the local uniqueness condition. So, the two authors could start the section "Applications" of [17] by the following sentence (my translation) :

"Let us mention first that the existence theorems proved by the Arzelá-Schmidt method [16] are all special cases of the fundamental theorem stated above."

In the Introduction of [21], I wrote (my translation)

"The method introduced in this work allows to find, under more general mathematical conditions, the existence theorems proved by the author with the help of Cesari's method in [20]."

It is always nice, for any mathematician, to meet with great ones, even in an anecdotal way.

References

1. Cesari, L., Periodic solutions of nonlinear differential systems, in *Proc. Symp. Active Networks and Feedback Systems,* pages 545–560, Polytechnic Press of Polytechn. Inst. Brooklyn, Brooklyn, 1960.
2. Cesari, L., Functional analysis and periodic solutions of nonlinear differential equations, in *Contributions to Differential Equations Vol. 1,* pages 149–187, Wiley, New York, 1963.

3. Cesari, L., *Asymptotic Behavior and Stability Problems in Ordinary Differential Equations,* 2nd ed., Springer, Berlin, 1963.

4. Cronin, J., *Fixed Points and Topological Degree in Nonlinear Analysis,* American Mathematical Society, Providence, RI, 1964.

5. Derwidué, L., Sur l'existence des solutions périodiques des équations quasi-linéaires à coefficients périodiques, *Acad. Roy. Belgique, Bull. Cl. Sci.* **(5) 49** (1963), 346–351, 461–469.

6. Ezeilo, J.O.C., On the existence of periodic solutions of a certain third-order differential equation, *Proc. Cambridge Phil. Soc.* **56** (1960), 381–389.

7. Faure, R., Solutions périodiques d'équations différentielles et méthode de Leray-Schauder (Cas des vibrations forcées), *Ann. Inst. Fourier (Grenoble)* **14** (1964), 195–204.

8. Faure, R., Sur l'existence de certaines solutions périodiques et méthode de Leray-Schauder (cas des vibrations forcées), in *Les vibrations forcées dans les systèmes non linéaires,* pages 261–266, CNRS, Paris, 1965.

9. Gaines, R.E., and Mawhin, J., *Coincidence Degree and Nonlinear Differential Equations,* Springer, Berlin, 1977.

10. Gomory, R., Critical points at infinity and forced oscillations, in *Contributions to the theory of nonlinear oscillations Vol. 3,* pages 85–126, Ann. Math. Studies 36, Princeton Univ. Press, Princeton, 1956.

11. Güssefeldt, G., Der topologische Abbildungsgrad für vollstatige Vectorfelder zum Nachweis von periodische Lösungen, *Math. Nachr.* **36** (1968), 231–233.

12. Hale, J.K., *Oscillations in Nonlinear Systems,* McGraW-Hill, New York, 1963.

13. Hale, J.K., *Ordinary Differential Equations,* Wiley-Interscience, 1969.

14. Knobloch, H.W., Remarks on a paper of Cesari on functional analysis and nonlinear differential equations, *Michigan Math. J.* **9** (1963), 303–309.

15. Krasnosel'skii, M.A., *The Operator of Translation along the Trajectories of Differential Equations,* American Mathematical Society, Providence, RI, 1968.

16. Leray, J., Etude de diverses équations intégrales non linéaires et de quelques problèmes que pose l'hydrodynamique, *J. Math. Pures Appl.* **12** (1933), 1–82.

17. Leray, J., and Schauder, J., Topologie et équations fonctionnelles, *Ann. Ec. Norm. Sup.* **51** (1934), 45–78.

18. Lefschetz, S., *Differential Equations: Geometric Theory,* 2nd ed., Wiley-Interscience, New York, 1959.

19. Mawhin, J., Application directe de la méthode de Cesari à l'étude des solutions périodiques de systèmes différentiels faiblement non linéaires, *Bull. Soc. Roy. Sci. Liège* **36** (1967), 193–210.

20. Mawhin, J., *Le problème des Solutions Périodiques en Mécanique non Linéaire,* Thèse de doctorat en sciences, Université de Liège, 1969.

21. Mawhin, J., Degré topologique et solutions périodiques des systèmes différentiels non linéaires, *Bull. Soc. Roy. Sci. Liège* **38** (1969), 308–398.

22. Mawhin, J., Equations intégrales et solutions périodiques des systèmes différentiels non linéaires, *Acad. Roy. Belgique, Bull. Cl. Sci.* **55** (1969), 934–947.

23. Mawhin, J., Equivalence theorems for nonlinear operator equations and coincidence degree theory for some mappings in locally convex topological vector spaces, *J. Diff. Eqns.* **12** (1972), 610–636.

24. Mawhin, J., *Topological Degree Methods in Nonlinear Boundary Value Problems*, American Mathematical Society, Providence, RI, 1979. Second printing 1981.

25. Minorsky, N., *Nonlinear Oscillations*, van Nostrand, Princeton, 1962.

26. Reissig, R., Ein funktionalanalytischer Existenzbeweis für periodische Lösungen, *Monatsber. Deutsche Akad. Wiss. Berlin* **6** (1964), 407–413.

27. Reissig, R., Funktionanalytischer Existenzbeweis für periodische Lösungen, *ZAMM* **45** (1965), T72–T73.

28. Reissig, R., Periodische Lösungen nichtlinearer Differentialgleichungen, *Monatsber. Deutsche Akad. Wiss. Berlin* **8** (1966), 779–782.

29. Reissig, R., Ueber die Existenz periodischer Lösungen bei einer nichtlinearen Differentialgleichung dritter Ordnung, *Math. Nachr.* **32** (1966), 83–88.

30. Reissig, R., Sansone, G., and Conti, R., *Nichtlineare Differentialgleichungen höherer Ordnung*, Cremonese, Roma, 1969. English transl.: Noordhoff, Leyden, 1974.

31. Rouche, N., and Mawhin, J., *Equations différentielles ordinaires. II. Stabilité et solutions périodiques*, Masson, Paris, 1973. English transl.: Pitman, Boston, 1980.

32. Sansone, G., and Conti, R., *Equazioni differenziali non lineari*, Cremonese, Roma, 1956. English transl.: Pergamon, Oxford, 1964.

33. Schmidt, E., Zur Theorie des linearen und nichtlinearen Integralgleichungen, II, *Math. Ann.* **64** (1907), 161–174.

34. Sedziwy, S., On periodic solutions of a certain third-order nonlinear differential equation, *Ann. Polon. Mat.* **17** (1965), 147–154.

35. Stoppelli, F., Su un'equazione differenziale della meccanica dei fili, *Rend. Accad. Sci. Fiz. Mat. Napoli* **(4) 19** (1952), 109–114.

36. Villari, G., Contributi allo studio dell'esistenza di soluzioni periodiche per i sistemi di equazioni differenziali ordinarie, *Ann. Mat. Pura Appl.* **(4) 69** (1965), 171–190.

37. Villari, G., Soluzioni periodiche di una classe di sistemi di tre equazioni differenziali ordinarie, *Ann. Mat. Pura Appl.* **(4) 69** (1965), 191–200.

38. Villari, G., Soluzioni periodiche di una classe di equazioni differenziali, *Ann. Mat. Pura Appl.* **(4) 73** (1966), 103–110.

Ten Mathematical Essays on Approximation in Analysis and Topology
J. Ferrera, J. López-Gómez, F. R. Ruiz del Portal, Editors

The stability of the equilibrium: a search for the right approximation

R. Ortega

Depto. de Matemática Aplicada, Universidad de Granada,
18071-Granada, Spain

Abstract

The stability of the equilibrium of a pendulum of variable length can be characterized in terms of the third approximation. In contrast, the traditional linearization procedure (first approximation) is not always faithful. Alternative characterizations of stability are also presented. They are based on degree theory and on the algebraic structure of the symplectic group.

Key words: stability, pendulum, Duffing equation, index, symplectic group

1. Introduction

In some textbooks in Mechanics, the phenomenon of parametric resonance is illustrated with the pendulum of variable length. After a change of the time variable, this class of pendula is modeled by the equation

$$\ddot{\theta} + \alpha(t) \sin \theta = 0 \tag{1}$$

where

$$\alpha(t) = g\ell(t)^3$$

and $\ell(t) > 0$ is the length at an instant t. It is traditional to assume that α is periodic, say of period $T > 0$. This model leads to suggestive examples of resonance because the equilibrium $\theta = 0$ becomes unstable if $\alpha(t)$ oscillates in an appropriate way. Although the system has only one degree of freedom [1], the study of the stability of

$$\theta = 0$$

---

[1] or one and a half if the dependence on time is counted

is not elementary. This probably explains why it is customary to substitute the original
equation by its linear approximation

$$\ddot{\theta} + \alpha(t)\theta = 0. \tag{2}$$

The main theme of these notes will be the validity of this procedure. It will turn out that the
linearization principle leads to the right conclusions in most cases, but there are exceptions.
Sometimes the equation (2) can be unstable while the equilibrium $\theta = 0$ is stable for (1).
In contrast, we shall find that the third order approximation (of Duffing type)

$$\ddot{\theta} + \alpha(t)\theta - \frac{1}{3!}\alpha(t)\theta^3 = 0 \tag{3}$$

is faithful. This means that the equilibrium $\theta = 0$ is stable for (1) if and only if the same
holds for (3). It must be noticed that the positivity of α is crucial for this result. Indeed,
if $\alpha(t)$ can change sign, probably none of the approximations obtained by truncating the
expansion of the sine function is faithful.

The idea of replacing a complicated equation by an approximation is central in Stability Theory. The first Lyapunov's method is the simplest instance. It can be applied to our
equation to prove instability in the easiest cases but it does not help in the proofs of stability. This is so because the notion of asymptotic stability (considered in Lyapunov's first
method) is strange to Hamiltonian mechanics. The study of the stability of the equilibrium
requires sophisticated techniques (KAM theory) which use the information on nonlinear
approximations. We refer to [2,6] for the perturbative case, where

$$\alpha(t) = \omega^2 + \epsilon\beta(t),$$

and to [26,23] for the general case. On the other hand, the results on instability also use
nonlinear approximations but are of a more elementary nature. Already in [15], Levi-Civita
obtained instability criteria using the quadratic approximation. His results were presented
for abstract mappings and applied to the study of a three body problem. The basic technique
in [15] is a detailed analysis of the dynamics around the equilibrium and it could be adapted
to the pendulum of variable length. Also, it would be possible to employ Lyapunov functions as in [32]. In these notes we shall show how to obtain instability criteria using a less
standard approach. Topological degree will be employed to reduce instability proofs to the
computation of certain indexes (localized versions of the degree). The rest of these notes
is organized in six sections. The notion of stability and its connection with the dynamics
of planar mappings is discussed in §2. The next section, §3, analyzes the linearized equation and the symplectic group $Sp(\mathbb{R}^2)$. In particular, the conjugacy classes in this group
are found. The basic facts about degree theory are collected in §4. The degree is useful to
define the index of the equilibrium of our differential equation, as shown in §5. Some links
between stability and index can be found in §6. Finally, in §7, several characterizations of
the stability of the equilibrium of the pendulum are presented. They are obtained in terms
of the index, the third approximation or the conjugacy classes of $Sp(\mathbb{R}^2)$. The notes are
concluded with some discussions about equations with more degrees of freedom.

2. Perpetual stability and discrete dynamical systems

We shall work with the class of differential equations

$$\ddot{\theta} = f(t, \theta) \tag{4}$$

where f is defined around $\theta = 0$, say $f : \mathbb{R} \times (-\epsilon, \epsilon) \to \mathbb{R}$ with $\epsilon > 0$. The function f satisfies

$$f(t, 0) = 0 \quad \forall t \in \mathbb{R}$$

and so $\theta = 0$ is an equilibrium of the equation. In addition, f is continuous, T-periodic in t and there is uniqueness for the initial value problem associated to (4).

Given a point

$$(\theta_0, \omega_0) \in (-\epsilon, \epsilon) \times \mathbb{R},$$

the solution satisfying

$$\theta(0) = \theta_0 \quad \text{and} \quad \dot{\theta}(0) = \omega_0$$

will be denoted by $\theta(t; \theta_0, \omega_0)$. In general one cannot say that this solution is defined in the whole real line but it is at least defined in a large interval for small values of $|\theta_0|$ and $|\omega_0|$.

The equilibrium $\theta = 0$ is said to be stable if given any neighborhood of the origin in $\mathbb{R}^2$, say $\mathcal{U}$, there exists another neighborhood $\mathcal{V}$ such that if (θ_0, ω_0) belongs to $\mathcal{V}$, then the solution $\theta(t; \theta_0, \omega_0)$ is defined in $(-\infty, \infty)$ and

$$(\theta(t; \theta_0, \omega_0), \dot{\theta}(t; \theta_0, \omega_0)) \in \mathcal{U} \quad \forall t \in \mathbb{R}.$$

This is the notion of perpetual stability, often employed in Hamiltonian dynamics (see Chapter 3 of [32]). The reader who is familiar with stability theory will notice that it means Lyapunov stability for the future and the past. Two simple examples are the equations

$$\ddot{\theta} + \theta = 0 \quad \text{and} \quad \ddot{\theta} - \theta = 0.$$

The equilibrium is stable only for the first.

Let us now consider the difference equation

$$\xi_{n+1} = M(\xi_n) \tag{5}$$

where

$$M : \mathcal{D} \subset \mathbb{R}^2 \to \mathbb{R}^2$$

is a one-to-one and continuous mapping defined in an open set $\mathcal{D}$. It is also assumed that the origin lies in $\mathcal{D}$ and it is a fixed point of M. Given an initial condition $\xi_0 \in \mathcal{D}$, the solution

$$\{\xi_n\}_{n \in I}, \quad \xi_n = M^n(\xi_0),$$

is defined on some subset I of $\mathbb{Z}$. The fixed point $\xi = 0$ is said to be stable if for each neighborhood $\mathcal{U}(0)$, there exists another neighborhood $\mathcal{V}(0)$ such that if $\xi_0 \in \mathcal{V}$ then $\{\xi_n\}$ is defined in $\mathbb{Z}$ and

$$\xi_n \in \mathcal{U} \quad \forall n \in \mathbb{Z}.$$

To practice with this definition the reader can consider the linear mappings M defined by the matrices

$$R[\Theta] = \begin{pmatrix} \cos\Theta & \sin\Theta \\ -\sin\Theta & \cos\Theta \end{pmatrix}, \qquad H_+[\Theta] = \begin{pmatrix} \cosh\Theta & \sinh\Theta \\ \sinh\Theta & \cosh\Theta \end{pmatrix}, \qquad \Theta \neq 0.$$

In the first case $\xi = 0$ is stable while in the second it is unstable.

There is of course a complete analogy between the definitions of stability for the continuous and discrete situations. Now we are going to immerse the study of stability for differential equations in the theory of difference equations. This is a central idea in dynamical systems that goes back to Poincaré.

The mapping

$$P(\theta_0, \omega_0) = (\theta(T; \theta_0, \omega_0), \dot{\theta}(T; \theta_0, \omega_0))$$

is well defined in a neighborhood of

$$\theta_0 = \omega_0 = 0$$

and, due to the uniqueness for the initial value problem, it is one-to-one and continuous. Moreover, the iterates P^n are obtained by evaluating the solutions at time $t = nT$. This property is crucial to prove that the equilibrium $\theta = 0$ is stable for (4) if and only if the fixed point $\theta_0 = \omega_0 = 0$ is stable for the mapping $M = P$. The mapping P is usually called the Poincaré map associated to the equation (4) and it has an important property: it preserves area and orientation. For smooth equations this is equivalent to the identity

$$\det P'(\theta_0, \omega_0) = 1$$

and it is a consequence of Liouville's theorem in Hamiltonian mechanics. The general case can be treated with the techniques in [31], Chapter IX.

To finish this section we notice that the notion of stability is invariant under changes of variables. For example, if φ is a local homeomorphism fixing the origin, the change

$$\xi = \varphi(\eta)$$

transforms

$$\xi_{n+1} = M(\xi_n)$$

into

$$\eta_{n+1} = M^\star(\eta_n)$$

with

$$M^\star = \varphi^{-1} \circ M \circ \varphi,$$

and the stability of $\xi = 0$ and $\eta = 0$ are equivalent.

3. The linear equation and the symplectic group

The linear equation

$$\ddot{\theta} + \alpha(t)\theta = 0, \tag{6}$$

where $\alpha(t)$ is continuous and T-periodic, is called Hill's equation and there are many studies about it. The book by Magnus and Winkler [20] is a classical reference. After passing to a first order system

$$\dot{\xi} = A(t)\xi, \qquad \xi = \begin{pmatrix} \theta \\ \omega \end{pmatrix}, \qquad A(t) = \begin{pmatrix} 0 & 1 \\ -\alpha(t) & 0 \end{pmatrix},$$

we find the matrix solution $X(t)$ satisfying

$$X(0) = I$$

(I is the 2×2 identity matrix). The Poincaré map associated to (6) is linear, namely

$$P\begin{pmatrix} \theta_0 \\ \omega_0 \end{pmatrix} = L\begin{pmatrix} \theta_0 \\ \omega_0 \end{pmatrix}, \qquad L = X(T).$$

We present two examples. For the harmonic oscillator ($\alpha \equiv 1$) and a fixed period T, L is the rotation $R[T]$ defined in the previous section. For the repulsive case ($\alpha \equiv -1$), L is the matrix $H_+[T]$.

Liouville's theorem implies that the matrix solution $X(t)$ always satisfies

$$\det X(t) = 1.$$

This property motivates our interest in the symplectic group. The group of 2×2 matrices with nonzero determinant will be denoted by $Gl(\mathbb{R}^2)$. The subgroup of $Gl(\mathbb{R}^2)$ composed by the matrices satisfying

$$\det L = 1$$

is the symplectic group, denoted by $Sp(\mathbb{R}^2)$. Given a matrix L in $Sp(\mathbb{R}^2)$, the eigenvalues μ_1, μ_2 satisfy

$$\mu_1 \mu_2 = 1$$

and one can distinguish three cases:

- *elliptic:* $\mu_1 = \overline{\mu_2}$, $|\mu_1| = 1$, $\mu_1 \neq \pm 1$
- *hyperbolic:* $\mu_1, \mu_2 \in \mathbb{R}$, $0 < |\mu_1| < 1 < |\mu_2|$
- *parabolic:* $\mu_1 = \mu_2 = \pm 1$

The conjugacy classes in the group $Sp(\mathbb{R}^2)$ can be described according to this classification. For an elliptic matrix L there exists $Q \in Sp(\mathbb{R}^2)$ such that $Q^{-1}LQ$ is a rotation

$$Q^{-1}LQ = R[\Theta], \quad \Theta \in (0, \pi) \cup (\pi, 2\pi).$$

A hyperbolic matrix is conjugate to a matrix in one of the two families

$$H_\pm[\Theta] = \begin{pmatrix} \pm\cosh\Theta & \sinh\Theta \\ \sinh\Theta & \pm\cosh\Theta \end{pmatrix}, \qquad \Theta \in (0, \infty).$$

Finally, a parabolic matrix will be conjugate to one of the six matrices

$$I, -I, P_+, P_-, -P_+, -P_- \quad where \ P_\pm = \begin{pmatrix} 1 & \pm 1 \\ 0 & 1 \end{pmatrix}.$$

All these facts can be proven from the theory of Jordan canonical forms. In fact that theory can be seen as the classification of the conjugacy classes in $Gl(\mathbb{R}^2)$. There is a more subtle point which does not follow from Jordan canonical form. From the point of view of the group $Gl(\mathbb{R}^2)$, the rotations $R[\Theta]$ and $R[2\pi - \Theta]$ are conjugate. This is not true in the symplectic group, for if $Q \in Gl(\mathbb{R}^2)$ satisfies

$$Q^{-1}R[\Theta]Q = R[2\pi - \Theta]$$

then

$$\det Q < 0.$$

In view of this property we can say that the angle Θ is a symplectic invariant. Similar situations appear in the parabolic case for the matrices P_+ and P_- (or $-P_+$ and $-P_-$). More details and geometric insights about this group can be found in the paper by Broer and Levi [5]. The reader can deduce from the previous discussions that the origin $\theta = 0$ is stable for (6) if and only if the monodromy matrix $X(T)$ is elliptic or parabolic with

$$X(T) = \pm I.$$

Hill's equation is invariant under translation and rescaling of time. This means that the change

$$t = \lambda(s + \tau), \quad x = x(s),$$

with $\lambda > 0$ and $\tau \in \mathbb{R}$, transforms the Hill's equation in another equation of the same type, namely

$$\frac{d^2\theta}{ds^2} + \alpha^\star(s)\theta = 0 \tag{7}$$

with

$$\alpha^\star(s) = \lambda^2\alpha(\lambda(s + \tau)).$$

The new period is

$$T^\star = \frac{T}{\lambda}.$$

We have made reference to this class of changes because they have a remarkable property, they are sufficient to arrive at the canonical form of monodromy matrices. More precisely,

Proposition 1. *Given $\alpha(t)$, continuous and T-periodic, there exists $\tau \in \mathbb{R}$ and $\lambda > 0$ such that the monodromy matrix associated to (7) is one of the matrices:*

- $R[\Theta]$, $\Theta \in (0, \pi) \cup (\pi, 2\pi)$ *(elliptic case)*
- $H_\pm[\Theta]$, $\Theta \neq 0$ *(hyperbolic case)*
- $I, -I, P_+, P_-, -P_+, -P_-$ *(parabolic case).*

The proof of this result can be seen in [25], Proposition 8, for the elliptic case and in [23], Lemma 2.1, for the parabolic case. Recently Yan and Zhang have found in [33] new applications of this result in the elliptic case. For the reader interested in details a proof in the hyperbolic case is presented.

Proof. Assume that the eigenvalues are μ_1 and μ_2. We find Floquet solutions associated to these eigenvalues. These are non-trivial solutions satisfying

$$\varphi(t+T) = \mu_1 \varphi(t), \quad \psi(t+T) = \mu_2 \psi(t).$$

The product

$$\Pi = \varphi \psi$$

is T-periodic and so there exists $\tau \in \mathbb{R}$ with

$$\dot{\Pi}(\tau) = 0.$$

The linear independence of φ and ψ implies that $\varphi(\tau)$ and $\psi(\tau)$ do not vanish. We select

$$\varphi(\tau) = \psi(\tau) = 1$$

and define

$$u = \frac{1}{2}(\varphi + \psi), \quad v = \frac{1}{2}(\varphi - \psi).$$

Then

$$u(\tau) = 1, \quad v(\tau) = 0, \quad \dot{u}(\tau) = 0, \quad \text{and} \quad \dot{v}(\tau) \neq 0.$$

Here one uses the definition of τ. From now on we shall assume

$$\dot{v}(\tau) > 0.$$

If this derivative is negative we exchange the roles of φ and ψ. The function

$$u^2 - v^2 = \Pi$$

is T-periodic and so

$$u(\tau + T)^2 - v(\tau + T)^2 = 1.$$

From

$$\dot{\Pi}(\tau + T) = 0,$$

we find that

$$\dot{u}(\tau + T)u(\tau + T) - \dot{v}(\tau + T)v(\tau + T) = 0.$$

The Wronskian formula implies that

$$\dot{v}(\tau + T)u(\tau + T) - \dot{u}(\tau + T)v(\tau + T) = \dot{v}(\tau).$$

From these equations one obtains

$$\dot{u}(\tau + T) = \dot{v}(\tau)v(\tau + T), \quad \dot{v}(\tau + T) = \dot{v}(\tau)u(\tau + T).$$

After the change

$$s = t + \tau,$$

the monodromy matrix takes the form $Q^{-1}MQ$ with

$$M = \begin{pmatrix} u(\tau + T) & v(\tau + T) \\ v(\tau + T) & u(\tau + T) \end{pmatrix}, \qquad Q = \begin{pmatrix} \dot{v}(\tau)^{1/2} & 0 \\ 0 & \dot{v}(\tau)^{-1/2} \end{pmatrix}.$$

We notice that M is of the type $H_{\pm}[\Theta]$ with

$$u(\tau + T) = \pm \cosh \Theta, \qquad v(\tau + T) = \sinh \Theta.$$

The matrix Q is eliminated with a change of scale. $\qquad \square$

4. Degree theory and index of zeros

Let us fix Ω, bounded and open subset of $\mathbb{R}^d$, $d \geq 1$. The degree is defined for continuous mappings from $\overline{\Omega}$ into $\mathbb{R}^d$ which do not vanish on the boundary. More precisely, given

$$F \in C(\overline{\Omega}, \mathbb{R}^d), \qquad F(\xi) \neq 0 \;\; \forall \xi \in \partial\Omega, \tag{8}$$

we can assign to it an integer which will be denoted by $\deg(F, \Omega)$. Among many other properties of degree we mention:

- *Existence.* If $\deg(F, \Omega) \neq 0$ then $F(\xi) = 0$ has at least one solution in Ω.
- *Invariance by homotopy.* If

$$\mathcal{F} : \overline{\Omega} \times [0, 1] \to \mathbb{R}^d, \qquad \mathcal{F} = \mathcal{F}(\xi, \lambda),$$

  is continuous and $\mathcal{F}(\cdot, \lambda)$ satisfies (8) for each $\lambda \in [0, 1]$, then

$$\deg(\mathcal{F}(\cdot, \lambda), \Omega) \quad \text{is independent of } \lambda.$$

- *Excision.* If K is a compact subset of $\overline{\Omega}$ and $F(\xi) \neq 0$ for $\xi \in K$, then

$$\deg(F, \Omega) = \deg(F, \Omega \setminus K).$$

In the properties of existence and excision it was assumed that F satisfied (8). There are many books about degree theory and we refer to [18] or [30] for more details.

Given an open set $\mathcal{U} \subset \mathbb{R}^d$ and $F \in C(\mathcal{U}, \mathbb{R}^d)$, let us assume that $\xi_\star \in \mathcal{U}$ is an isolated root of $F(\xi) = 0$. This means that

$$F(\xi_\star) = 0$$

and, for some $\delta > 0$,

$$F(\xi) \neq 0 \;\; \text{if } 0 < |\xi - \xi_\star| \leq \delta.$$

We define the index of F at $\xi_\star$ by

$$\text{ind}[F, \xi_\star] = \deg(F, B_\delta(\xi_\star)),$$

where $B_\delta(\xi_\star)$ is the ball of radius δ centered at $\xi_\star$. The property of excision shows that this ball could be replaced by any small neighborhood of $\xi_\star$.

In dimension one ($d = 1$), the index can only take the values ± 1 and 0. Namely,

$$\text{ind}[F, \xi_\star] = 1 \quad \text{if } F(\xi_\star - \delta) < 0 < F(\xi_\star + \delta),$$

,

$$\text{ind}[F, \xi_\star] = -1 \quad \text{if } F(\xi_\star - \delta) > 0 > F(\xi_\star + \delta)$$

and

$$\text{ind}[F, \xi_\star] = 0 \quad \text{otherwise.}$$

In dimension two ($d = 2$) the index can take any integer value. The prototypes are (in complex notation) $z \mapsto z^n$ for positive index n and $z \mapsto \bar{z}^n$ for index $-n$. The simplest procedure to compute the index is by linearization. Given $F \in C^1(\mathcal{U}, \mathbb{R}^d)$ and $\xi_\star \in \mathcal{U}$ with $F(\xi_\star) = 0$, if the Jacobian matrix $F'(\xi_\star)$ is non-singular then

$$\text{ind}[F, \xi_\star] = \text{sign}\{\det F'(\xi_\star)\}.$$

The linearization technique is also useful for degenerate zeros ($\det F'(\xi_\star) = 0$) as long as the Jacobian matrix is not identically zero. In such a case the computation of the index is not direct, but at least one can reduce the dimension. This idea can be found in the book [11]. Next we describe the simplest situation.

Assume that $d = 2$ and $F = (F_1, F_2)$ is a C^1 mapping with

$$F_1(0,0) = F_2(0,0) = 0 \quad \text{and} \quad \frac{\partial F_1}{\partial \xi_2}(0,0) \neq 0.$$

We can apply the Implicit Function Theorem to

$$F_1(\xi_1, \xi_2) = 0$$

and solve in ξ_2, say

$$\xi_2 = \varphi(\xi_1).$$

Define the function

$$\Phi(\xi_1) = F_2(\xi_1, \varphi(\xi_1))$$

and assume that $\xi_1 = 0$ is an isolated zero of this function. Then

$$\xi_1 = \xi_2 = 0$$

is an isolated zero of F and

$$\text{ind}_{\mathbb{R}^2}[F, 0] = -\sigma \, \text{ind}_{\mathbb{R}}[\Phi, 0]$$

where

$$\sigma = \text{sign}\{\frac{\partial F_1}{\partial \xi_2}(0,0)\}.$$

To exercise with the properties of degree we sketch a proof. Assume that we are in the case

$$\frac{\partial F_1}{\partial \xi_2}(0,0) > 0$$

and, for $\lambda \in [0, 1]$, consider the system of equations

$$\begin{cases} \lambda F_1(\xi_1, \xi_2) + (1 - \lambda)(\xi_2 - \varphi(\xi_1)) = 0, \\ \lambda F_2(\xi_1, \xi_2) + (1 - \lambda)\Phi(\xi_1) = 0. \end{cases}$$

This is well defined in a neighborhood of the origin and the only solutions of the first equation are $\xi_2 = \varphi(\xi_1)$. This is a consequence of the uniqueness of the implicit function since

$$\frac{\partial \mathcal{F}_1(0,0,\lambda)}{\partial \xi_2} = \lambda \frac{\partial F_1}{\partial \xi_2}(0,0) + (1-\lambda) > 0.$$

Once the first equation is solved, we substitute into the second and deduce that this last equation becomes equivalent to

$$\Phi(\xi_1) = 0.$$

In consequence the only solution of the system is $\xi_1 = \xi_2 = 0$ because $\xi_1 = 0$ was isolated for $\Phi = 0$. The invariance by homotopies leads us to the computation of the index for

$$(\xi_1, \xi_2) \mapsto (\xi_2 - \varphi(\xi_1), \Phi(\xi_1)).$$

When $\Phi'(0) \neq 0$ the computation of this index can be done by linearization. In the case $\Phi'(0) = 0$ the reader could try a proof or read more about the computation of indexes in [12].

We finish this section with an example on how to compute the index using the third approximation. Consider a planar and smooth mapping F with Taylor expansion

$$(\xi_1, \xi_2) \mapsto (k\xi_2 + \alpha\xi_1^3 + \beta\xi_1^2\xi_2 + \gamma\xi_1\xi_2^2 + \delta\xi_2^3 + \cdots, a\xi_1^3 + b\xi_1^2\xi_2 + c\xi_1\xi_2^2 + d\xi_2^3 + \cdots).$$

Then, if $k \neq 0$ and $a \neq 0$,

$$\mathrm{ind}[F, (0,0)] = -\mathrm{sign}(ka).$$

To prove this we first solve $F_1 = 0$ and find

$$\xi_2 = \varphi(\xi_1) = O(\xi_1^3).$$

Thus,

$$\Phi(\xi_1) = a\xi_1^3 + O(\xi_1^4).$$

5. The index of an equilibrium

Again we consider the differential equation

$$\ddot{\theta} = f(t, \theta), \tag{9}$$

in the same conditions as in Section 2. The equilibrium $\theta = 0$ is isolated (period T) if there exists $\delta > 0$ such that the equation (9) has no T-periodic solutions satisfying

$$0 < |\theta(t)| + |\dot{\theta}(t)| \leq \delta, \quad \forall t \in \mathbb{R}.$$

As an example consider the equation

$$\ddot{\theta} + \theta = 0,$$

then $\theta = 0$ is isolated (period T) if T is not a multiple of 2π. In general, if $\theta = 0$ is isolated (period T), the origin of $\mathbb{R}^2$ is an isolated root of the equation

$$(I - P)(\xi) = 0,$$

where I is the identity and P is the Poincaré map. We define the index of $\theta = 0$ as

$$\gamma_T(0) = \mathrm{ind}[I - P, 0].$$

The differential equation is periodic in time and we have fixed the period as T. The multiples nT, $n \geq 2$, are also admissible as periods of (9) and so one can consider iterated indexes

$$\gamma_{nT}(0) = \mathrm{ind}[I - P^n, 0]$$

when $\theta = 0$ is isolated (period nT).

To understand this definition one must recall that the Poincaré map for period nT is just the iteration

$$P^n = P \circ^{(n} \cdots \circ P.$$

Also, we notice that if $\gamma_{nT}(0)$ is well defined then the same happens for $\gamma_{kT}(0)$ if k is a divisor of n. In this section we shall concentrate on the first index $\gamma_T(0)$ and describe some methods to compute it. To this end we shall assume that the force $f(t, \theta)$ is smooth in θ. This will be understood in the following sense: the partial derivatives $\frac{\partial^n f}{\partial \theta^n}(t, \theta)$ exist everywhere in $\mathbb{R} \times (-\epsilon, \epsilon)$ and the functions

$$(t, \theta) \mapsto \frac{\partial^n f}{\partial \theta^n}(t, \theta)$$

are continuous for each $n \geq 1$. The first approximation to (9) is

$$\ddot{\theta} + a(t)\theta = 0, \qquad a(t) = \frac{\partial f}{\partial \theta}(t, 0). \tag{10}$$

The general theory of differential equations says that P is smooth and the matrix $P'(0)$ is precisely the monodromy matrix $X(T)$ which was defined in Section 3. If $X(T)$ is elliptic, hyperbolic or parabolic with

$$\mu_1 = \mu_2 = -1,$$

the index can be computed by linearization. Namely,

$$\gamma_T(0) = \mathrm{sign}\{\det(I - X(T))\} = \mathrm{sign}\{(1 - \mu_1)(1 - \mu_2)\},$$

where μ_1, μ_2 are the eigenvalues of $X(T)$, often called the Floquet multipliers. The computation of the index in the degenerate case

$$\mu_1 = \mu_2 = 1$$

is more delicate and requires information on nonlinear approximations. Assume now that (9) can be expanded as

$$\ddot{\theta} + a(t)\theta + c(t)\theta^p + \cdots = 0,$$

where $c(t)$ is not identically zero. From now on, the dots in an expansion will refer to terms with an order higher than the order of those explicitly mentioned. The nonlinear expansion of P up to the order p is

$$\begin{pmatrix} \theta_0 \\ \omega_0 \end{pmatrix} \mapsto X(T) \begin{pmatrix} \theta_0 + \partial_{\omega_0} H(\theta_0, \omega_0) + \cdots \\ \omega_0 - \partial_{\theta_0} H(\theta_0, \omega_0) + \cdots \end{pmatrix}$$

where

$$H(\theta_0, \omega_0) = \frac{1}{p+1} \int_0^T c(t)(\phi_1(t)\theta_0 + \phi_2(t)\omega_0)^{p+1} dt$$

and $\phi_1(t)$, $\phi_2(t)$ are the solutions of the linear equation (10) satisfying

$$\phi_1(0) = \dot{\phi}_2(0) = 1, \qquad \dot{\phi}_1(0) = \phi_2(0) = 0.$$

The way to obtain this formula is rather standard but we outline a proof. First we observe that the solution of

$$\ddot{\delta} + a(t)\delta = b(t), \qquad \delta(0) = \dot{\delta}(0) = 0,$$

is

$$\delta(t) = \int_0^t [\phi_1(s)\phi_2(t) - \phi_1(t)\phi_2(s)]b(s)ds.$$

Since the solution of (9) satisfies

$$\theta(t; \theta_0, \omega_0) = \phi_1(t)\theta_0 + \phi_2(t)\omega_0 + \cdots$$

we can apply the previous formula with

$$\delta(t) = \theta(t; \theta_0, \omega_0) - \phi_1(t)\theta_0 - \phi_2(t)\omega_0$$

and

$$b(t) = -c(t)\theta(t; \theta_0, \omega_0)^p + \cdots$$

to obtain

$$\theta(t; \theta_0, \omega_0) = \phi_1(t)\theta_0 + \phi_2(t)\omega_0$$
$$- \int_0^t [\phi_1(s)\phi_2(t) - \phi_1(t)\phi_2(s)]c(s)(\phi_1(s)\theta_0 + \phi_2(s)\omega_0)^p ds + \cdots$$

A similar formula is valid for the derivative and the conclusion follows for $t = T$.

Once we have a nonlinear expansion of P, we can employ the methods of the previous section to compute the index. Let us assume first that $X(T)$ is one of the matrices P_+, P_-. The function

$$F = I - P, \qquad F = F(\theta_0, \omega_0),$$

satisfies

$$\frac{\partial F_1}{\partial \omega_0}(0,0) = \mp 1.$$

By solving $F_1 = 0$ one obtains

$$\omega_0 = \varphi(\theta_0) = O(\theta_0^p).$$

Hence,

$$\Phi(\theta_0) = F_2(\theta_0, \varphi(\theta_0)) = \partial_{\theta_0} H(\theta_0, \varphi(\theta_0)) + \cdots$$
$$= (\int_0^T c(t)\phi_1(t)^{p+1} dt)\theta_0^p + \cdots$$

Assume now that the integral appearing in the previous formula is not zero, then $\theta_0 = 0$ is isolated for Φ and so $\theta = 0$ is isolated (period T). Moreover,

$$\gamma_T(0) = \begin{cases} 0 & \text{if } p \text{ is even} \\ \nu \operatorname{sign}\{\int_0^T c(t)\phi_1(t)^{p+1} dt\} & \text{if } p \text{ is odd.} \end{cases}$$

Here

$$\nu = \begin{cases} 1 & \text{if } X(T) = P_+, \\ -1 & \text{if } X(T) = P_-. \end{cases}$$

The computation of the index when

$$X(T) = I$$

is based on different ideas. Assume that the origin is the only critical point of H, we shall prove that $\theta = 0$ is isolated (period T) and

$$\gamma_T(0) = \operatorname{ind}[\nabla H, 0].$$

To prove this statement we notice that, since $X(T)$ is the identity, the map

$$F = I - P$$

has the form

$$F = J\nabla H + R,$$

where

$$J = \begin{pmatrix} 0 & -1 \\ 1 & 0 \end{pmatrix}$$

and $R = R(\theta_0, w_0)$ is a remainder of order higher than p. Consider the system of equations

$$J\nabla H(\theta_0, w_0) + \lambda R(\theta_0, w_0) = 0, \qquad \lambda \in [0, 1].$$

We are going to prove that, in a small neighborhood of the origin, there are no solutions different from

$$\theta_0 = w_0 = 0.$$

This is done by a contradiction argument. Assume that there is a sequence of solutions

$$\xi_n = (\theta_{0n}, w_{0n}), \qquad \lambda_n \in [0, 1],$$

with

$$\xi_n \neq 0 \quad \text{and} \quad |\xi_n| \to 0.$$

Define

$$\eta_n = \frac{\xi_n}{|\xi_n|}$$

and extract a convergent subsequence, say

$$\eta_n \to \eta_*, \qquad \lambda_n \to \lambda_*.$$

Dividing the equations by $|\xi_n|^p$ and passing to the limit we conclude that η_* is a critical point of H. This is impossible because $|\eta_*| = 1$ and 0 was the only critical point. The invariance by homotopies implies that

$$\gamma_T(0) = \mathrm{ind}[J\nabla H, 0].$$

It remains to prove that

$$\mathrm{ind}[J\nabla H, 0] = \mathrm{ind}[\nabla H, 0].$$

To do this we notice that J has positive determinant and so J and I are in the same component of $Gl(\mathbb{R}^2)$. Let J_λ be a continuous arc in $Gl(\mathbb{R}^2)$ joining I and J. The equation

$$J_\lambda \nabla H = 0$$

is equivalent to $\nabla H = 0$ and this homotopy proves the identity between the two indexes.

We sum up the discussions of this section. If the Floquet multipliers are different from 1, the index $\gamma_T(0)$ can be computed from the variational equation (10). In the critical case $\mu_1 = \mu_2 = 1$ one can use the first nonlinear approximation to compute the index in many cases.

6. Stability and index

The connections between stability and index are delicate and some interesting questions remain unanswered. For an autonomous system

$$\dot{x} = X(x), \qquad x \in \mathbb{R}^d$$

having $x = 0$ as an isolated and stable equilibrium, we would like to know the index of the vector field X. For dimension $d = 2$ it is known that

$$\mathrm{ind}[X, 0] = 1.$$

For dimension $d \geq 3$ the index can be any integer (at least for C^∞ vector fields). We refer to the interesting book by Krasnosel'skii and Zabreiko [12] for more information. See also the papers by Erle [9] and Bonati and Villadelprat [4]. In the recent paper [4] there is a construction which contradicts some of the assertions in [12].

Given a T-periodic equation

$$\dot{x} = X(t, x), \qquad x \in \mathbb{R}^d$$

having $x = 0$ as an isolated (period T) and stable equilibrium, the interest is in the index $\gamma_T(0)$. For dimensions $d \geq 3$, the construction for the autonomous case can be adapted and so the index can take any value (see [27] for more details). In [10] Krasnosel'skii stated that the index is always 1 in dimension $d = 2$. He also said that this fact could be proved easily, but he did not present the proof. Dancer and I found a proof which probably cannot be called elementary. It is based on the arc translation lemma due to Brouwer. In particular we adapted the proof of this lemma given by M. Brown in [7]. Our equation (9) can be transformed, in the usual way, into a periodic system in $\mathbb{R}^2$. Hence, as a consequence of the results in [8] we have,

Theorem 2. *Assume that $\theta = 0$ is an equilibrium of (9) which is isolated (period T) and stable. Then $\gamma_T(0) = 1$.*

Equations of period T are also of period nT, $n \geq 2$. We are lead to the following consequence,

Corollary 3. *Assume that $\theta = 0$ is an equilibrium of (9) which is isolated (period nT, $n \geq 2$) and stable. Then $\gamma_{nT}(0) = 1$.*

This result is of practical value. It allows to obtain instability criteria via degree theory. In fact, if one of the indexes is different from 1, we can say that $\theta = 0$ is unstable. As an example consider the equation

$$\ddot{\theta} + \theta + c(t)\theta^2 = 0 \tag{11}$$

and assume that $c(t)$ has period

$$T = \frac{2\pi}{3}.$$

The linearization ($\ddot{\theta} + \theta = 0$) has monodromy matrices (for periods T, $2T$ and $3T$),

$$X(T) = R[\frac{2\pi}{3}], \qquad X(2T) = R[\frac{4\pi}{3}], \qquad X(3T) = R[2\pi] = I.$$

This implies that, for periods T and $2T$, it is possible to compute the index by linearization. Namely, $\theta = 0$ is isolated (period $2T$) and

$$\gamma_T(0) = \text{sign}\{\det(I - X(T))\} = 1,$$
$$\gamma_{2T}(0) = \text{sign}\{\det(I - X(2T))\} = 1.$$

To compute the third index we must employ the discussions about the degenerate case in Section 5. Since

$$X(3T) = I$$

and

$$\phi_1(t) = \cos t, \qquad \phi_2(t) = \sin t,$$

for period $3T$ the function H becomes

$$H(\theta_0, \omega_0) = \frac{1}{3} \int_0^{2\pi} c(t)(\theta_0 \cos t + \omega_0 \sin t)^3 dt.$$

In complex notation,

$$\xi = \theta_0 + i\omega_0, \qquad \bar{\xi} = \theta_0 - i\omega_0,$$

equals

$$H(\xi, \bar{\xi}) = \frac{1}{24} \int_0^{2\pi} c(t)(\xi e^{-it} + \bar{\xi} e^{it})^3 dt.$$

The function $c(t)$ has period $\frac{2\pi}{3}$ and this implies that

$$\int_0^{2\pi} c(t) e^{it} dt = 0.$$

Some computations lead to

$$H(\xi, \overline{\xi}) = \frac{1}{8}\left(\gamma\overline{\xi}^3 + \overline{\gamma}\xi^3\right)$$

where

$$\gamma = \int_0^{\frac{2\pi}{3}} c(t)e^{3it}\,dt. \tag{12}$$

The derivative

$$H_{\overline{\xi}} = \frac{1}{2}(H_{\theta_0} + iH_{\omega_0})$$

is $\frac{3}{8}\gamma\overline{\xi}^2$ and so, if $\gamma \neq 0$, the only critical point of H is the origin. It follows that $\theta = 0$ is isolated (period $3T$) and

$$\gamma_{3T}(0) = \text{ind}[\nabla H, 0] = -2.$$

The conclusion is that $\theta = 0$ is unstable as soon as the quantity defined by (12) does not vanish.

This example shows that the linearization procedure is not valid for a general equation of the type (9). In this example $\theta = 0$ was stable for the linearization and unstable for the original equation. The reader who is familiar with hamiltonian dynamics will have recognized the phenomenon of resonance at the roots of the unity. In this case it was the third root

$$\omega = e^{\frac{2\pi i}{3}}$$

and we refer to [32,19] for more details.

7. The pendulum of variable length

Consider again the equation

$$\ddot{\theta} + a(t)\sin\theta = 0 \tag{13}$$

where $a(t)$ is continuous, T-periodic and positive. We shall compute the second index $\gamma_{2T}(0)$. Let us start with the linearization principle. If μ_1 and μ_2 are the Floquet multipliers of the linearized equation (period T), the eigenvalues of

$$X(2T) = X(T)^2$$

are μ_1^2 and μ_2^2. In the elliptic case,

$$\mu_1 = \overline{\mu_2}, \qquad \mu_1 \neq \pm 1,$$

and

$$\gamma_{2T}(0) = \text{sign}\{(1 - \mu_1^2)(1 - \mu_2^2)\} = \text{sign}|1 - \mu_1^2|^2 = 1.$$

In the hyperbolic case,

$$|\mu_1| < 1 < |\mu_2|$$

and

$$\gamma_{2T}(0) = \text{sign}\{(1 - \mu_1^2)(1 - \mu_2^2)\} = -1.$$

In the parabolic case
$$\mu_1 = \mu_2 = \pm 1,$$
we notice that $X(2T)$ must be conjugate in $Sp(\mathbb{R}^2)$ to one of the matrices I, P_+, P_-. Going back to the methods of computation in the degenerate case and considering the third order approximation
$$\ddot{\theta} + a(t)\theta - \frac{1}{3!}a(t)\theta^3 = 0 \tag{14}$$
we obtain an expansion of the Poincaré map like in Section 5, with
$$H(\theta_0, \omega_0) = -\frac{1}{24}\int_0^{2T} a(t)(\phi_1(t)\theta_0 + \phi_2(t)\omega_0)^4 dt.$$

Since H has a strict maximum at the origin $\xi = 0$, we can use Euler's theorem for homogeneous functions to deduce that
$$\xi \cdot \nabla H(\xi) = 4H(\xi) < 0 \qquad \forall \xi \neq 0.$$
This inequality implies that $\xi = 0$ is the only critical point of H and so we can discuss the case
$$X(2T) = I.$$
More precisely, $\theta = 0$ is isolated (period $2T$) with
$$\gamma_{2T}(0) = \text{ind}[\nabla H, 0] = 1.$$
Here we have used a typical property of the index for gradient operators (see [12] or [1]).

Let us now assume that $X(2T)$ is conjugate to P_+ or P_-. We apply the proposition in Section 3 and find a change of independent variable
$$t = \lambda(s + \tau)$$
such that the equation (13) becomes
$$\frac{d^2\theta}{ds^2} + a^\star(s)\theta - \frac{a^\star(s)}{3!}\theta^3 + \cdots = 0, \qquad a^\star(s) = \lambda^2 a(\lambda(s + \tau))$$
and the monodromy matrix $X^\star(2T^\star)$ of the linearization is precisely P_+ or P_-. This transformation in time does not alter the index of $\theta = 0$. For if $P^\star$ is the Poincaré map of the new equation, then
$$L^{-1}P^\star L = P$$
with
$$L(\theta_0, \omega_0) = (\theta_0, \lambda\omega_0).$$
The commutativity theorem for degree shows that the indexes of $I - P$ and $I - P^\star$ coincide. Incidentally we notice that the stability of $\theta = 0$ is also preserved. We apply once again the discussions of Section 5 and conclude that
$$\gamma_{2T}(0) = \nu \text{sign}\{-\frac{1}{3!}\int_0^{T^\star} a^\star(s)\phi_1^\star(s)^4 ds\} = -\nu$$
with $\nu = 1$ if $X(2T) \sim P_+$ and $\nu = -1$ if $X(2T) \sim P_-$.

At this point the reader may think that the computation of other indexes $\gamma_{kT}(0)$ could lead to more instability criteria. However this is not the case, as can be seen after computing all indexes. The next step is to discuss the stability of $\theta = 0$. This can be done but the techniques which are required go beyond the scope of these notes. The details can be seen in [26,23], the second paper in collaboration with Núñez. Summing up the previous discussions and the results in these papers one obtains

Theorem 4. *The following statements are equivalent:*

(i) $\theta = 0$ *is stable for* (13)

(ii) $\theta = 0$ *is isolated (period 2T) and* $\gamma_{2T}(0) = 1$

(iii) $\theta = 0$ *is stable for the Duffing equation* (14)

(iv) *the monodromy matrix* $X(2T)$ *is conjugate in* $Sp(\mathbb{R}^2)$ *to* $R[\Theta]$, *for some* $\Theta \in \mathbb{R}$, *or to* P_-.

We notice that the assertion (iv) is the answer to the question posed in the introduction of these notes. The linearization procedure is valid for the pendulum of variable length excepting when

$$X(2T) \sim P_-.$$

In this situation $\theta = 0$ is stable for the original equation (13) but unstable for the linearization. The analysis leading to Theorem 4 is not exclusive of the pendulum and can be applied to other equations. The crucial property is that the coefficient of the cubic term has a sign. Other results about stability using the third approximation can be found in [28,16,17,22,24,29,34,13,14].

A natural question about Theorem 4 is its possible extension to more degrees of freedom. To fix the ideas consider the system

$$\ddot{\theta} + A\theta + \alpha(t)S(\theta) = 0, \qquad \theta = \begin{pmatrix} \theta_1 \\ \vdots \\ \theta_N \end{pmatrix}$$

where A is the $N \times N$ tridiagonal matrix, coming from the discretization of the Laplacian,

$$A = \epsilon \begin{pmatrix} -2 & 1 & 0 & \cdots & 0 & 0 \\ 1 & -2 & 1 & \cdots & 0 & 0 \\ 0 & 1 & -2 & \cdots & 0 & 0 \\ \vdots & \vdots & \vdots & \ddots & \vdots & \vdots \\ 0 & 0 & 0 & \cdots & -2 & 1 \\ 0 & 0 & 0 & \cdots & 1 & -2 \end{pmatrix}, \qquad \epsilon > 0,$$

and

$$S(\theta) = \begin{pmatrix} \sin\theta_1 \\ \vdots \\ \sin\theta_N \end{pmatrix}.$$

It is not clear if the approach in these notes can be extended. In principle one cannot expect a result like Theorem 2 because we are in more dimensions and the index of a stable equilibrium can be any number. However, our system is analytic and Hamiltonian and we are in a rather special situation. Is there a version of Theorem 2 applicable to this example? In any case it is probable that one can obtain instability criteria for the third approximation using Lyapunov functions. The stability is more delicate. Probably the notion of perpetual stability is too demanding as to obtain reasonable results. There is the notion of formal stability, associated to the Birkhoff normal form [3], which seems easier to study. This formal stability implies (via KAM theory) the notion of stability introduced by Moser in his conference in ICM Berlin 99 [21]. We finish these notes by recalling Moser's definition of *stability in measure*: instead of requiring that all orbits of a certain neighborhood are bounded for all times, one asks that most orbits (in the sense of measure) are bounded.

References

1. Amann, H., A note on degree theory for gradient mappings, *Proc. Amer. Math. Soc.* **85** (1982), 591–595.
2. Arnold, V.I., and Avez, A., *Ergodic Problems of Classical Mechanics*, Addison Wesley, 1989.
3. Birkhoff, G.D., *Dynamical Systems*, Amer. Math. Soc., 1927.
4. Bonati, C., and Villadelprat, J., The index of stable critical points, *Topology Appl.* **126** (2002), 263–271.
5. Broer, H.W., and Levi, M., Geometrical aspects of stability theory for Hill's equation, *Arch. Rat. Mech. Anal.* **131** (1995), 225–240.
6. Broer, H.W., and Vegter, G., Bifurcational aspects of parametric resonance, *Dynamics Reported (N.S.)* **1** (1992), 1–53.
7. Brown, M., A new proof of Brouwer's lemma on translation arcs, *Houston Math. J.* **10** (1984), 35–41.
8. Dancer, E.N., and Ortega, R., The index of Lyapunov stable fixed points in two dimensions, *J. Dynam. Diff. Eqns.* **6** (1994), 631–637.
9. Erle, D., Stable equilibria and vector field index, *Topol. Appl.* **49** (1993), 231–235.
10. Krasnosel'skii, M.A., *Translations Along Trajectories of Differential Equations*, Trans. Math. Monographs 19, Amer. Math. Soc., 1968.
11. Krasnosel'skii, M.A., Perov, A.I., Povolotskiy, A.I., and Zabreiko, P.P., *Plane Vector Fields*, Academic Press, 1966.
12. Krasnosel'skii, M.A., and Zabreiko, P.P., *Geometrical Methods of Nonlinear Analysis*, Springer-Verlag, 1984.
13. Lei, J., and Zhang, M., Twist property of periodic motion of an atom near a charged wire, *Letters in Math. Phys.* **60** (2002), 9–17.

14. Lei, J., Li, X., Yan, P., and Zhang, M., Twist character of the least amplitude periodic solution of the forced pendulum, *SIAM J. Math. Anal.* **35** (2003), 844–867.

15. Levi-Civita, T., Sopra alcuni criteri di instabilita, *Annali di Matematica* **V** (1901), 221–307.

16. Liu, B., The stability of the equilibrium of a conservative system, *J. Math. Anal. App.* **202** (1996), 133–149.

17. Liu, B., The stability of the equilibrium of reversible systems, *Trans. Am. Math. Soc.* **351** (1999), 515–531.

18. Lloyd, N.G., *Degree Theory*, Cambridge Univ. Press, 1978.

19. Meyer, K., Counter-examples in dynamical systems via normal form theory, *SIAM Rev.* **28** (1986), 41–51.

20. Magnus, W., and Winkler, S., *Hill's Equation*, Dover, 1979.

21. Moser, J., Dynamical systems-past and present, *Doc. Math. J., DMV Extra Volume* **ICM I** (1998), 381–402.

22. Núñez, D., The method of lower and upper solutions and the stability of periodic oscillations, *Nonl. Anal.* **51** (2002), 1207–1222.

23. Núñez, D., and Ortega, R., Parabolic fixed points and stability criteria for nonlinear Hill's equation, *Z. Angew. Math. Phys.* **51** (2000), 890–911.

24. Núñez, D., and Torres, P., Periodic solutions of twist type of an earth satelite equation, *Discr. Con. Dyn. Syst.* **7** (2001), 303–306.

25. Ortega, R., The twist coefficient of periodic solutions of a time-dependent Newton's equation, *J. Dynam. Diff. Eqns.* **4** (1992), 651–665.

26. Ortega, R., The stability of the equilibrium of a nonlinear Hill's equation, *SIAM J. Math. Anal.* **25** (1994), 1393–1401.

27. Ortega, R., Some applications of the topological degree to stability theory, in *Topological Methods in Differential Equations and Inclusions* (Granas, A., and Frigon, M., Eds.), pages 377–409, Kluwer Academic, 1995.

28. Ortega, R., Periodic solutions of a newtonian equation: stability by the third approximation, *J. Diff. Eqns.* **128** (1996), 491–518.

29. Torres, P., Twist solutions of a Hill's equation with singular term, *Adv. Nonl. Studies* **2** (2002), 279–287.

30. Rothe, E.H., *Introduction to Various Aspects of Degree Theory in Banach Spaces*, Math. Surveys 23, Amer. Math. Soc., 1986.

31. Sell, G., *Topological Dynamics and Ordinary Differential Equations*, Van Nostrand-Reinhold, 1971.

32. Siegel, C., and Moser, J., *Celestial Mechanics*, Springer-Verlag, 1971.

33. Yan, P., and Zhang, M., Higher order nonresonance for differential equations with singularities, *Math. Meth. in App. Sci.*, **26** (2003), 1067–1074.

34. Zhang, M., The best bound on the rotations in the stability of periodic solutions of a newtonian equation, *J. London Math. Soc.* **67** (2003), 137–148.

Ten Mathematical Essays on Approximation in Analysis and Topology
J. Ferrera, J. López-Gómez, F. R. Ruiz del Portal, Editors

The Bishop–Phelps Theorem

R. R. Phelps

Department of Mathematics, University of Washington,
Box 354-350, Seattle, WA 98195, USA

Abstract

What follows is a history of the Bishop-Phelps theorem on the density of support function-als, together with a sketch of its proof. This is followed by descriptions of a number of extensions and applications, plus two basic open questions.

Key words: Banach spaces, convex sets, support functionals, support points, Bishop-Phelps

1. Introduction

The theorem of the title is an elementary but fundamental result about convex sets and continuous linear functionals on real Banach spaces. I will test the reader's patience by starting with a personal narrative, giving some historical background before describing various extensions and applications, including two open questions. The ideas underlying the proof of the original result will be presented, but most of this article will be descriptive, with appropriate references.

To start, suppose that E is a Banach space and that

$$B = \{x \in E \colon \|x\| \leq 1\}$$

is its unit ball. A corollary to the Hahn–Banach theorem states that to each $x \in E$ with $\|x\| = 1$ there corresponds at least one continuous linear functional $x^* \in E^*$ of norm one which attains its supremum on the unit ball B at x. That is, x^* satisfies

$$\langle x^*, x \rangle = 1 = \sup\{\langle x^*, y \rangle \colon \|y\| \leq 1\} \equiv \|x^*\|.$$

Since B lies in the closed half-space

$$\{y \in E \colon \langle x^*, y \rangle \leq 1\}$$

defined by the closed hyperplane

$$H = \{y \in E \colon \langle x^*, y \rangle = 1\}$$

and since $x \in H \cap B$, we also say that H (or, equivalently, x^*) *supports B at x*. Since there are many points of norm one in E, there should be "many" such *support functionals* of B in the unit sphere S^* of E^*. Another corollary of the Hahn–Banach theorem (the separation theorem) is that *every closed convex set is the intersection of all the closed half–spaces which contain it* and —for reasons which are a bit obscure to me now, some 43 years later, but which arose out of a question in best approximation— it seemed desirable to know that, for bounded closed convex sets, it would suffice to use only those half–spaces defined by functionals which attain their norm. Here is the connection with best approximation: Suppose that C is closed and convex and that a point $x \notin C$ has $y \in C$ as a *nearest point*; that is y satisfies

$$\|x - y\| = \inf\{\|x - z\| \colon z \in C\} \equiv d(x, C).$$

For simplicity, assume that

$$d(x, C) = 1,$$

say, so that

$$(x + \mathrm{int}\,B) \cap C = \varnothing.$$

Any functional of norm one which separates C from $x + \mathrm{int}\,B$ (there is at least one) simultaneously attains its supremum on C at y *and* its supremum on B at $x - y$ and, of course, separates x from C. It is not hard to show that *if every bounded closed convex set is the intersection of "norm–attaining" half–spaces, then the support functionals of B must be norm dense in S^** (and conversely).

Now, Robert James' deep theorem states that if *every* functional in S^* supports the unit ball B, then the space is necessarily reflexive. Thus, it seemed reasonable to give the name "subreflexive" to the larger class of normed linear spaces for which the support functionals are merely dense in S^*. I proceeded to investigate the subreflexive spaces for my thesis ([21,22]), showing, among other things, that there exists an incomplete normed linear space which is not subreflexive. Of course, I looked at examples, namely, the classical non–reflexive Banach spaces, finding that indeed, spaces such as

$$c_0, \quad \ell_1, \quad C(X), \quad \cdots$$

are all subreflexive. In each case, it was an enjoyable exercise to characterize the norm–attaining functionals and show that they were dense in S^*, but unfortunately there was no general pattern to the proofs. After getting my Ph.D. and moving to the Institute for Advanced Study in Princeton for two years, I worked on other questions. I kept returning, however, to the intriguing problem as to whether *every* Banach space must necessarily be subreflexive. When I moved to Berkeley in 1960, I thought of a function–space approach (start by embedding E isometrically in a $C(X)$) and it seemed to me that Errett Bishop would be the ideal person to ask about it. He was. Although the embedding approach was a dead end, Bishop came up with an original idea which, combined with a result I had proved earlier in a different context [23], led us to an affirmative answer [1].

Theorem 1. (Bishop–Phelps) *In any real Banach space E, the linear functionals in E^* which attain their supremum on the unit ball of E are norm–dense in E^*.*

After submitting our paper, we realized that the same technique proved the density of support functionals for an *arbitrary* bounded closed convex subset of a Banach space. That result solved in the affirmative a question that Victor Klee [11] had posed three years earlier:

Must a bounded closed convex subset C of a Banach space necessarily have any support points?

That is, points where a nontrivial continuous linear functional attains its supremum on the set. This having been settled, it seemed time to drop the subject and move on. Nevertheless, I couldn't stop thinking about the proof. Getting up much too sleepily one morning, it seemed likely that my habitual morning work period would be a waste of time, but I sat out on our deck in the California sunshine anyway, with the usual cup of coffee and pad of paper. I can still remember the "Eureka! moment" when I realized what should have been obvious all along: The partial ordering we had used could be defined by means of a convex cone. This geometric viewpoint turned out to be illuminating and fruitful, leading in [2] to both a simpler exposition and substantial generalizations of the original result, including a proof that the support points of C are, in fact, always dense in the boundary of C. A number of expositions of these matters have since appeared (see, e.g. [4,8,31]); the density of support points has even showed up as an exercise in Bourbaki [3, pp. 138].

The cone referred to above is defined as follows: If $x^* \in S^*$ and $0 < k < 1$, define

$$K(x^*; k) = \{x \in E: \ k\|x\| \le \langle x^*, x \rangle\}.$$

Since the function

$$f = k\|\cdot\| - x^*$$

is continuous, convex and positive–homogeneous and

$$K = \{f \le 0\},$$

the latter is closed, convex and preserved under multiplication by positive scalars. To visualize it, sketch the slice of the unit ball B defined by

$$A \equiv \{x \in B: \ \langle x^*, x \rangle \ge k\}.$$

Since $0 < k < 1$, this is nonempty —in fact, has nonempty interior— and misses the origin. It is easy to show that the cone

$$R^+ A \equiv \{ra: \ r \ge 0, \ a \in K\}$$

that it generates is precisely $K(x^*; k)$. Here is a rough description of the way to use such cones to produce support points and support functionals for any proper closed convex set C.

Given a point x in the boundary of C, choose a nearby point z outside of C and separate z from C by a linear functional x^* of norm one. Choosing $0 < k < 1$, form the cone

$$K \equiv K(x^*; k).$$

Using induction (or Zorn's lemma) and completeness, produce a point y in $(x + K) \cap C$ which is a *maximal point* in this set with respect to the partial ordering defined by K, that is, which satisfies

$$y \in (x + K) \cap C \quad \text{and} \quad C \cap (y + K) = \{y\}.$$

(One says that *the cone $y + K$ supports C at y.*) The interior of the cone $y + K$ necessarily misses C and hence can be separated from C by a norm–one functional y^*. The latter must support C at y. If k is sufficiently close to 1 (so that the cone K is rather narrow), then y will be close to x. If k is close to 0, (so that K is rather wide) then y^* will be close to x^*, this can be used to prove the density of support functionals.

Once the density of the sets of support points and support functionals had been established, it was natural to look at related questions: Are there density theorems for linear *operators* which attain their norms? Is the Bishop–Phelps theorem valid in *complex* Banach spaces? What is the topological or set–theoretic nature of the sets of support points or functionals? Is the method of proof of the Bishop–Phelps theorems useful in other contexts? Are any of the results valid in locally convex spaces? We'll give a brief survey of these matters below.

2. Complex Bishop–Phelps theorem

At a 1985 conference at Kent State University, Gilles Godefroy raised the question as to whether there is a valid version of the Bishop–Phelps theorem in complex Banach spaces. The answer is trivially "yes" if one restricts oneself to the real scalars and the real parts of linear functionals; the precise question was this:

Suppose that C is a bounded closed convex subset of the complex Banach space E. Must the functionals $x^* \in E^*$ which satisfy

$$\sup\{|\langle x^*, y \rangle| : y \in C\} = |\langle x^*, x \rangle|$$

for some $x \in C$ be dense in E^*?

This is easily seen to be true if C is the unit ball or, more generally, if C is closed under multiplication by complex scalars of modulus 1, or is a translate of such a set. It is also true for arbitrary bounded closed convex sets in any Banach space having the Radon–Nikodým property (RNP) [30]. (Spaces with the RNP have been discussed at length in [4,8] and [9]; in addition to all reflexive space, they contain, e.g. all separable dual spaces.) The question remained open until 1999, when Victor Lomonosov [13,14] constructed an ingenious counter–example, using spaces of analytic functions in the unit disk.

3. Operators which attain their norm

The question was raised in [1] whether linear *operators* T which attain their norms are necessarily dense in the space of all bounded operators from one Banach space to another.

That is, one seeks the density of those T for which

$$\|T\| = \|Tx\|$$

for some x of norm 1. Joram Lindenstrauss [12] showed that while this is not true in general, it *is* true if the domain space is reflexive. He obtained a number of results showing that the question has an affirmative answer in other special cases, and he went on to show intimate connections between such questions and the notion of strongly exposed points of the unit ball. Jean Bourgain [3] carried this work further. He started by defining a bounded absolutely convex set $C \subset E$ (that is, $\lambda C \subset C$ for all $|\lambda| \leq 1$) to have the *Bishop–Phelps property* if every bounded operator from E into another Banach space can be approximated in norm by operators T for which

$$\sup\{\|Tx\|: \ x \in C\}$$

is attained at some point of C. He then defined a Banach space E to have the Bishop–Phelps property if this was the case for each of its bounded closed absolutely convex subsets. This led to the beautiful theorem that *a space E has the Bishop–Phelps property if and only if it has the Radon–Nikodým property.*

4. Topological and set–theoretic properties of support points

The topological nature of support points was first considered in [24]. It was shown that the support points of a closed convex subset of a metrizable locally convex space E is always an F_σ–set (and nonempty, if E a Banach space) and, under additional hypotheses, the set of support points was shown to have certain connectedness properties. For instance, in a weakly compact convex set, the support points are weakly connected, or in a reflexive Banach space, the support points of any closed convex set which contains no hyperplane are arcwise connected. George Luna [15,16] has improved upon these results in several directions. The question of connectedness of the normalized (to have supremum 1) support *functionals* of a closed convex set was also considered in [24]; it was shown that if C has nonempty interior and contains no hyperplane, then the normalized support functionals are weak* connected. In particular, the functionals in S^* which attain their norm are weak* connected, but the following question remains open. (It has an affirmative answer in all the concrete examples we have examined.)

Problem 2. *Are the norm–attaining functionals in S^* arcwise connected in the norm topology?*

A set–theoretic question is the following:

Can the (normalized) support functionals of a bounded closed convex set be a countable subset of S^*?

Surprisingly, this natural question was only raised relatively recently, by Ludek Zajiček in January, 1999, and is still open.

Problem 3. *Suppose that C is a bounded nonempty closed convex subset of the Banach space E, with $\dim E > 1$. Must the set $\Sigma(C)$ of normalized support functionals of C be an uncountable subset of S^*?*

Suppose that the answer is "No", that is, *suppose that there exists such a set C for which $\Sigma(C)$ is countable*. Since, by the Bishop–Phelps theorem, $\Sigma(C)$ is dense in S^*, it follows that E^* must be separable, while if E were reflexive, then $\Sigma(C)$ would coincide with the uncountable set S^*. Thus, E *must be a nonreflexive Banach space with separable dual.* Also, *the interior of C must be empty*: Indeed, assume without loss of generality that

$$0 \in \text{int} C.$$

Choose a two-dimensional subspace M of E (any reflexive subspace will suffice). The functionals $f \in S_{M^*}$ which support $M \cap C$ form an uncountable set, and the Hahn-Banach theorem yields an extension of each one of them to an element

$$\tilde{f} \in \Sigma(C).$$

Moreover, the map

$$\Sigma(M \cap C) \ni f \to \tilde{f} \in \Sigma(C)$$

is one–one, so $\Sigma(C)$ would be uncountable.

These two problems are slightly related, since any nontrivial arcwise connected set is necessarily uncountable.

5. Non-support points

If a closed convex set C has interior points, then the support points coincide with its boundary, so the set $N(C)$ of all points of C which are *not* support points is the same as $\text{int}\, C$, the interior of C. Now, there are a number of theorems dealing with the differentiability of a real–valued convex continuous function f defined on the interior of a convex set. For instance, in a reflexive space (or, more generally, in an Asplund space), the set of points of Fréchet differentiability of f form a dense G_δ subset of $\text{int}\, C$. (See, e.g., [31].) If C has empty interior, then it is still the case (as noted above) that $S(C)$ is an F_σ set and hence

$$N(C) \text{ is a } G_\delta \text{ subset of } C.$$

It is easily seen to be convex and either empty (e.g., if C lies in a closed hyperplane) or *dense* in C. It is always nonempty in a separable Banach space. Maria Elena Verona [33] was the first to show that $N(C)$ could be a substitute for the interior of C when considering differentiability questions, proving that *in a separable Banach space, a locally Lipschitzian convex function on $N(C)$ is Gâteaux differentiable at the points of a dense G_δ subset of $N(C)$*. This generalizes Mazur's classical theorem on Gâteaux differentiability of continuous convex functions on separable Banach spaces. (Note that while continuous convex functions on $\text{int}\, C$ are automatically locally Lipschitzian (see e.g., [31]), simple examples show that this need not be true for $N(C)$ [32].) Verona's theorem was subsequently extended in several ways by John Rainwater [32].

6. Generalizations of the Bishop–Phelps proof

Suppose that f is a proper convex lower semicontinuous extended real–valued function on a Banach space E. By *proper* we mean that $-\infty < f \leq \infty$ and that $f(x) < \infty$ for at least one point x. The *epigraph* of f is the set

$$\text{epi}(f) = \{(x,r) \in E \times \mathcal{R} \colon f(x) \leq r\}$$

of all points in $E \times \mathcal{R}$ lying above or on the graph of f. This a closed and convex subset of $E \times \mathcal{R}$. By the Bishop–Phelps theorem, $\text{epi}(f)$ will have "many" supporting hyperplanes. Of course, these are necessarily defined by elements

$$(x^*, s) \in E^* \times \mathcal{R} \equiv (E \times \mathcal{R})^*.$$

If one of these hyperplanes is "non–vertical", that is, if $s \neq 0$, and if (x^*, s) attains its supremum on epi$f$ at $(x, f(x))$, say, then, setting

$$y^* = s^{-1} x^*,$$

it follows that the affine function

$$y^* + f(x) - \langle y^*, x \rangle$$

is dominated by f and equals it at x. This fact can be written as

$$\langle y^*, y - x \rangle \leq f(y) - f(x) \text{ for all } y \in E.$$

Any such functional y^* is said to be a *subdifferential* of f at x. Prior to the Bishop–Phelps theorem, it was not known whether subdifferentials necessarily existed for arbitrary proper lower semicontinuous convex functions. On the other hand, the argument described above had a qualifier; note that it assumed that $s \neq 0$. It is somewhat complicated to produce subdifferentials using this approach; the original – and better – approach was developed by Arne Brøndsted and Terry Rockafellar [5], who modified the Bishop–Phelps proof to show that, given x_0 with $f(x_0) < \infty$ and a functional x_0^* which is *almost* a subdifferential of f at x_0, that is, which for some $\epsilon > 0$ satisfies

$$\langle x_0^*, y - x_0 \rangle \leq f(y) - f(x_0) + \epsilon \text{ for all } y \in E,$$

there exists a subdifferential x^* of f at a point x for which x^* is near x_0^* and x is near x_0. "Near" in this context can be taken to be within $\sqrt{\epsilon}$. Ivar Ekeland [10] generalized the Bishop–Phelps argument even further, proving a *variational principle* for arbitrary proper lower semicontinuous functions on a complete metric space. This led him to unified proofs of not only the Brøndsted–Rockafellar and Bishop–Phelps theorems but of a number of other results in nonlinear analysis. In addition to [10], an exposition of all the above may be found in [31].

A different modification of the Bishop–Phelps proof was used by Felix Browder [6,7]. Recall that the cones $K(x^*, k)$ can be written in the form $\mathcal{R}^+ A$, where A is a certain bounded closed convex set having nonempty interior and missing the origin. He showed that locally, at least, one can obtain maximal points for an *arbitrary* closed set (not necessarily convex) with respect to the ordering defined by *any* cone of the form

$$K = \mathcal{R}^+ B$$

where B has the same properties as A (above). Specifically, he proved the following. (Here, $B_\delta(x)$ is the ball of radius δ centered at x.)

Lemma 4. (F. E. Browder [6]) *Suppose that S is a proper closed subset of E, that z is a point in the boundary of S and that $\epsilon > 0$. Then there exists $\delta > 0$, a cone K as above and a point $x \in S$ such that*

$$\|x - z\| < \epsilon \quad \text{and} \quad S \cap (K + x) \cap B_\delta(x) = \{x\}.$$

This was used by Browder to obtain theorems about "normal solvability" of certain nonlinear operators connected with partial differential equations. One takes $S = f(F)$ where f is a certain nonlinear mapping from another Banach space F into E. In addition to [6,7], see [26] for an exposition and additional references.

7. Locally convex spaces

Tenney Peck [18] dashed any hopes for finding support points without the hypothesis that E be a Banach space by proving the following result.

Theorem 5. *If E is the product space of an infinite sequence of non-reflexive Banach spaces, then E contains a bounded closed convex nonempty subset which has no support points.*

Such a product space is, of course, a locally convex Fréchet space. For the special locally convex space consisting of the dual space E^* of a Banach space E, provided with its weak* topology, it is possible to obtain some norm–density theorems; see e.g., [17] and [27].

8. Miscellany

As noted in the introductory section, a closed convex subset C of a Banach space can always be represented as the intersection of the closed half–spaces which support it. The question as to which sets S of support points can be removed from C and still have C represented by the half spaces supporting it at $C \setminus S$ has been investigated in [28]. The question as to whether *vector–valued* lower semicontinuous convex functions need have subdifferentials was answered in the negative in [29]; this is a consequence of an example of two proper lower semicontinuous functions on ℓ_2 with no common point of subdifferentiability.

References

1. Bishop, E., and Phelps, R.R., A proof that every Banach space is subreflexive, *Bull. Amer. Math. Soc.* **67** (1961), 97–98.

2. Bishop, E., and Phelps, R.R., The support functionals of a convex set, in *Convexity, Proc. Symp. Pure Math. VII*, pages 27–35, Amer. Math. Soc., 1963.
3. Bourbaki, N., *Espaces Vectoriels Topologiques*, 2nd Ed., Hermann, Paris, 1966.
4. Bourgin, R.D., *Geometric Aspects of Convex Sets with the Radon–Nikodym Property*, Lecture Notes in Maths. 993, Springer-Verlag, 1983.
5. Brøndsted, A., and Rockafellar, R.T., On the subdifferentiability of convex functions, *Proc. Amer. Math. Soc.* **16** (1965), 605–611.
6. Browder, F.E., Normal solvability and the Fredholm alternative for mappings into infinite dimensional manifolds, *J. Funct. Anal.* **8** (1971), 250–274.
7. Browder, F.E., Normal solvability and ϕ–accretive mappings of Banach spaces, *Bull. Amer. Math. Soc.* **78** (1972), 186–192.
8. Diestel, J., *Geometry of Banach Spaces. Selected Topics*, Lecture Notes in Maths. 485, Springer-Verlag, 1975.
9. Diestel, J., and Uhl, J.J., *Vector Measures*, Math. Surveys 15, Amer. Math. Soc., 1977.
10. Ekeland, I., Nonconvex minimization problems, *Bull. Amer. Math. Soc.* (N.S.) **1** (1979), 443–474.
11. Klee, V.L., Extremal structure of convex sets II, *Math. Zeit.* **69** (1958), 90–104.
12. Lindenstrauss, J., On operators which attain their norm, *Israel J. Math.* **3** (1963), 139–148.
13. Lomonosov, V., A counterxample to the Bishop–Phelps theorem in complex spaces, *Israel. J. Math.* **115** (2000), 25–28.
14. Lomonosov, V., On the Bishop–Phelps theorem in complex spaces, *Quaest. Math.* **23** (2000), 187–191.
15. Luna, G., Connectedness properties of support points of convex sets, *Rocky Mount. J. Math.* **16** (1986), 147–151.
16. Luna, G., Local connectedness of support points, *Rocky Mount. J. Math.* **18** (1988), 179–184.
17. Luna, G., The dual of a theorem of Bishop and Phelps, *Proc. Amer. Math. Soc.* **47** (1975), 171–174.
18. Peck, N.T., Support points in locally convex spaces, *Duke Math. J.* **38** (1971), 271–278.
19. Phelps, R.R., Convex Sets and Nearest Points, *Proc. Amer. Math. Soc.* **8** (1957), 780–797.
20. Phelps, R.R., Convex Sets and Nearest Points II, *Proc. Amer. Math. Soc.* **9** (1958), 867–873.
21. Phelps, R.R., Subreflexive normed linear spaces, *Archiv der Math.* **8** (1957), 444–450.
22. Phelps, R.R., Some subreflexive Banach spaces, *Archiv der Math.* **10** (1959), 162–169.
23. Phelps, R.R., A representation theorem for bounded convex sets, *Proc. Amer. Math. Soc.* **11** (1960), 976–983.
24. Phelps, R.R., Some topological properties of support points, *Israel J. Maths.* **13** (1972), 327–336.
25. Phelps, R.R., Support cones and their generalizations, *Proc. Symp. Pure Math.* **7** (1962), 393–401.
26. Phelps, R.R., Support cones in Banach spaces and their applications, *Adv. in Math.* **13** (1974), 1–19.
27. Phelps, R.R., Weak* Support Points of Convex Sets in E^*, *Israel J. Maths.* **2** (1964), 177–182.

28. Phelps, R.R., Removable sets of support points of convex sets in Banach spaces, *Proc. Amer. Math. Soc.* **99** (1987), 319–322.

29. Phelps, R.R., Counterexamples concerning support theorems for convex sets in Hilbert space, *Canadian Math. Bull.* **31** (1988), 121–128.

30. Phelps, R.R., The Bishop–Phelps theorem in complex spaces: An open problem, in *Function Spaces* (Jarosz, K., Ed.), pages 337-340, Lect. Notes in Pure and Appl. Math. 136, Dekker, N. Y., 1992.

31. Phelps, R.R., *Convex Functions, Monotone Operators and Differentiability* (2nd Edition), Lecture Notes in Maths. 1364, Springer–Verlag, 1993.

32. Rainwater, J., Yet more on the differentiability of convex functions, *Proc. Amer. Math. Soc.* **103** (1988), 773–778.

33. Verona, M.E., More on the differentiability of convex functions, *Proc. Amer. Math. Soc.* **103** (1988), 137–140.

An essay on some problems of approximation theory

A. G. Ramm

Mathematics Department, Kansas State University,
Manhattan, KS 66506-2602, USA

Abstract

Several questions of approximation theory are discussed:

1) can one approximate stably in L^∞ norm f' given approximation f_δ, $\| f_\delta - f \|_{L^\infty} < \delta$, of an unknown smooth function f such that $\| f'(x) \|_{L^\infty} \leq m_1$?

2) can one approximate an arbitrary $f \in L^2(D)$, where $D \subset \mathbb{R}^n$, $n \geq 3$, is a bounded domain, by linear combinations of the products $u_1 u_2$, where $u_m \in N(L_m)$, $m = 1, 2$, L_m is a formal linear partial differential operator, and $N(L_m)$ is the null-space of L_m in D,

$$N(L_m) := \{w : L_m w = 0 \text{ in } D\} \text{ ?}$$

3) can one approximate an arbitrary $L^2(D)$ function by an entire function of exponential type whose Fourier transform has support in an arbitrary small open set? Is there an analytic formula for such an approximation?

Key words: stable differentiation, approximation theory, Property C, elliptic equations, Runge-type theorems, scattering solutions

1. Introduction

In this essay I describe several problems of approximation theory which I have studied and which are of interest both because of their mathematical significance and because of their importance in applications.

1.1. *Approximation of the derivative of noisy data*

The first question I have posed around 1966. The question is: suppose that $f(x)$ is a smooth function, say $f \in C^\infty(\mathbb{R})$, which is T-periodic (just to avoid a discussion of its behavior near the boundary of an interval), and which is not known; assume that its δ-approximation $f_\delta \in L^\infty(\mathbb{R})$ is known,

$$\|f_\delta - f\|_\infty < \delta,$$

where $\| \cdot \|_\infty$ is the $L^\infty(\mathbb{R})$ norm. Assume also that

$$\|f'\|_\infty \leq m_1 < \infty.$$

Can one approximate stably in $L^\infty(\mathbb{R})$ the derivative f', given the above data $\{\delta, f_\delta, m_1\}$?

By a possibility of a stable approximation (estimation) I mean the existence of an operator L_δ, linear or nonlinear, such that

$$\sup_{\substack{f \in C^\infty(\mathbb{R}) \\ \|f-f_\delta\|_\infty \leq \delta,\ \|f'\|_\infty \leq m_1}} \|L_\delta f_\delta - f'\|_\infty \leq \eta(\delta) \to 0 \quad \text{as } \delta \to 0, \tag{1}$$

where $\eta(\delta) > 0$ is some continuous function, $\eta(0) = 0$. Without loss of generality one may assume that $\eta(\delta)$ is monotonically growing.

In 1962-1966 there was growing interest to ill-posed problems. Variational regularization was introduced by D. L. Phillips [2] in 1962 and a year later by A. N. Tikhonov [33]. It was applied in [1] in 1966 to the problem of stable numerical differentiation. The method for stable differentiation proposed in [1] was complicated.

I then proposed and published in 1968 [3] the idea to use a divided difference for stable differentiation and to use the stepsize $h = h(\delta)$ as a regularization parameter. If

$$\|f''\| \leq m_2,$$

then

$$h(\delta) = \sqrt{\frac{2\delta}{m_2}},$$

and if one defines ([3]):

$$L_\delta f_\delta := \frac{f_\delta(x + h(\delta)) - f_\delta(x - h(\delta))}{2h(\delta)}, \tag{2}$$

then

$$\|L_\delta f_\delta - f'\|_\infty \leq \sqrt{2m_2\delta} := \varepsilon(\delta). \tag{3}$$

It turns out that the choice of L_δ, made in [3], that is, L_δ defined in (2), is the best possible among all linear and nonlinear operators T which approximate $f'(x)$ given the information $\{\delta, m_2, f_\delta\}$. Namely, if

$$\mathcal{K}(\delta, m_j) := \{f : f \in C^j(\mathbb{R}),\ m_j < \infty,\ \|f - f_\delta\|_\infty \leq \delta\} \quad \text{and} \quad m_j = \|f^{(j)}\|_\infty,$$

then

$$\inf_T \sup_{f \in \mathcal{K}(\delta, m_2)} \|T f_\delta - f'\| \geq \sqrt{2m_2\delta}. \tag{4}$$

One can find a proof of this result and more general ones in [4], [5], [8], [28], [31], [32] and various applications of these results in [4]-[7].

The idea of using the stepsize h as a regularization parameter became quite popular after the publication of [3] and was used by many authors later.

In [25] formulas are given for a simultaneous approximation of f and f'.

1.2. *Property C: completeness of the set of products of solutions to homogeneous PDE*

The second question that I will discuss is the following one:

can one approximate, with an arbitrary accuracy, an arbitrary function $f \in L^2(D)$, or in $L^p(D)$ with $p \geq 1$, by a linear combination of the products $u_1 u_2$, where

$$u_m \in N(L_m), \quad m = 1, 2,$$

L_m is a formal linear partial differential operator, and $N(L_m)$ is the null-space of L_m in D,

$$N(L_m) := \{w : L_m w = 0 \text{ in } D\} \ ?$$

This question has led me to the notion of Property C for a pair of linear formal partial differential operators $\{L_1, L_2\}$.

Let us introduce some notations. Let $D \subset \mathbb{R}^n$, $n \geq 3$, be a bounded domain,

$$L_m u(x) := \sum_{|j| \leq J_m} a_{jm}(x) D^j u(x), \quad m = 1, 2,$$

j is a multiindex, $J_m \geq 1$ is an integer, $a_{jm}(x)$ are some functions whose smoothness properties we do not specify at the moment,

$$D^j u = \frac{\partial^j u}{\partial x_1^{j_1} \dots \partial x_n^{j_n}}, \qquad |j| = j_1 + \dots + j_n.$$

Define

$$N_m := N_D(L_m) := \{w : L_m w = 0 \text{ in } D\},$$

where the equation is understood in the sense of distribution theory. Consider the set of products $\{w_1 w_2\}$, where $w_m \in N_m$ and we use all the products which are well-defined. If L_m are elliptic operators and $a_{jm}(x) \in C^\gamma(\mathbb{R}^n)$, then, by elliptic regularity, the functions w_m are smooth and, therefore, the products $w_1 w_2$ are well defined.

Definition 1. A pair $\{L_1, L_2\}$ has Property C if and only if the set $\{w_1 w_2\}_{\forall w_m \in N_m}$ is total in $L^p(D)$ for some $p \geq 1$.

In other words, if $f \in L^p(D)$, then

$$\left\{\int_D f(x) w_1 w_2 dx = 0, \quad \forall w_m \in N_m\right\} \Rightarrow f = 0, \tag{5}$$

where $\forall w_m \in N_m$ means for all w_m for which the products $w_1 w_2$ are well defined.

Definition 2. If the pair $\{L, L\}$ has Property C then we say that the operator L has this property.

From the point of view of approximation theory Property C means that any function $f \in L^p(D)$ can be approximated arbitrarily well in $L^p(D)$ norm by a linear combination of the set of products $w_1 w_2$ of the elements of the null-spaces N_m.

For example, if $L = \nabla^2$ then $N(\nabla^2)$ is the set of harmonic functions, and the Laplacian has Property C if the set of products $h_1 h_2$ of harmonic functions is total (complete) in $L^p(D)$.

The notion of Property C has been introduced in [9]. It was developed and widely used in [9] - [20]. It proved to be a very powerful tool for a study of inverse problems [14]-[17], [19]-[20], [28]-[29].

Using Property C the author has proved in 1987 the uniqueness theorem for 3D inverse scattering problem with fixed-energy data [11], [12], [16], uniqueness theorems for inverse problems of geophysics [11], [15], [17], and for many other inverse problems [17]. The above problems have been open for several decades.

1.3. *Approximation by entire functions of exponential type*

The third question that I will discuss, deals with approximation by entire functions of exponential type. This question is quite simple but the answer was not clear to engineers in the fifties. It helped to understand the problem of resolution ability of linear instruments [21], [22], and later it turned to be useful in tomography [30]. This question in applications is known as spectral extrapolation.

To formulate it, let us assume that $D \subset \mathbb{R}^n_x$ is a known bounded domain,

$$\widetilde{f}(\xi) := \int_D f(x) e^{i\xi \cdot x} dx := \mathcal{F}f, \quad f \in L^2(D), \tag{6}$$

and assume that $\widetilde{f}(\xi)$ is known for $\xi \in \widetilde{D}$, where $\widetilde{D}$ is a domain in $\mathbb{R}^n_\xi$. The question is:

can one recover $f(x)$ from the knowledge of $\widetilde{f}(\xi)$ in $\widetilde{D}$?

Uniqueness of $f(x)$ with the data $\{\widetilde{f}(\xi), \xi \in \widetilde{D}\}$ is immediate: $\widetilde{f}(\xi)$ is an entire function of exponential type and if $\widetilde{f}(\xi) = 0$ in $\widetilde{D}$, then, by analytic continuation, $\widetilde{f}(\xi) \equiv 0$, and therefore $f(x) = 0$. Is it possible to derive an analytic formula for recovery of $f(x)$ from $\{\widetilde{f}(\xi), \xi \in \widetilde{D}\}$? It turns out that the answer is yes ([23] - [24]). Thus we give an analytic formula for inversion of the Fourier transform $\widetilde{f}(\xi)$ of a compactly supported function $f(x)$ from a compact set $\widetilde{D}$.

From the point of view of approximation theory this problem is closely related to the problem of approximation of a given function $h(\xi)$ by entire functions of exponential type whose Fourier transform has support inside a given convex region. This region is fixed but can be arbitrarily small.

In Sections 2,3 and 4 the above three questions of approximation theory are discussed in more detail, some of the results are formulated and some of them are proved.

2. Stable approximation of the derivative from noisy data.

In this section we formulate an answer to Question 1.1. Denote

$$\|f^{(1+a)}\| := m_{1+a},$$

where $0 < a \le 1$, and

$$\|f^{(1+a)}\| = \|f'\|_\infty + \sup_{x,y\in\mathbb{R}} \frac{|f'(x) - f'(y)|}{|x - y|^a}. \tag{7}$$

Theorem 3. *There does not exist an operator T such that*

$$\sup_{f\in\mathcal{K}(\delta,m_j)} \|Tf_\delta - f'\|_\infty \le \eta(\delta) \to 0 \quad as\ \delta \to 0, \tag{8}$$

if $j = 0$ or $j = 1$. There exists such an operator if $j > 1$. For example, one can take $T = L_{\delta,j}$ where

$$L_{\delta,j}f_\delta := \frac{f_\delta(x + h_j(\delta)) - f_\delta(x - h_j(\delta))}{2h_j(\delta)}, \quad h_j(\delta) := \left(\frac{\delta}{m_j(j - 1)}\right)^{\frac{1}{j}}, \tag{9}$$

and then

$$\sup_{f\in\mathcal{K}(\delta,m_j)} \|L_{\delta,j}f_\delta - f'\|_\infty \le c_j\delta^{\frac{j-1}{j}}, \quad 1 < j \le 2, \tag{10}$$

where

$$c_j := \frac{j}{(j - 1)^{\frac{j-1}{j}}} m_j^{\frac{1}{j}}. \tag{11}$$

Proof.
1. Nonexistence of T for $j = 0$ and $j = 1$.

Let

$$f_\delta(x) = 0, \quad f_1(x) := -\frac{mx(x - 2h)}{2}, \quad 0 \le x \le 2h.$$

Extend $f_1(x)$ to $\mathbb{R}$ so that

$$\|f_1^{(j)}\|_\infty = \sup_{0\le x\le 2h} \|f_1^{(j)}\|, \quad j = 0, 1, 2,$$

and set

$$f_2(x) := -f_1(x).$$

Also, denote

$$(Tf_\delta)(0) := b.$$

One has

$$\|T f_\delta - f_1'\| \geq |(T f_\delta)(0) - f_1'(0)| = |b - mh|, \tag{12}$$

and

$$\|T f_\delta - f_2'\| \geq |b + mh|. \tag{13}$$

Thus, for $j = 0$ and $j = 1$, one has:

$$\gamma_j := \inf_T \sup_{f \in \mathcal{K}(\delta, m_j)} \|T f_\delta - f'\| \geq \inf_{b \in \mathbb{R}} \max\left[|b - mh|, |b + mh|\right] = mh. \tag{14}$$

Since

$$\|f_s - f_\delta\|_\infty \leq \delta, \quad s = 1, 2,$$

and $f_\delta = 0$, one gets

$$\|f_s\|_\infty = \frac{mh^2}{2} \leq \delta, \quad s = 1, 2. \tag{15}$$

Take

$$h = \sqrt{\frac{2\delta}{m}}. \tag{16}$$

Then

$$m_0 = \|f_s\|_\infty = \delta, \quad m_1 = \|f_s'\|_\infty = \sqrt{2\delta m},$$

so (14) yields

$$\gamma_0 = \sqrt{2\delta m} \to \infty \quad \text{as } m \to \infty, \tag{17}$$

and (8) does not hold if $j = 0$.

If $j = 1$, then (14) yields

$$\gamma_1 = \sqrt{2\delta m} = m_1 > 0, \tag{18}$$

and, again, (8) does not hold if $j = 1$.

2. Existence of T for $j > 1$.

If $j > 1$, then the operator $T = L_{\delta,j}$, defined in (9) yields estimate (10), so (8) holds with

$$\eta(\delta) = c_j \delta^{\frac{j-1}{j}}$$

and c_j is defined in (11). Indeed,

$$\|L_{\delta,j} f_\delta - f'\|_\infty \leq \|L_{\delta,j}(f_\delta - f)\|_\infty + \|L_{\delta,j} f - f'\|_\infty \leq \frac{\delta}{h} + m_j h^{j-1}. \tag{19}$$

Minimizing the right-hand side of (19) with respect to $h > 0$ for a fixed $\delta > 0$, one gets (10) and (11). Theorem 2.1 is proved. $\qquad\square$

Remark 4. If $j = 2$, then $m_2 = m$, where m is the number introduced in the beginning of the proof of Theorem 3, and using the Taylor formula one can get a better estimate in the right-hand side of (19) for $j = 2$, namely

$$\|L_{\delta,2} f_\delta - f'\|_\infty \leq \frac{\delta}{h} + \frac{m_2 h}{2}.$$

Minimizing with respect to h, one gets

$$h(\delta) = \sqrt{\frac{2\delta}{m_2}}$$

and

$$\min_{h>0} \left(\frac{\delta}{h} + \frac{m_2 h}{2} \right) := \varepsilon(\delta) = \sqrt{2\delta m_2}.$$

Now (14), with $m = m_2$, yields

$$\gamma_2 \geq \sqrt{2\delta m_2} = \varepsilon(\delta). \tag{20}$$

Thus, we have obtained:

Corollary 5. *Among all linear and nonlinear operators T, the operator*

$$Tf = L_\delta f := \frac{f(x + h(\delta)) - f(x - h(\delta))}{2h(\delta)}, \quad h(\delta) = \sqrt{\frac{2\delta}{m_2}},$$

yields the best approximation of f', $f \in \mathcal{K}(\delta, m_2)$, and

$$\gamma_2 := \inf_{T} \sup_{f \in \mathcal{K}(\delta, m_2)} \|Tf_\delta - f'\|_\infty = \varepsilon(\delta) := \sqrt{2\delta m_2}. \tag{21}$$

Proof. We have proved that $\gamma_2 \geq \varepsilon(\delta)$. If $T = L_\delta$ then

$$\|L_\delta f_\delta - f'\|_\infty \leq \varepsilon(\delta),$$

as follows from the Taylor's formula: if

$$h = \sqrt{\frac{2\delta}{m_2}},$$

then

$$\|L_\delta f_\delta - f'\| \leq \frac{\delta}{h} + \frac{m_2 h}{2} = \varepsilon(\delta), \tag{22}$$

so $\gamma_2 = \epsilon(\delta)$, which concludes the proof. $\qquad\square$

3. Property C

3.1. *Genericity of Property C*

In the introduction we have defined Property C for PDE. Is this property generic or is it an exceptional one?

Let us show that this property is generic: a linear formal partial differential operator with constant coefficients, in general, has Property C. In particular, the operators

$$L = \nabla^2, \quad L = i\partial_t - \nabla^2, \quad L = \partial_t - \nabla^2, \quad \text{and } L = \partial_t^2 - \nabla^2,$$

all have Property C.

A necessary and sufficient condition for a pair $\{L_1, L_2\}$ of partial differential operators to have Property C was found in [9] and [26] (see also [17]).

Let us formulate this condition and use it to check that the four operators, mentioned above, have Property C.

Let

$$L_m u := \sum_{|j| \leq J_m} a_{jm} D^j u(x), \quad m = 1, 2,$$

where $a_{jm} = $ const, $x \in \mathbb{R}^n$, $n \geq 2$, and $J_m \geq 1$. Define the algebraic varieties

$$\mathcal{L}_m := \{z : z \in \mathbb{C}^n, \ L_m(z) := \sum_{|j| \leq J_m} a_{jm} z^j = 0\}, \quad m = 1, 2.$$

Definition 6. Let us write

$$\mathcal{L}_1 \nmid \mathcal{L}_2$$

if and only if there exist

$$z^{(1)} \in \mathcal{L}_1 \quad \text{and} \quad z^{(2)} \in \mathcal{L}_2$$

such that T_m, the tangent spaces in $\mathbb{C}^n$ to $\mathcal{L}_m$ at the points $z^{(m)}$, $m = 1, 2$, are transversal, that is

$$T_1 \nmid T_2.$$

Remark 7. A pair $\{\mathcal{L}_1, \mathcal{L}_2\}$ fails to have the property $\mathcal{L}_1 \nmid \mathcal{L}_2$ if and only if $\mathcal{L}_1 \cup \mathcal{L}_2$ is a union of parallel hyperplanes in $\mathbb{C}^n$.

Theorem 8. *A pair $\{L_1, L_2\}$ of formal linear partial differential operators with constant coefficients has Property C if and only if $\mathcal{L}_1 \nmid \mathcal{L}_2$.*

Example 9. Let $L = \nabla^2$, then

$$L = \{z : z \in \mathbb{C}^n, \ z_1^2 + z_2^2 + \cdots + z_n^2 = 0\}.$$

It is clear that the tangent spaces to $\mathcal{L}$ at the points $(1, 0, 0)$ and $(0, 1, 0)$ are transversal. So the Laplacian $L = \nabla^2$ does have Property C (see Definition 2 in Section 1). In other words, given an arbitrary bounded domain $D \subset \mathbb{R}^n$ and an arbitrary function $f \in L^p(D)$, $p \geq 1$, for example, $p = 2$, one can approximate f in the norm of $L^p(D)$ by linear combinations of the product $h_1 h_2$ of harmonic in $L^2(D)$ functions.

Example 10. One can check similarly that the operators

$$\partial_t - \nabla^2, \quad i\partial_t - \nabla^2 \quad \text{and} \quad \partial_t^2 - \nabla^2$$

have Property C.

Proof of Theorem 8. We prove only the sufficiency and refer to [17] for the necessity. Assume that

$$\mathcal{L}_1 \nmid \mathcal{L}_2.$$

Note that

$$e^{x \cdot z} \in N(L_m) := N_m$$

if and only if

$$L_m(z) = 0, \quad z \in \mathbb{C}^n,$$

that is

$$z \in \mathcal{L}_m.$$

Suppose

$$\int_D f(x) w_1 w_2 dx = 0 \quad \forall w_m \in N_m.$$

Then

$$F(z_1 + z_2) := \int_D f(x) e^{x \cdot (z_1 + z_2)} dx = 0 \quad \forall z_m \in \mathcal{L}_m. \tag{23}$$

The function $F(z)$ defined in (23) is entire. It vanishes identically if it vanishes on an open set in $\mathbb{C}^n$ (or $\mathbb{R}^n$). The set

$$\{z_1 + z_2\}_{\forall z_m \in N_m}$$

contains a ball in $\mathbb{C}^n$ if (and only if)

$$\mathcal{L}_1 \nparallel \mathcal{L}_2.$$

Indeed, if z_1 runs though the set $\mathcal{L}_1 \cap B(z^{(1)}, r)$, where

$$B(z^{(1)}, r) := \{z : z \in \mathbb{C}^n, \, |z^{(1)} - z| < r\},$$

and z_2 runs through the set $\mathcal{L}_2 \cap B(z^{(2)}, r)$, then, for all sufficiently small $r > 0$, the set $\{z_1 + z_2\}$ contains a small ball $B(\zeta, \rho)$, where

$$\zeta = z^{(1)} + z^{(2)},$$

and $\rho > 0$ is a sufficiently small number. To see this, note that T_1 has a basis $h_1, \ldots h_{n-1}$, which contains $n - 1$ linearly independent vectors of $\mathbb{C}^n$, and T_2 has a basis such that at least one of its vectors, call it h_n, has a non-zero projection onto the normal to T_1, so that $\{h_1, \ldots, h_n\}$ are n linearly independent vectors in $\mathbb{C}^n$. Their linear combinations fill in a ball $B(\zeta, \rho)$. Since the vectors in T_m approximate well the vectors in $\mathcal{L}_m \cap B(z^{(m)}, r)$ if $r > 0$ is sufficiently small, the set

$$\{z_1 + z_2\}_{\forall z_m \in \mathcal{L}_m}$$

contains a ball $B(\zeta, \rho)$ if $\rho > 0$ is sufficiently small. Therefore, condition (23) implies $F(z) \equiv 0$, and so $f = 0$. This means that the pair $\{L_1, L_2\}$ has Property C. We have proved:

$$\{\mathcal{L}_1 \nparallel \mathcal{L}_2\} \Longrightarrow \{\{L_1, L_2\} \text{ has Property } C\}.$$

In [17] one can find the proof of the converse implication. □

How does one prove Property C for a pair

$$\{L_1, L_2\} := \{\nabla^2 + k^2 - q_1(x), \nabla^2 + k^2 - q_2(x)\}$$

of the Schrödinger operators, where $k = \text{const} \geq 0$ and $q_m(x) \in L^2_{\text{loc}}(\mathbb{R}^n)$ are some real-valued, compactly supported functions?

One way to do it [17] is to use the existence of the elements

$$\psi_m \in N_m := \{w : [\nabla^2 + k^2 - q_m(x)]w = 0 \text{ in } \mathbb{R}^n\}$$

which are of the form

$$\psi_m(x, \theta) = e^{ik\theta \cdot x}[1 + R_m(x, \theta)], \quad k > 0, \tag{24}$$

where

$$\theta \in M := \{\theta : \theta \in \mathbb{C}^n, \ \theta \cdot \theta = 1\}.$$

Here

$$\theta \cdot w := \sum_{j=1}^{n} \theta_j w_j,$$

(note that there is no complex conjugation above w_j), the variety M is noncompact, and [17], [19]:

$$\|R_m(x, \theta)\|_{L^\infty(D)} \leq c \left(\frac{\ln |\theta|}{|\theta|}\right)^{\frac{1}{2}}, \quad \theta \in M, \quad |\theta| \to \infty, \quad m = 1, 2, \tag{25}$$

where $c = \text{const} > 0$ does not depend on θ, c depends on D and on $\|q\|_{L^\infty(B_a)}$, where $q = \bar{q}, q = 0$ for $|x| > a$, and $D \subset \mathbb{R}^n$ is an arbitrary bounded domain. Also

$$\|R_m(x, \theta)\|_{L^2(D)} \leq \frac{c}{|\theta|}, \quad \theta \in M, \quad |\theta| \to \infty, \quad m = 1, 2. \tag{26}$$

It is easy to check that for any $\xi \in \mathbb{R}^n, n \geq 3$, and any $k > 0$, one can find (many) θ_1 and θ_2 such that

$$k(\theta_1 + \theta_2) = \xi, \quad |\theta_1| \to \infty, \quad \theta_1, \theta_2 \in M, \quad n \geq 3. \tag{27}$$

Therefore, using (27) and (25), one gets:

$$\lim_{\substack{|\theta_1| \to \infty \\ \theta_1 + \theta_2 = \xi, \ \theta_1, \theta_2 \in M}} \psi_1 \psi_2 = e^{i\xi \cdot x}. \tag{28}$$

Since the set $\{e^{i\xi \cdot x}\}_{\forall \xi \in \mathbb{R}^n}$ is total in $L^p(D)$, it follows that the pair $\{L_1, L_2\}$ of the Schrödinger operators under the above assumptions does have Property C.

3.2. Approximating by scattering solutions

Consider the following problem of approximation theory [20].

Let $k = 1$ (without loss of generality), $\alpha \in S^2$ (the unit sphere in $\mathbb{R}^3$), and $u := u(x, \alpha)$ be the scattering solution that is, an element of $N(\nabla^2 + 1 - q(x))$ which solves the problem:

$$[\nabla^2 + 1 - q(x)] u = 0 \quad \text{in } \mathbb{R}^3, \tag{29}$$

$$u = \exp(i\alpha \cdot x) + A(\alpha', \alpha)\frac{e^{i|x|}}{|x|} + o\left(\frac{1}{|x|}\right), \quad |x| \to \infty, \quad \alpha' := \frac{x}{|x|}, \tag{30}$$

where $\alpha \in S^2$ is given.

Let
$$w \in N(\nabla^2 + 1 - q(x)) := N(L)$$
be arbitrary, $w \in H^2_{loc}$, H^l is the Sobolev space. The problem is:

Is it possible to approximate w in $L^2(D)$ with an arbitrary accuracy by a linear combination of the scattering solutions $u(x, \alpha)$?

In other words, given an arbitrary small number $\varepsilon > 0$ and an arbitrary fixed, bounded, homeomorphic to a ball, Lipschitz domain $D \subset \mathbb{R}^n$, can one find $\nu_\varepsilon(\alpha) \in L^2(S^2)$ such that

$$\|w - \int_{S^2} u(x, \alpha)\nu_\varepsilon(\alpha)d\alpha\|_{L^2(D)} \leq \varepsilon \ ? \tag{31}$$

If yes,

What is the behavior of $\|\nu_\varepsilon(\alpha)\|_{L^2(S^2)}$ as $\varepsilon \to 0$, if $w = \psi(x, \theta)$ where ψ is the special solution (24)-(25), $\theta \in M$, $\mathrm{Im}\,\theta \neq 0$?

The answer to the first question is yes. A proof [17] can go as follows. If (31) is false, then one may assume that $w \in N(L)$ exists such that

$$\int_D \overline{w} \left(\int_{S^2} u(x, \alpha)\nu(\alpha)d\alpha \right) dx = 0 \quad \forall \nu \in L^2(S^2). \tag{32}$$

This implies

$$\int_D \overline{w}u(x, \alpha)dx = 0 \quad \forall \alpha \in S^2. \tag{33}$$

From (33) and formula (5) in [17, pp. 46] one concludes:

$$v(y) := \int_D \overline{w}G(x, y)dx = 0 \quad \forall y \in D' := \mathbb{R}^3 \setminus D, \tag{34}$$

where $G(x, y)$ is the Green function of L:

$$\left[\nabla^2 + 1 - q(x)\right] G(x, y) = -\delta(x, y) \text{ in } \mathbb{R}^3, \tag{35}$$

$$\lim_{r \to \infty} \int_{|x|=r} \left| \frac{\partial G}{\partial |x|} - iG \right|^2 ds = 0. \tag{36}$$

Note, that formula (5) in [17, pp. 46] is:

$$G(x, y, k) = \frac{e^{i|y|}}{4\pi|y|}u(x, \alpha) + o\left(\frac{1}{|y|}\right), \quad |y| \to \infty, \quad \frac{y}{|y|} = -\alpha. \tag{37}$$

From (34) and (35) it follows that

$$\left[\nabla^2 + 1 - q(x)\right] v(x) = -\overline{w}(x) \quad \text{in } D, \tag{38}$$

$$v = v_N = 0 \quad \text{on } S := \partial D, \tag{39}$$

where N is the outer unit normal to S, and (39) holds because $v = 0$ in D' and $v \in H^2_{loc}$, by elliptic regularity, if $q \in L^2_{loc}$. Multiply (38) by w, integrate over D and then, on the left, by parts, using (39), and get:

$$\int_D |w|^2 dx = 0. \tag{40}$$

Thus, $w = 0$ and (31) is proved. Consequently, the answer to the first question is affirmative.

Let us prove that

$$\lim_{\varepsilon \to 0} \|\nu_\varepsilon(\alpha)\|_{L^2(S^2)} = \infty. \tag{41}$$

We proceed by contradiction. Assuming

$$\|\nu_\varepsilon(\alpha)\|_{L^2(S^2)} \le c \quad \forall \varepsilon \in (0, \varepsilon_0),$$

where $\varepsilon_0 > 0$ is some number, one can select a weakly convergent in $L^2(S^2)$ sequence

$$\nu_n(\alpha) \to \nu(\alpha), \quad n \to \infty,$$

and pass to the limit in (31) with

$$w = \psi(x, \theta),$$

to get

$$\left\| \psi(x, \theta) - \int_{S^2} u(x, \alpha) \nu(\alpha) d\alpha \right\|_{L^2(D)} = 0. \tag{42}$$

The function

$$U(x) := \int_{S^2} u(x, \alpha) \nu(\alpha) d\alpha \in N(L)$$

and

$$\|U(x)\|_{L^\infty(\mathbb{R}^n)} < \infty,$$

because

$$\sup_{x \in \mathbb{R}^n, \alpha \in S^2} |u(x, \alpha)| < \infty.$$

Since (42) implies that

$$\psi(x, \theta) = U(x) \quad \text{in } D,$$

and both $\psi(x, \theta)$ and $U(x)$ solve equation (27), the unique continuation principle for the solutions of the elliptic equation (27) implies

$$U(x) = \psi(x, \theta) \quad \text{in } \mathbb{R}^3.$$

This is a contradiction since $\psi(x, \theta)$ grows exponentially as $|x| \to \infty$ in certain directions because $\text{Im}\,\theta \ne 0$ (see formula (24)). Therefore, we have proved that if

$$w = \psi(x, \theta), \quad \text{Im}\,\theta \ne 0,$$

then (41) holds, where $v_\varepsilon(\alpha)$ is the function from (31).

For example, if $q(x) = 0$ and $k = 1$ then

$$\psi(x, \theta) = e^{i\theta \cdot x} \quad \text{and} \quad u(x, \alpha) = e^{i\alpha \cdot x}.$$

So, if

$$\text{Im}\,\theta \ne 0, \quad \theta \in M, \quad \text{and} \quad \left\| e^{i\theta \cdot x} - \int_{S^2} \nu_\varepsilon(\alpha) e^{i\alpha \cdot x} d\alpha \right\|_{L^2(D)} < \varepsilon,$$

then

$$\lim_{\varepsilon \to 0} \|\nu_\varepsilon(\alpha)\|_{L^2(S^2)} = \infty.$$

It is interesting to estimate the rate of growth of

$$\inf_{\substack{\nu \in L^2(S^2) \\ \|e^{i\theta \cdot x} - \int_{S^2} \nu(\alpha)e^{i\alpha \cdot x} d\alpha\|_{L^2(D)} < \varepsilon}} \|\nu(\alpha)\|_{L^2(S^2)}.$$

This is done in [20] (and [27]).

Property C for ordinary differential equations is defined, proved and applied to many inverse problems in [18], [28]-[29].

4. Approximation by entire functions of exponential type.

Let

$$f \in L^2(B_a), \quad B_a := \{x \, : \, x \in \mathbb{R}^n_x, \, |x| \le a\},$$

$a > 0$ is a fixed number, $f(x) = 0$ for $|x| > a$, and

$$\widetilde{f}(\xi) = \int_{B_a} f(x)e^{i\xi \cdot x} dx. \tag{43}$$

Assume that $\widetilde{f}(\xi)$ is known for all $\xi \in \widetilde{D} \subset \mathbb{R}^n_\xi$, where $\widetilde{D}$ is a (bounded) domain.

The problem is to find $\widetilde{f}(\xi)$ for all $\xi \in \mathbb{R}^n_\xi$, (this is called spectral extrapolation), or, equivalently, to find $f(x)$ (this is called inversion of the Fourier transform $\widetilde{f}(\xi)$ of a compactly supported function $f(x)$, supp$f \subset B_a$, from a compact $\widetilde{D}$).

In applications the above problem is also of interest in the case when $\widetilde{D}$ is not necessarily bounded. For example, in tomography $\widetilde{D}$ may be a union of two infinite cones (the limited-angle data).

In the fifties and sixties there was an extensive discussion in the literature concerning the resolution ability of linear instruments. According to the theory of the formation of optical images, the image of a bright point, which one obtains when the light, issued by this point, is diffracted on a circular hole in a plane screen, is the Fourier transform $\widetilde{f}(\xi)$ of the function $f(x)$ describing the light distribution on the circular hole B_a, the two-dimensional ball. This Fourier transform is an entire function of exponential type. One says that the resolution ability of a linear instrument (system) can be increased without a limit if the Fourier transform of the function $f(x)$, describing the light distribution on B_a, can approximate the delta-function $\delta(\xi)$ with an arbitrary accuracy. The above definition is not very precise, but it is usual in applications and can be made precise: it is sufficient to specify the metric in which the delta-function is approximated. For our purposes, let us take a delta-type sequence $\delta_j(\xi)$ of continuous functions which is defined by the requirements:

$$\left| \int_{\widetilde{D}} \delta_j(\xi) d\xi \right| \le c,$$

where the constant c does not depend on $\widetilde{D}$ and j, and

$$\lim_{j \to \infty} \int_{\widetilde{D}} \delta_j(\xi) d\xi = \begin{cases} 1 \text{ if } 0 \in \widetilde{D}, \\ 0 \text{ if } 0 \notin \widetilde{D}. \end{cases}$$

The approximation problem is:

Can one approximate an arbitrary continuous (or $L^1(\widetilde{D})$) function $g(\xi)$ in an arbitrary fixed bounded domain $\widetilde{D} \subset \mathbb{R}^n_\xi$ by an entire function of exponential type

$$\widetilde{f}(\xi) = \int_{B_a} f(x) e^{i\xi \cdot x} dx,$$

where $a > 0$ is an arbitrary small number?

The engineers discussed this question in a different form:

Can one transmit with an arbitrary accuracy a high-frequency signal $g(\xi)$ by using low-frequency signals $\widetilde{f}(\xi)$? The smallness of a means that the "spectrum" $f(x)$ of the signal $\widetilde{f}(\xi)$ contains only "low spatial frequencies".

From the mathematical point of view the answer is nearly obvious: yes. The proof is very simple: if an approximation with an arbitrary accuracy were impossible, then

$$0 = \int_{\widetilde{D}} g(\xi) \left(\int_{B_a} e^{i\xi \cdot x} f(x) dx \right) d\xi \quad \forall f \in L^2(B_a).$$

This implies the relation

$$0 = \int_{\widetilde{D}} g(\xi) e^{i\xi \cdot x} d\xi \quad \forall x \in B_a.$$

Since $\widetilde{D}$ is a bounded domain, the integral above is an entire function of $x \in \mathbb{C}^n$ which vanishes in a ball B_a. Therefore this function vanishes identically and consequently

$$g(\xi) \equiv 0.$$

This contradiction proves that the approximation of an arbitrary $g(\xi) \in L^2(\widetilde{D})$ by the entire functions

$$\widetilde{f}(\xi) = \int_{B_a} f(x) e^{i\xi \cdot x} dx$$

is possible with an arbitrary accuracy in $L^2(\widetilde{D})$.

Now let us turn to another question:

How does one derive an analytic formula for finding $f(x)$ if $\widetilde{f}(\xi)$ is given in $\widetilde{D}$?

In other words,

How does one invert analytically the Fourier transform $\widetilde{f}(\xi)$ of a compactly supported function $f(x)$, $\operatorname{supp} f \subset B_a$, from a compact $\widetilde{D}$?

We discuss this question below, but first let us discuss the notion of apodization, which was a hot topic at the end of the sixties. Apodization is a method to increase the resolution ability of a linear optical system (instrument) by putting a suitable mask on the outer pupil of the instrument. Mathematically one deals with an approximation problem: by choosing a mask $g(x)$, which transforms the function $f(x)$ on the outer pupil of the instrument into a function $g(x)f(x)$, one wishes to change the image $\tilde{f}(\xi)$ on the image plane to an image $\delta_j(\xi)$ which is close to the delta-function $\delta(\xi)$, and therefore increase the resolution ability of the instrument. That the resolution ability can be increased without a limit (only in the absence of noise!) follows from the above argument: one can choose $g(x)$ so that

$$\widetilde{g(x)f(x)}$$

will approximate arbitrarily accurately $\delta_j(\xi)$, which, in turn, approximates $\delta(\xi)$ arbitrarily accurately.

This conclusion contradicts to the usual intuitive idea according to which one cannot resolve details smaller than the wavelength.

In fact, if there is no noise, one can, in principle, increase resolution ability without a limit (superdirectivity in the antenna theory), but since the noise is always present, in practice there is a limit to the possible increase of the resolution ability.

Let us turn to the analytic formula for the approximation by entire functions and for the inversion of the Fourier transform of a compactly supported function from a compact $\tilde{D}$.

Multiply (43) by

$$(2\pi)^{-n}\tilde{\delta}_j(\xi)e^{-i\xi \cdot x}$$

and integrate over $\tilde{D}$ to get

$$f_j(x) = \int_{B_a} f(y)\delta_j(x-y)dy = \frac{1}{(2\pi)^n}\int_{\tilde{D}} \tilde{f}(\xi)\tilde{\delta}_j e^{-i\xi \cdot x}d\xi \qquad (44)$$

where $\tilde{\delta}_j(\xi)$ is the Fourier transform of $\delta_j(x)$.

Let us choose $\delta_j(x)$ so that it will be a delta-type sequence (in the sense defined above). In this case $f_j(x)$ approximates $f(x)$ arbitrarily accurately:

$$\lim_{j\to\infty} \|f - f_j\| = 0, \qquad (45)$$

where the norm $\|\cdot\|$ is $L^2(B_a)$ norm if $f \in L^2(B_a)$, and $C(B_a)$-norm if $f \in C(B_a)$.

If $\|f\|_{C^1(B_a)} \le m_1$, then

$$\|f - f_j\|_{C(B_a)} \le cm_1 j^{-\frac{1}{2}}. \qquad (46)$$

The conclusions (45) and (46) hold if, for example,

$$\delta_j(x) := P_j(|x|^2)\left(\mathcal{F}^{-1}\tilde{h}\right)(x), \qquad (47)$$

where

$$P_j(r) := \left(\frac{j}{4\pi a_1^2}\right)^{\frac{n}{2}}\left(1 - \frac{r}{4a_1^2}\right)^j, \quad 0 \le r \le a, \quad a_1 > a, \qquad (48)$$

$$\widetilde{h}(\xi) \in C_0^\infty(\widetilde{D}), \quad \frac{1}{(2\pi)^n} \int_{\widetilde{D}} \widetilde{h}(\xi) d\xi = 1. \tag{49}$$

Theorem 11. *If* (47)-(49) *hold, then the sequence* $f_j(x)$, *defined in* (44), *satisfies* (45) *and* (46).

Thus, formula (44):

$$f(x) = \mathcal{F}^{-1}\left[\widetilde{f}(\xi)\widetilde{\delta}_j(\xi)\right], \tag{50}$$

where $\widetilde{\delta}_j(\xi) := \mathcal{F}\delta_j(x)$, and $\delta_j(x)$ is defined by formulas (4.5)-(4.7), is an inversion formula for the Fourier transform $\widetilde{f}(\xi)$ of a compactly supported function $f(x)$ from a compact $\widetilde{D}$ in the sense (45). A proof of a result similar to Theorem 11 has been originally published in [23].

In [21], [22] apodization theory and resolution ability are discussed. In [24] a one-dimensional analog of Theorem 11 is given. In this analog one can choose analytically explicitly a function similar to $\widetilde{h}(\xi)$.

Proof of Theorem 11 is given in [30, pp. 260-263].

Let us explain a possible application of Theorem 11 to the limited-angle data in tomography. The problem is: let

$$\widehat{f}(\alpha, p) := \int_{l_{\alpha p}} f(x) ds,$$

where $\widehat{f}(\alpha, p)$ is the Radon transform of $f(x)$, and

$$l_{\alpha p} = \{x : \alpha \cdot x = p\}$$

is a plane.

Given $\widehat{f}(\alpha, p)$ *for all* $p \in \mathbb{R}$ *and all* $\alpha \in K$, *where* K *is an open proper subset of* S^2, *find* $f(x)$, *assuming* $\mathrm{supp} f \subset B_a$.

It is well known [30], that

$$\int_{-\infty}^{\infty} \widehat{f}(\alpha, p) e^{ipt} dp = \widetilde{f}(t\alpha), \quad t \in \mathbb{R}, \quad t\alpha := \xi.$$

Therefore, if one knows $\widehat{f}(\alpha, p)$ for all $\alpha \in K$ and all $p \in \mathbb{R}$, then one knows $\widetilde{f}(\xi)$ for all ξ in a cone $K \times \mathbb{R}$. Now Theorem 11 is applicable for finding $f(x)$ given $\widehat{f}(\alpha, p)$ for $\alpha \in K$ and $p \in \mathbb{R}$.

References

1. Dolgopolova, T., and Ivanov, V., On numerical differentiation, *J. of Comput. Math. and Phys.* **6** (1966), 570–576.

2. Phillips, D.L., A technique for numerical solution of certain integral equations of the first kind, *J. Assoc. Comput. Math.* **9** (1962), 84–97.

3. Ramm, A.G., On numerical differentiation, *Math. Izvestija Vuzov* **11** (1968), 131–135.

4. Ramm, A.G., *Scattering by Obstacles*, D. Reidel, Dordrecht, 1986.

5. Ramm, A.G., *Random fields estimation theory*, Longman Scientific and Wiley, New York, 1990.

6. Ramm, A.G., Stable solutions of some ill-posed problems, *Math. Meth. in the Appl. Sci.* **3** (1981), 336–363.

7. Ramm, A.G., Simplified optimal differentiators, *Radiotech.i Electron.* **17** (1972), 1325–1328 (English Translation pp.1034–1037).

8. Ramm, A.G., Inequalities for the derivatives, *Math. Ineq. and Appl.* **3** (2000), 129–132.

9. Ramm, A.G., On completeness of the products of harmonic functions, *Proc. Amer. Math. Soc.* **99** (1986), 253–256.

10. Ramm, A.G., Inverse scattering for geophysical problems, *Phys. Letters* **99A** (1983), 258–260.

11. Ramm, A.G., Completeness of the products of solutions to PDE and uniqueness theorems in inverse scattering, *Inverse Problems* **3** (1987), L77–L82.

12. Ramm, A.G., Recovery of the potential from fixed energy scattering data, *Inverse Problems* **4** (1988), 877–886 (**5** (1989) 255).

13. Ramm, A.G., Multidimensional inverse problems and completeness of the products of solutions to PDE, *J. Math. Anal. Appl.* **134** (1988), 211–253 (and **139** (1989), 302).

14. Ramm, A.G., Multidimensional inverse scattering problems and completeness of the products of solutions to homogeneous PDE, *Zeitschr. f. Angew. Math. u. Mech.* **69** (1989), T13–T22.

15. Ramm, A.G., Completeness of the products of solutions of PDE and inverse problems, *Inverse Problems* **6** (1990), 643–664.

16. Ramm, A.G., Uniqueness theorems for multidimensional inverse problems with unbounded coefficients, *J. Math. Anal. Appl.* **136** (1988), 568–574.

17. Ramm, A.G., *Multidimensional inverse scattering problems*, Longman/Wiley, New York, 1992.

18. Ramm, A.G., Property C for ODE and applications to inverse problems, in *Operator Theory and Its Applications* (Ramm, A.G., Shivakumar, P.N., and Strauss, A.V., Eds.), pages 15-75, Fields Institute Communications 25, Amer. Math. Soc., Providence, 2000.

19. Ramm, A.G., Stability of solutions to inverse scattering problems with fixed-energy data, *Milan J. of Math.* **70** (2002), 97–161.

20. Ramm, A.G., Stability estimates in inverse scattering, *Acta Appl. Math.* **28** (1992), 1–42.

21. Ramm, A.G., Apodization theory II, *Opt. and Spectroscopy* **29** (1970), 390–394.

22. Ramm, A.G., Increasing of the resolution ability of the optical instruments by means of apodization, *Opt. and Spectroscopy* **29** (1970), 594–599.

23. Ramm, A.G., Approximation by entire functions, *Math. Izv. Vusov* **10** (1978), 72–76.

24. Ramm, A.G., Signal estimation from incomplete data, *J. Math. Anal. Appl.* **125** (1987), 267–271.

25. Ramm, A.G., On simultaneous approximation of a function and its derivative by interpolation polynomials, *Bull. Lond. Math. Soc.* **9** (1977), 283–288.

26. Ramm, A.G., Necessary and sufficient condition for a PDE to have Property C, *J. Math. Anal. Appl.* **156** (1991), 505–509.
27. Ramm, A.G., *Multidimensional inverse scattering problems*, Mir Publishers, Moscow, 1994.
28. Ramm, A.G., *Inverse Problems*, Springer, New York, 2004.
29. Ramm, A.G., One-dimensional inverse scattering and spectral problems, *Cubo. A Math. J.* **6** (2004), 313–426.
30. Ramm, A.G., and Katsevich, A.I., *The Radon Transform and Local Tomography*, CRC Press, Boca Raton, 1996.
31. Ramm, A.G., and Smirnova, A.B., On stable numerical differentiation, *Math. of Comp.* **70** (2001), 1131–1153.
32. Ramm, A.G., and Smirnova, A.B., Stable numerical differentiation: When is it possible? *J. Korean SIAM* **7** (2003), 47–61.
33. Tikhonov, A.N., Solving ill-posed problems and regularization, *Doklady Acad. Sct. USSR* **157** (1963), 501–504.

Index